AF167477

The Synergy Between Dynamics and Reactivity at Clusters and Surfaces

NATO ASI Series

Advanced Science Institutes Series

A Series presenting the results of activities sponsored by the NATO Science Committee, which aims at the dissemination of advanced scientific and technological knowledge, with a view to strengthening links between scientific communities.

The Series is published by an international board of publishers in conjunction with the NATO Scientific Affairs Division

A Life Sciences	Plenum Publishing Corporation
B Physics	London and New York
C Mathematical and Physical Sciences	Kluwer Academic Publishers
D Behavioural and Social Sciences	Dordrecht, Boston and London
E Applied Sciences	
F Computer and Systems Sciences	Springer-Verlag
G Ecological Sciences	Berlin, Heidelberg, New York, London,
H Cell Biology	Paris and Tokyo
I Global Environmental Change	

PARTNERSHIP SUB-SERIES

1. **Disarmament Technologies**	Kluwer Academic Publishers
2. **Environment**	Springer-Verlag / Kluwer Academic Publishers
3. **High Technology**	Kluwer Academic Publishers
4. **Science and Technology Policy**	Kluwer Academic Publishers
5. **Computer Networking**	Kluwer Academic Publishers

The Partnership Sub-Series incorporates activities undertaken in collaboration with NATO's Cooperation Partners, the countries of the CIS and Central and Eastern Europe, in Priority Areas of concern to those countries.

NATO-PCO-DATA BASE

The electronic index to the NATO ASI Series provides full bibliographical references (with keywords and/or abstracts) to more than 50000 contributions from international scientists published in all sections of the NATO ASI Series.
Access to the NATO-PCO-DATA BASE is possible in two ways:

– via online FILE 128 (NATO-PCO-DATA BASE) hosted by ESRIN,
Via Galileo Galilei, I-00044 Frascati, Italy.

– via CD-ROM "NATO-PCO-DATA BASE" with user-friendly retrieval software in English, French and German (© WTV GmbH and DATAWARE Technologies Inc. 1989).

The CD-ROM can be ordered through any member of the Board of Publishers or through NATO-PCO, Overijse, Belgium.

Series C: Mathematical and Physical Sciences – Vol. 465

The Synergy Between Dynamics and Reactivity at Clusters and Surfaces

edited by

Louis J. Farrugia

Department of Chemistry,
University of Glasgow,
Glasgow, Scotland

Springer-Science+Business Media, B.V.

Proceedings of the NATO Advanced Research Workshop on
The Synergy Between Dynamics and Reactivity at Clusters and Surfaces
Drymen, Scotland
July 3–8, 1994

A C.I.P. Catalogue record for this book is available from the Library of Congress

ISBN 978-0-7923-3522-1   ISBN 978-94-011-0133-2 (eBook)
DOI 10.1007/978-94-011-0133-2

Printed on acid-free paper

All Rights Reserved
© 1995 Springer Science+Business Media Dordrecht
Originally published by Kluwer Academic Publishers in 1995

No part of the material protected by this copyright notice may be reproduced or
utilized in any form or by any means, electronic or mechanical, including photo-
copying, recording or by any information storage and retrieval system, without written
permission from the copyright owner.

CONTENTS

vi

PREFACE

The analogy between the chemistry of molecular transition metal clusters, and the processes of chemisorption and catalysis at metal surfaces (the **Cluster Surface** analogy) has, for a number of years, provided an interplay between experimental and theoretical Inorganic and Physical chemists. The attraction of this analogy lies in the fact that it is far easier to examine ligand stereogeometry and reactivity in molecular clusters, than to tackle the (often formidable) problem of characterising absorbate interactions on metal surfaces. To date, the most productive area of this collaborative approach has been the use of the well defined modes of metal-ligand *bonding* in discrete molecular clusters, as *models* for metal-ligand *binding* on surfaces. Crystallographic and spectroscopic (mainly IR) data, obtained from molecular clusters containing for example carbonyl, alkyne, vinylidene and benzene ligands, have been used by surface scientists to define binding modes of these ligands on metal surfaces.

The dynamic properties of ligands in molecular clusters have been studied for a number of years. Although it is well known that the *energetics* of ligand migrations in clusters and on surfaces are of similar magnitudes, there has been remarkably little effort to correlate ligand dynamic *mechanisms* in clusters and surfaces. In part this is due to the great difficulties in studying ligand mobility on surfaces. Only recently has it been possible to examine in detail ligand dynamics on surfaces through such techniques as ESDIAD and solid state NMR. Advances in solution NMR such as 2D EXSY have also made it possible to obtain accurate quantitative information about the complex multisite ligand exchange mechanisms which occur in molecular clusters. It was therefore decided that it was particularly timely to organise a NATO Advanced Research Workshop on this topic. The lectures and discussions at this meeting form the basis of this book.

The Workshop took place in July 1994 at the Buchanan Arms Hotel in Drymen, a small village near Glasgow not far from the shores of Loch Lomond in western Scotland. About forty scientists from west and east Europe, North America and Taiwan attended. Many commented that the cool Scottish "summer" was such a pleasant change from the heat of their home countries, and it proved very condusive for scientific discussions! The meeting brought together chemists studying experimental and theoretical aspects of molecular cluster dynamics and reactivity, with surface scientists studying the dynamics of metal-ligand interactions.

The key lecturers were Heinrich Vahrenkamp, Arthur Carty, Raymond Roulet, Julius Jellinek, Tom Fehlner, Alex Bradshaw and John Yates (who very kindly agreed to step into these shoes at extremely short notice), all of whom provided very stimulating talks. Some of the key topics discussed at this meeting included:

1.   Mechanisms of fluxional behaviour in clusters in the liquid phase and the connections with diffusion processes on extended surfaces. The role of metal-metal bond breaking in diffusion.

2.   Analogies in the structure of chemisorbed species and related ligands on metallic clusters.

3.   Analogies between benzene surface chemistry on extended metal surfaces and on metal faces in molecular cluster compounds, with particular reference to structural distortions.

4.   The role of mobile precursors for dissociation or chemisorption on extended metals and on clusters. Are there analogies in the ligand attachment during cluster compound synthesis?

5.   The role of defect sites on metal surfaces in catalyzing chemical reactions and the connection to the special bonding properties of sites on metal clusters having lowest metal-metal coordination.

6.   The size of metal clusters needed to mimic surface phenomena on bulk metal surfaces. Different sites needed for different phenomena.

The discussion sessions of the meeting were held at the end of the formal lectures, but due to an oversight on the part of the main Organiser (Louis Farrugia) the Workshop took place right in the middle of the World Cup matches. This resulted in a serious conflict of interests for several participants regarding their contributions to these sessions. However, it was all worth it to see the look on Carlo Mealli's face as he strode into the conference dining rooms, after Italy had won their match on the Wednesday evening!

A meeting of this nature involves a considerable effort on the part of many people. The Director and Co-Director on the Organising Committee were joined by Professors Heinrich Vahrenkamp, Dario Braga, Arthur Carty, and Brian Johnson, and also Pierre Braunstein who afforded   several valuable suggestions, but was unfortunately unable to attend due to pressing engagments in the Antipodes. We thank NATO and the NATO Science Committee for their help and their substantial funding, without which this meeting would not have been possible. We also thank the University of Glasgow, and the Department of Chemistry for much help in kind, particularly Liz Hughes and Bob Munro for their patience in trying times. Glasgow City Corporation is thanked for the opportunity to explore the recesses of Glagow City Halls, and we are also grateful for the financial contributions from Procter & Gamble Ltd. and Plenum Publishing Company Ltd. The International Science Foundation is thanked for their generous support for the participants from Russia and Lithuania. Finally we thank Michaela Ruff and all the staff of the Buchanan Arms Hotel for their hospitality, and for a most memorable and extremely well fed sojourn in Drymen.

Louis J. Farrugia
John T. Yates Jr.
Glasgow, Scotland,
January, 1995

CAN WE PUT THE CLUSTER-SURFACE ANALOGY ON A SOUND STRUCTURAL BASIS?

A. M. BRADSHAW

Fritz-Haber-Institut der Max-Planck-Gesellschaft
Faradayweg 4 - 6, D-14195 Berlin, Germany

A thorough examination of the analogy between the adsorption complexes formed by molecules and molecular fragments adsorbed on metal surfaces and the corresponding co-ordination compounds has hitherto not been possible because of a lack of accurate structural information on surface systems. X-ray diffraction, which gives bondlengths and bond angles to very high accuracy in inorganic and organometallic systems, is not readily applicable to problems of molecular chemisorption. Nevertheless, some progress is now being made using techniques based on electron scattering phenomena. The few data so far available indicate that there are strong similarities between the two situations, at least as far as ligand structure and metal-ligand bondlengths are concerned.

1. Introduction

Although Muetterties is generally credited with first drawing attention to the so-called cluster-surface analogy [1], many authors had previously pointed out the similarity between molecular chemisorption and the metal-ligand interaction in co-ordination chemistry. Eischens and Pliskin [2], for example, having measured infrared absorption spectra of CO adsorbed on supported metal catalysts, assigned the C-O stretching frequencies on the basis of IR data for the metal carbonyls. Blyholder [3] later applied the back-donation model for metal-ligand bonding to produce a qualitative, but generally consistent picture of CO chemisorption. The success of this approach was such that it was

1

L. J. Farrugia (ed.), The Synergy Between Dynamics and Reactivity at Clusters and Surfaces, 1–20.
© 1995 *Kluwer Academic Publishers*.

extended to nitric oxide adsorption on metals, with the nitrosyl complexes in co-ordination chemistry serving as analogons for the assignment of the various N-O stretches [4]. Similarly, early consideration of the bonding of unsaturated hydrocarbons to metal surface were dictated by the known structure of the corresponding organometallic compounds [5,6]. However, substantial progress in the application of these ideas first became possible with the development of high resolution electron energy loss spectroscopy (EELS) by Ibach and co-workers in the 1970's [7]. Its routine application to the measurement of vibrational spectra of atoms and molecules adsorbed on compact solid surfaces has been arguably the most important single development in "dry" surface chemistry since the work of Langmuir. This wider data base convinced Albert and Yates in their recent monograph [8] that the cluster-surface analogy should be extended to include mononuclear co-ordination compounds and focus on the structure and properties of the ligands.

The major problem which confronts us in any comparative study of the two fields is the lack of precise structural information on molecular chemisorption systems. Various semi-quantitative techniques have been available for some 10 - 15 years, but methods that reliably give adsorption site, orientation and bondlengths are practised by only a few groups and have so far been applied to comparatively few systems. In this situation qualitative arguments based on "chemical intuition" have assumed an important rôle. The structural models they derived have often been confirmed in subsequent, mostly semi-quantitative studies, although there are exceptions. The cyanide anion, for example, was originally thought to be bonded to metal surfaces via the C atom in a terminal or bridging configuration as in co-ordination chemistry. Even quantum chemical cluster calculations were found to be consistent with this bonding geometry. Later work has shown, however, that in those systems so far investigated with structure-sensitive techniques the C-N axis is actually oriented parallel to the surface [9,10]. Similarly, the assignment of CO adsorption sites solely on the basis of the C-O stretching frequency has recently been questioned following quantitative structural studies of the system Ni{111}-CO [11] (see below) and Pd{110}-CO [12]. In short, the need for accurate structural information in surface chemistry has recently become very apparent and is a necessary pre-requisite for putting the analogy with co-ordination chemistry on a solid footing.

In the next section the various quantitative and semi-quantitative techniques for surface structural research are briefly discussed. In the third section, some recently published structure determinations for adsorbed carbon monoxide, trifluorophosphine, the formate and acetate species, ethylene and benzene will be described and compared with the structure of certain co-ordination compounds.

2. The techniques

Since the various methods in surface science are well described in several general texts (e. g. [13]), this account is very brief. Beginning with the *semi-quantitative* methods, we should first mention the use of selection rules in both vibrational spectroscopy and photoelectron spectroscopy. On a metal surface, only those vibrational modes with a component of the dynamic dipole moment perpendicular to the surface can be excited in infrared reflection-absorption spectroscopy (IRAS) or in the dipole scattering mode of EELS. This is due to the so-called screening properties of metal electrons at IR frequencies. Thus the observation of certain vibrational frequencies and not of others provides us in favourable circumstances with a structural tool [7]. For example, in the system Ni{111}-PF$_3$, which will be encountered again below, only the symmetric P-F stretch and the symmetric P-F bend are observed in the spectral range 400 - 1000 cm^{-1} [14]. The corresponding antisymmetric modes are not present, implying that their associated dynamic dipoles are parallel to the surface. This in turn indicates that the point group of the surface molecule remains C$_{3v}$, i. e. that the threefold symmetry axis of the molecule is normal to the surface.

Selection rules can also be applied in photoelectron spectroscopy, since the promotion of an electron from a filled orbital of the molecule into a continuum state is also a dipole excitation. By selecting appropriate photoelectron emission directions and varying the orientation of the electric vector of the incident UV radiation, a "forbidden" geometry can be found, for which a particular orbital gives no emission peak. This condition depends, in turn, on the point group of the adsorbed molecule and thus on its orientation [15]. As an example, the disappearance of the 4a$_1$ feature in the photoelectron spectrum of the methoxy species on a Cu{111} surface under appropriate conditions tells us that the O-C axis is normal to the surface [16].

Selection rules only give "yes" or "no" answers, i. e., they enable us to decide for or against a particular (usually high symmetry) configuration, and are thus limited in their application. A technique which has provided somewhat more information in recent years is near-edge X-ray absorption spectroscopy (NEXAFS) [17], although some of the early data may have been over-interpreted. Monochromatic soft X-rays are used to excite, say, a C1s electron into the unfilled, antibonding orbitals of the surface molecule. Again, we are dealing with an optical excitation, and in this case the linear dichroism effect is exploited to obtain orientational information. Since a conventional absorption experiment is not possible, the electron yield from the surface (proportional to the absorption cross-section) is measured. A recent example illustrates the kind of application: for the case of the pyridine molecule adsorbed on a

Ni{111} surface, the ring plane is inclined at an angle of ~ 20° to the surface normal [18]. The accuracy of such determinations (usually estimated to be ±10°) is often open to question since oversimplified notions of molecular orbital theory are generally used and formal symmetry arguments ignored [19].

Two more techniques deserve mention. The ions or excited atoms which are desorbed from an adsorbate-covered surface as a result of electron beam-induced fragmentation may have a characteristic angular distribution (ESDIAD) which reflects the molecular orientation, and perhaps even the internal molecular structure [20]. Thus, in the case of PF_3 adsorbed on Ni{111}, the angular distribution of desorbed F^+ ions suggests that the molecule is adsorbed terminally via the P atom to a Ni atom with the PF bonds azimuthally oriented over neighbouring Ni atoms [21]. There are two such orientations on the Ni{111} surface which then give rise to the observed six-fold rotationally symmetric F^+ pattern (see 3.2 below). The method has proved particularly useful in obtaining orientational information of this sort for adsorbed PF_3, NH_3, H_2O and CO. The scanning tunnelling microscope, on the other hand, has been able to detect molecular adsorbates at low temperature [22], but so far has not delivered any precise information on adsorption site or orientation. This does not detract from its undisputedly spectacular achievements in studies of surface microtopography and adsorbate-induced reconstruction.

What is the present status of *quantitative* surface structure analysis? Since powerful computers and semi-automated procedures now allow fast, almost routine X-ray structure determination of co-ordination compounds (providing suitable crystals are available), X-ray diffraction might appear at first sight to be the method of choice. In fact, precise information on adsorption site, orientation and bondlengths of adsorbed molecules has so far been obtained exclusively from experiments based on the phenomenon of *electron* scattering. These derive their surface sensitivity from the low mean free path for inelastic electron scattering in solids. X-ray diffraction can be made sufficiently surface-sensitive if grazing-incidence geometries are used to minimise the penetration depth, but does require the high spectral brilliance of synchrotron radiation [23]. Although a few studies of low Z atomic adsorbates on metals have been reported, molecules have so far not been investigated, at least not as far as the author is aware. (Ordered overlayers would of course be a necessary pre-requisite.) X-ray standing wave analysis [24] has been used to study at least one molecular chemisorption system, however, and will be referred to in the next section.

In low energy electron diffraction (LEED) a monoenergetic beam of electrons typically in the energy range 30 - 300 eV is incident at a single crystal surface, see e. g. [13]. The elastically back-scattered electrons form an interference pattern which can be displayed two-dimensionally on a fluorescent screen. An ordered adsorbate layer produces additional, fractional order features; their relation to the integral order features enables the periodicity of the adsorbate mesh to be determined. However, the third Laue condition in the direction normal to the surface is not fulfilled and a kinematic analysis of the diffraction intensities is not possible. Structure determination with LEED thus consists of modelling the intensities as a function of energy using a dynamical theory, i.e. one which includes multiple electron scattering. Best agreement between theory and experiment for an optimised model structure produces the final result (the trial-and-error approach). The technique has been used with some success for ordered layers of molecules on metal surfaces, as will be shown below. Moreover, recent work has indicated that structure analysis is also possible when the intensity of the diffuse scattering contribution is measured [25]. This enables DLEED, as it is termed, to be applied to adsorbed layers not exhibiting long-range order, a situation frequently encountered in the chemisorption of molecules and molecular fragments.

A method related to LEED is photoelectron diffraction [26]. In a variant of conventional x-ray photoelectron spectroscopy, or ESCA, the intensity of an adsorbate core level photoelectron line is measured at a pre-selected emission angle as a function of photon energy, and thus of photoelectron energy. Clearly, synchrotron radiation is required for this experiment, since conventional soft x-ray sources for photoelectron spectroscopy have a fixed photon energy. The resulting plot of photoelectron intensity against kinetic energy is modulated by the interference between the directly emitted component of the photoelectron wave and the components that arise from elastic scattering at the substrate atoms. The modulations depend on the respective pathlength differences and thus provide information on local structure. The latter is extracted in the same way as in LEED by comparing experimental diffraction curves with calculated ones. One important advantage over conventional LEED is that again no ordered structure is necessary. The method has recently been applied to a series of molecular chemisorption systems, and even to relatively large molecules where the chemical shift can be utilised [27].

Finally, no brief account of this field would be complete without reference to surface EXAFS (= extended x-ray absorption fine structure) [28]. As in the case of solids ("bulk" EXAFS), the technique involves measuring the x-ray absorption spectrum in the range several hundred eV above the core level excitation threshold for an adsorbate atom. It is thus essentially the same experiment as near-edge absorption spectroscopy described above. In the extended energy range, however, there is structure due to the interference between the emitted photoelectron wave and waves backscattered from neighbouring substrate atoms towards the emitter. Although a simple Fourier transform procedure can reveal the distances to these nearest neighbours (assuming that the scattering phase shifts are known), more recent work utilises modelling procedures and takes into account multiple scatttering events. The technique has been mainly applied to atomic adsorbates, although some molecular systems

6

have been investigated, a recent example being Cu{110}-H₂O [29]. There are certain similarities to photoelectron diffraction, but a full discussion of the relative merits of the two methods is beyond the scope of this article. It suffices here to say that SEXAFS has the disadvantage of a smaller data base and cannot readily utilise the chemical shift information.

3. Some examples of surface molecular systems

3.1   Ni{111}-CO

Various infrared and EELS studies of the adsorption of CO on Ni{111} surface (see [30,31] and references therein) have shown the development of a band at 1816 cm^{-1} at low coverages due to the C-O stretching frequency. This is replaced by a band at 1831 cm^{-1} as a function of increasing coverage which eventually shifts to 1905 cm^{-1}, corresponding to a c(4x2) structure at a coverage of 0.5. (In the standard nomenclature [13] for describing ordered overlayers on single crystal surfaces the adsorbate unit mesh is just related to the substrate unit mesh in this simple way. Coverage refers to the number of adsorbed species per substrate atom in the outermost layer.) Conform with the classification scheme introduced by Eischens and Pliskin [2] and modified by

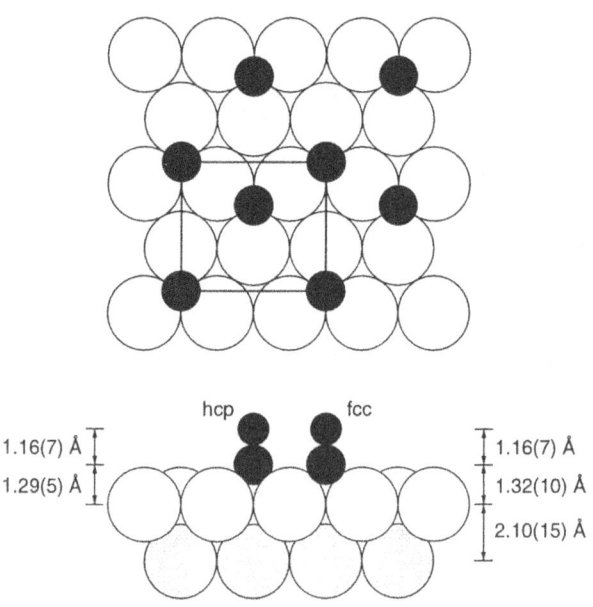

Figure 1. The structure of the Ni{111}c(4x2)-CO layer.
After Schindler *et al* [11].

later authors [32,33], the initial feature was assigned to CO in threefold symmetric hollow sites and the band shifting in the range 1831 - 1905 cm^{-1} to CO in bridge sites. The large coverage-dependent frequency shifts are known to be due to lateral interactions in the adlayer, including dipole-dipole coupling effects [34]. Later SEXAFS and C 1s photoelectron diffraction studies have shown that the site assignment is incorrect: in fact the CO remains in threefold hollow sites even though the band at 1905 cm^{-1} is firmly in the "bridging" region [11]. Figure 1 shows a schematic digram of the c(4x2) structure with some of the structural parameters. Note that both the inequivalent threefold hollow sites ("fcc" and "hcp") are occupied in this structure and that there is a 3% expansion (+ 0.07 Å) of the outermost Ni-Ni layer spacing.

For comparison, fig. 2 shows the structure of the molecule $(C_5H_5)_3Ni_3(\mu_3\text{-}CO)_2$ which contains two threefold co-ordinated CO ligands [35]. The C-Ni bondlength is almost the same as in the chemisorption case (1.932 Å compared to a mean value of 1.97 Å on the surface), but the (symmetric) C-O stretching frequency is considerable lower at 1741 cm^{-1}. As De la Cruz and

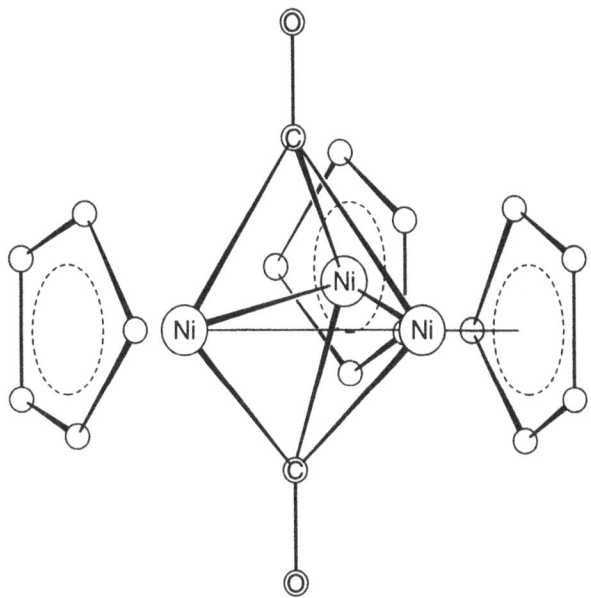

Figure 2. Schematic representation of the molecule $(Cp)_3Ni_3(CO)_2$. After Ref. [35].

Sheppard have recently discussed [36], the presence of cyclopentadienyl ligands is likely to lower the C-O stretch considerably compared to that in a similar *binary* carbonyl cluster compound due to the stonger electron-donating properties of this ligand. Thus in $Rh_6(CO)_{12}(\mu_3\text{-}CO)_4$ the C-O stretch of the threefold co-ordinated CO is at 1800 cm^{-1}, compared to 1690 cm^{-1} and 1667 cm^{-1} in $RuRh_3(CO)_3(\mu_3\text{-}CO)_2(C_5H_5)_2$ [37,38]. Allowing for the three cyclopentadienyl ligands in $(C_5H_5)_3Ni_3(\mu_3\text{-}CO)_2$, the C-O stretch of threefold co-ordinated CO in a hypothetical binary Ni carbonyl would probably be at about 1900 cm^{-1} which corresponds roughly to the frequency of an isolated CO molecule in a threefold hollow site on Ni{111}. This comparison should not be taken too far, however, since - apart from the obvious differences between the two systems - many other factors have not been taken into account, in particular the effect of the "image dipole" in the surface case.

Finally, we note that the C-O bondlength is 1.183 Å in $(C_5H_5)_3Ni_3(\mu_3\text{-}CO)_2$, which represents a small increase compared to the free molecule, consistent with the notion of an increased donation of charge into the π^* orbitals of the CO ligands and a concomitant weakening of the C-O bonds. For cluster compounds containing just one CO ligand there is indeed a correlation between the C-O stretching frequency and the C-O bondlength, as was also pointed out in the paper by De la Cruz and Sheppard [36]. Unfortunately, it is unlikely that the level of precision in structure determinations will ever be high enough to make such correlations useful in surface science: something better than ± 0.01 Å would probably be necessary to test, for example, Badger's rule [7] or to put the σ^* resonance correlation in soft x-ray absorption spectroscopy on a quantitative footing [17].

3.2 Ni{111}-PF$_3$

From the sixfold symmetric ESDIAD pattern Alvey and Yates [21] inferred that PF$_3$ adsorbs on the Ni{111} surface in atop sites via the P atom, as might be expected, and that at 85 K the molecule is azimuthally oriented so that the individual P-F bonds are directed over neighbouring Ni atoms. Since the barrier to rotation about the threefold axis is low (80 ± 20 cm^{-1} [39]), this azimuthal orientation disappears on warming the surface at low coverages to room temperature, but persists at higher coverages due to lateral interactions. The system has since been studied by photoelectron diffraction [40], SEXAFS [41] and the x-ray standing wave technique [41] to give a rather complete picture of the structure of the adsorption complex (fig. 3). The Ni-P bondlength is 2.07 Å and a 0.03 Å expansion of the outermost Ni-Ni layer spacing takes place. The P-F bondlength is 1.54 Å and the angle made by the P-F internuclear axis and the surface normal is 62°, showing that within the limits of the accuracy of the techniques employed there is no distortion compared to the free PF$_3$ molecule where the corresponding values are 1.535 Å and 62.2°, respectively. The polarisation dependence of the P 1s X-ray absorption spectrum has confirmed that the C$_3$ symmetry axis of the molecule is perpendicular to the surface [41].

In the analogous inorganic complex Ni(PF$_3$)$_4$, in which the central Ni atom is tetrahedrally co-ordinated, the ligands are also barely distorted relative to the free PF$_3$ molecule: the P-F bondlength is 1.561(± 0.003) Å and the angle between the P-F internuclear axis and the (former) C_3 axis 61.6° [42]. Even more interesting from the present point of view is the Ni-P bondlength of 2.099 (± 0.003) Å which is within three hundredths of an Å of the corresponding surface value. Thus, both Ni{111}-CO and Ni{111}-PF$_3$ not only show little change in the structural parameters of the adsorbate compared to the free molecule, but they also have metal-ligand bondlengths, which are very similar to those in the corresponding inorganic or organometallic systems. The comparison of the structural data for the two chemisorption systems produces, however, one small surprise: in view of the frequently discussed similarity in the metal-ligand bonding mechanism of CO and PF$_3$, e.g. Ref. [43], identical adsorption sites might have been expected. Vibrational spectroscopy suggests that atop adsorption of CO on Ni{111} is only observed at high coverages in the presence of CO in threefold hollow sites, although so far no quantitative structural information is available for this phase.

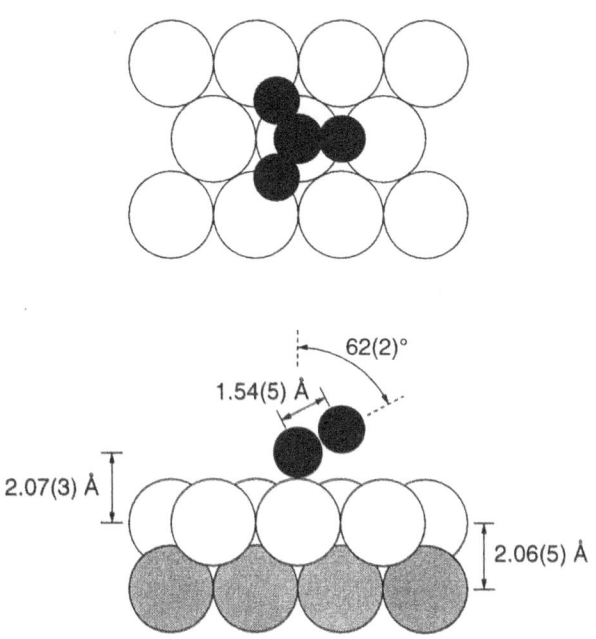

Figure 3. The structure of the adsorption complex formed by PF$_3$ on Ni{111}. After Alvey and Yates [21], Dippel et al [40] and Kerkar et al [41].

3.3 Cu{110}-OOCH and Cu{110}-OOCCH$_3$

The formate and acetate species are well-characterised surface intermediates in the catalytic decomposition of formic acid and acetic acid over copper surfaces. In particular, it is known from vibrational spectroscopy

[44,45] that in each case there are two equivalent C-O bonds and that the molecular plane is oriented perpendicular, or nearly perpendicular to the surface. Their structure has been investigated with both photoelectron diffraction and X-ray absorption spectroscopy on the <110> surface of copper [46,27]. The formate species is depicted in fig. 4. The molecular plane is indeed perpendicular to the surface and aligned in the <110> azimuth. A so-called aligned bridge site is occupied in which the two oxygen atoms are

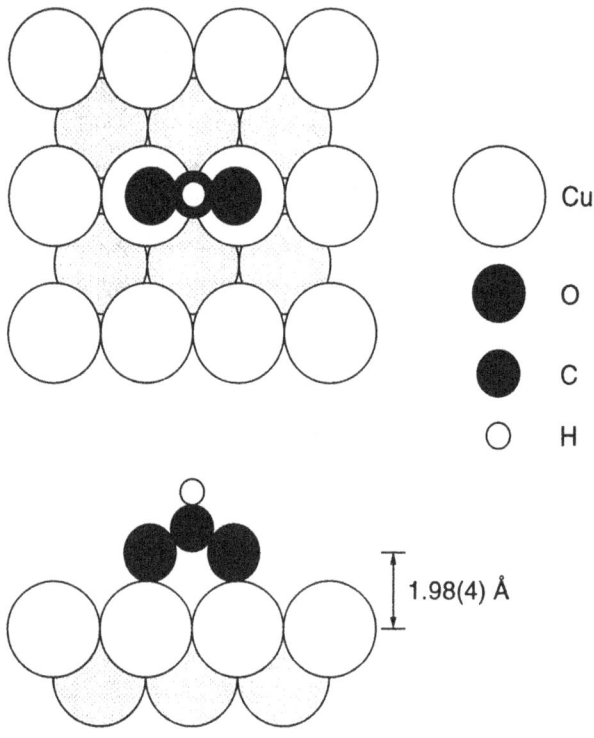

Figure 4. The structure of the surface formate species on Cu{110}. The position of the H atom is not known. After Woodruff et al. [46].

almost atop; the Cu-O nearest-neighbour distance is 1.98 ($\pm$ 0.04) Å. Assuming a C-O bondlength of 1.25 Å, the O-C-O bond angle is 135 ($\pm$ 5)°. (Since this was a rather early photoelectron diffraction experiment, the number of parameters determined was rather limited.) A similar aligned bridge site with almost identical structural parameters was found for the formate species on the Cu(100) surface [46]. As might perhaps be expected, the structure of the surface acetate species on Cu{110} is also very similar (fig. 5) with an oxygen-Cu nearest-neighbour bondlength of 1.91 ($\pm$ 0.04) Å [27]. In this case the chemical shift between the C 1s binding energies of the two carbons enables the position of each atom to be measured separately. This useful asset of the chemical shift in photoelectron diffraction may prove in future to be even more valuable in analysing the structure of different, co-adsorbed species on surfaces, e.g. molecular fragments resulting from a decomposition reaction [47].

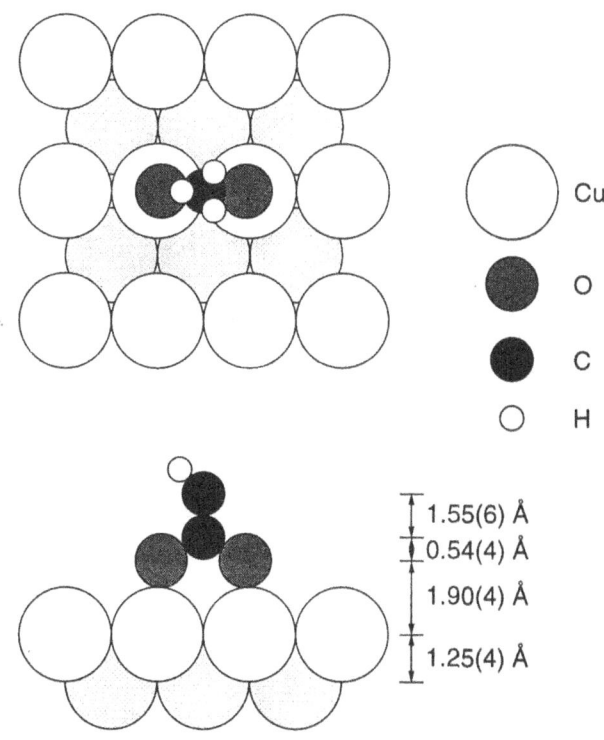

Figure 5. The structure of the surface acetate species on Cu{110}. The positions of the H atoms are not known. After Weiss et al [27].

There are several inorganic formates and acetates in which the ligand configuration is identical to the surface cases and the structural parameters are very similar. In the copper formate anion [(RCOO)$_4$Cu$_2$(NCS)$_2$] depicted in fig. 6 [48], where R = H, the Cu-Cu separation is slightly greater (2.716 ($\pm$ 0.002) Å) than for the nearest-neighbour distance on the surface (2.56 Å), but the Cu-O bondlength is identical (1.983 ($\pm$ 0.004) Å). The agreement is not quite so good for the corresponding acetate (R = CH$_3$): although the Cu-Cu separation is smaller at 2.643 ($\pm$ 0.003) Å (typical for dimeric cupric acetate complexes), the Cu-O bondlength (2.03 ($\pm$ 0.01) Å) is about one tenth of an Å greater than on the surface (1.91 ($\pm$ 0.02) Å). There is nonetheless impressive agreement between the inorganic and surface systems, despite the fact that the Cu-Cu separation in the complex probably precludes strong metal-metal bonding [48].

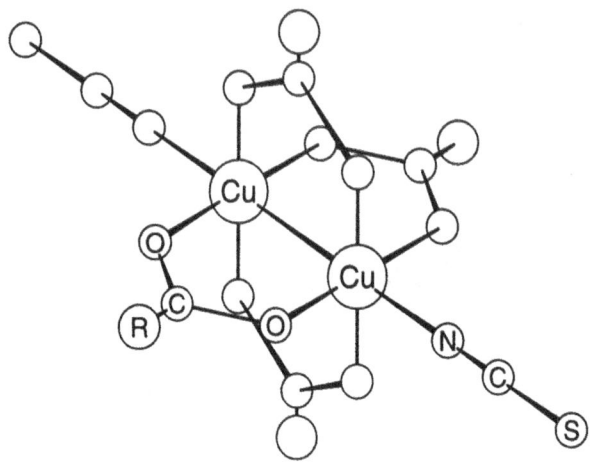

Figure 6. Schematic representation of the molecular anion
[(RCOO)$_4$Cu$_2$(NCS)$_2$], where R = H or CH$_3$.
After Goodgame *et al* [48].

3.4 Ni{111}-C$_2$H$_4$

Vibrational spectroscopy has been widely used to characterise the bonding of ethylene (ethene) to metal single crystal surfaces. Sheppard [49] has summarised much of this work and posed a classification scheme for the three different types of adsorption complex. Type I spectra are characterised by a symmetric C-H stretch in the range 2910 - 3000 cm^{-1} and a C-C stretch of the fully deuterated species which is distributed over two bands in the ranges 1100 - 1200 cm^{-1} and 850 - 950 cm^{-1}. Such a spectrum is assigned to a 1,2 di-σ or (μ-η^1, η^1-C$_2$H$_4$) species as found in the almost unique complex (C$_2$H$_4$)Os$_2$(CO)$_8$ [50-52]. Type II spectra are characterised by a symmetric C-H stretch in the range 2990 - 3075 cm^{-1} and a C-C stretch of the deuterated species in the range 1320 - 1420 cm^{-1}. Here, π bonding is thought to pertain as in the (η^2-C$_2$H$_4$) configuration of Zeise's salt K[(C$_2$H$_4$)PtCl$_3$] [53]. The bond order lies between 1.6 and 1.8, i.e., somewhat greater than in the Type I case. A third type of bonding with spectral characteristics intermediate between Types I and II has been identified and designated I' [49]. Recent comparisons with the complexes (C$_2$H$_4$)Os(CO)$_4$ and (C$_2$H$_4$)Fe(CO)$_4$ suggest a metallacyclopropane-type structure [54].

In a recent photoelectron diffraction study of C$_2$H$_4$ on Ni{111} the structural predictions of vibrational spectroscopy for this Type I system have been largely confirmed [55]. The molecular axis is parallel to the surface with the two carbon atoms in off-atop sites (fig. 7). The C-C bondlength is 1.60 (± 0.18) Å and the carbon-Ni first layer separation 1.90 (± 0.02) Å, giving a Ni-C bondlength of 1.95 Å. In addition, there is a surprisingly large Ni first/second layer expansion of 0.15 (± 0.10) Å. Because of the large error bars on the C-C bondlength it is difficult to compare this directly with the vibrational data which suggest a bond order approaching unity. A bondlength of 1.60 Å is actually greater than 1.54 Å in ethane (the C-C bondlength in

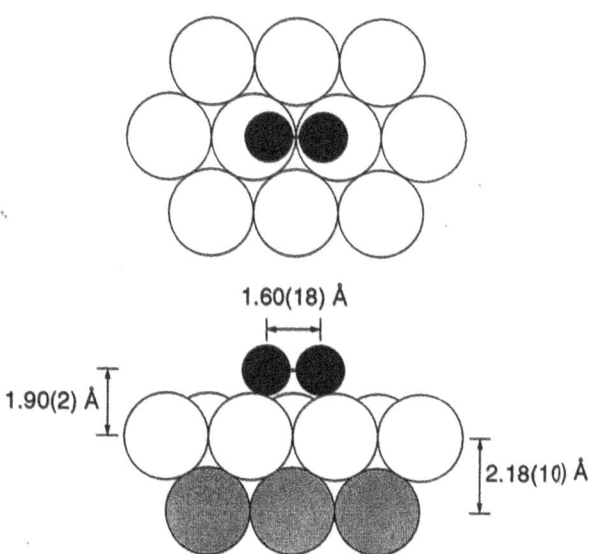

Figure 7. The structure of the adsorption complex formed by ethylene on Ni{111}. After Bao *et al* [55].

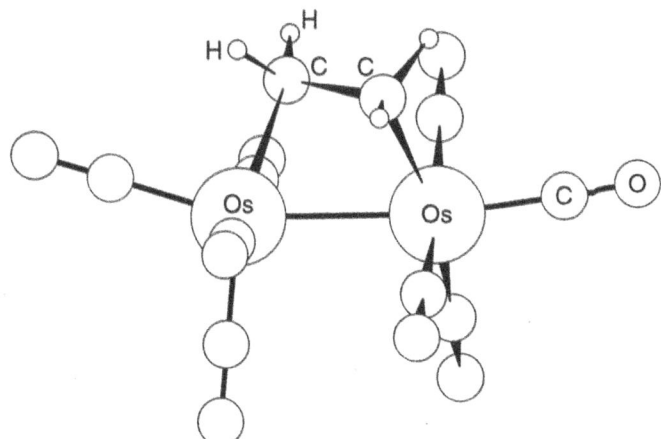

Figure 8. Schematic representation of the molecule $(C_2H_4)Os_2(CO)_8$. After Anderson *et al* [52].

in ethane (the C-C bondlength in ethylene is 1.34 Å). The positions of the hydrogen atoms are unknown, but due to the probable sp^3 hybridisation at the C atom, the molecule is certainly no longer planar.

In the complex $(C_2H_4)Os_2(CO)_8$ the C-C bondlength is 1.523 ($\pm$ 0.003) Å [52], but the local $(C_2H_4)Os_2$ symmetry is reduced from C_{2v} to C_2, in that the C-C axis is twisted by about 25° relative to the Os-Os axis (fig. 8). Although the C-C axis in the Ni{111}-C_2H_4 system appears to remain parallel to the Ni-Ni axis, some evidence for a twisted configuration has been found in the system Cu{110}-C_2H_4 [56].

3.5 Pt{111}-C_6H_6

A large number of experimental and theoretical investigations, particularly with photoelectron spectroscopy and EELS, have pointed to an adsorption geometry for benzene in which the molecular plane is parallel to the metal surface. Photoemission selection rules indicate that various adsorption sites can be occupied, giving rise to different local symmetries, e.g. Refs. [57,58]. There are, however, very few quantitative structural studies. Somorjai, Van Hove and co-workers have applied the LEED method to the ordered co-adsorption layers formed by benzene and CO on the {111} surfaces of Pd, Rh and Pt [59]. In the case of Rh{111}-C_6H_6/2CO, for example, the centre of the molecule is above the hcp threefold hollow site and a strong Kekulé distortion occurs. In the corresponding Pd system the distortion is less pronounced. Co-adsorption with CO apparently has the useful function of producing a higher degree of order in such overlayers, thus making structure determination easier with conventional LEED. Such data are harder to obtain for pure benzene layers, although the systems Pt{111}-C_6H_6 and Ni{111}-C_6H_6 have recently been studied with diffuse LEED [60] and photoelectron diffraction [56], respectively. The Pt case will be discussed in more detail here.

For disordered benzene on Pt{111} the molecule adsorbs such that its centre is above a twofold symmetric bridge site (fig. 9) [60]. Two C atoms, shown grey in fig. 9, lie almost directly over Pt atoms and are bent down slightly towards the surface to give short Pt-C bondlengths of 2.02 ($\pm$ 0.02) Å. The other four C atoms are in slightly non-symmetric bridge sites with Pt-C bondlengths of 2.58 ($\pm$ 0.02) Å and 2.61 ($\pm$ 0.02) Å. There is also a significant distortion giving rise to two C-C bondlengths of 1.63 ($\pm$ 0.05) Å and four of 1.45 ($\pm$ 0.10) Å, so that it is difficult to still refer to the molecule as adsorbed benzene. (In the free molecule the C-C bondlength is 1.397 Å.) This structure should be compared with the earlier result of the Somorjai group for benzene co-adsorbed with CO in the ordered array. The primary difference is a rotation of the molecule by 30° about the surface normal with its centre remaining above the bridge site. Essentially identical Pt-C bondlengths of 2.25 ($\pm$ 0.05) Å are associated with all six C atoms. The C-C bondlengths 1.65 (± 0.15) Å and 1.76 (± 0.15) Å) are greater than in the pure benzene layer, but no out-of-plane bending occurs.

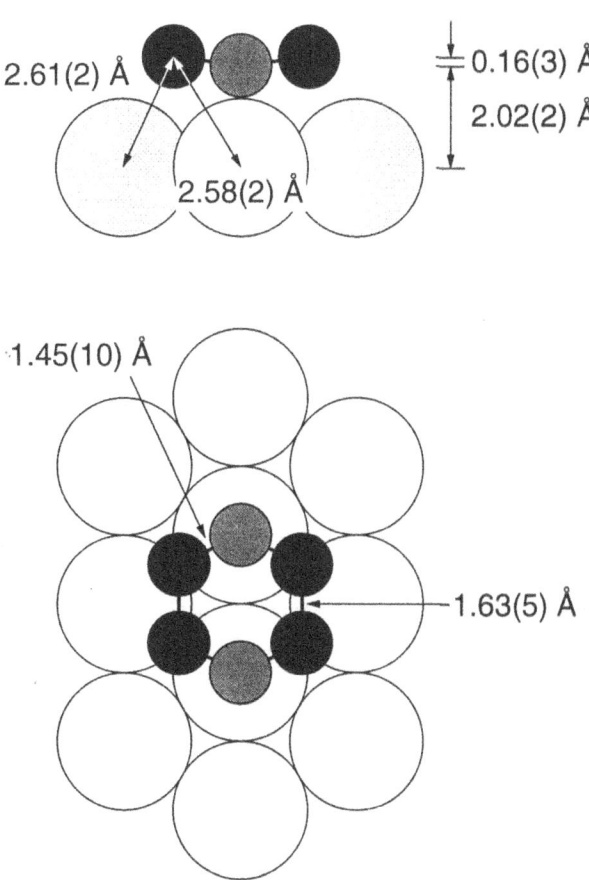

Figure 9. The structure of benzene adsorbed on Pt{111}.
After Wander *et al* [60].

In metal-arene complexes the η^6 bonding mode is usually found, although some bridging benzene ligands are known [61-63]. In the so-called face-bridging mode, for example, which is encountered in some multi-nuclear cluster compounds, the bonding geometry is characterised by an interaction between the hexagonal C atom ring and three metal atoms. This configuration is analogous to the adsorption complexes probably formed on several fcc(111) or hcp(0001) surfaces, although only in two cases - for the C_6H_6/CO co-adsorption layers on Pd{111} and Rh{111} - has a quantitative study been performed [64]. The bonding configuration most similar to that in the pure benzene layer on Pt{111} discussed here is probably that found in the dinuclear rhodium complex $(\eta_5\text{-}C_5H_5)^2Rh_2(\mu\text{-}\eta_3\text{:}\eta_3\text{-}C_6H_6)$ prepared by Müller and co-workers [63]. The μ_2-benzene ligand has a boat-like form (fig. 10) with an angle of 127° between the two η_3-enyl planes. Apart from the fact that the two apical C atoms are bent up away from the metal atoms, the main

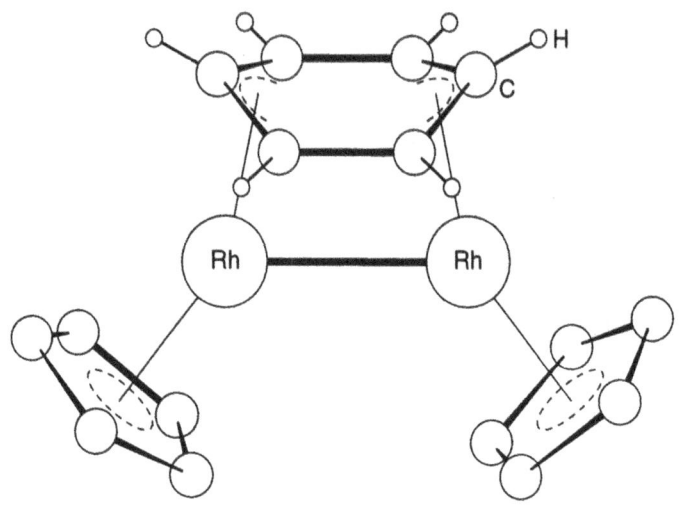

Figure 10. Schematic representation of the molecule $(Cp)_2Rh_2(C_6H_6)$. After Müller et al [62].

difference compared to the Pt adsorption complex is that there is very little distortion and the bondlengths are generally shorter. The two C-C bonds linking the enyl units are only marginally longer (1.46 Å) than the four C-C bonds within the enyl groups (1.42 Å). A similar small distortion is observed in the dinuclear ruthenium cluster $Ru_2(CO)_6(\mu\text{-}\eta_3\text{:}\eta_3\text{-}C_6H_{12})$ recently prepared by Johnson and co-workers [63]. This contains the same boat-like arene configuration; the corresponding bondlengths are 1.40 -1.44 Å and 1.48 - 1.49 Å. At the moment there too few examples of this kind of ligand and, indeed, insufficient surface structural data, to explore this question of distortion in any more detail. It should, however, be noted that not all the benzene adsorption systems for which data exist show the rather large overall increases in C-C bondlength.

4. Conclusions

This brief comparison between some molecular adsorption systems and their inorganic analogues shows that there is a surprising degree of similarity as far as the internal structure of the ligand and the metal-ligand bondlength are concerned. The data set is, however, certainly not large enough to establish any trends and there may still be many systems which do not conform to this simple picture. Considering the large number of papers in the past claiming to have established the surface-cluster analogy largely on the basis of qualitative evidence - at least from the surface side - the present conclusions may be regarded as a not inconsiderable success for surface spectroscopy and chemical intuition.

Acknowledgements

Useful discussions with L. J. Farrugia, G. Hess, Ph. Hofmann, J. Müller, O. Schaff, K.-M. Schindler, N. Sheppard and D. P. Woodruff are gratefully acknowledged. The photoelectron diffraction studies were performed by the author's group at the Berlin synchrotron radiation source BESSY in collaboration with D. P. Woodruff at the University of Warwick and V. Fritzsche at the Technical University Dresden. Financial support for this work was received from the German *Bundesministerium für Forschung und Technologie* under contract number 05 5EBFX 2 as well as from the SCIENCE and Human Capital and Mobility programmes of the European Union.

References

1.   Muetterties, E. L. (1975) *Bull. Soc. Chim. Belg.* **84**, 959; Muetterties, E. L. (1976) *Bull. Soc. Chem. Belg.* **85**, 451; Muetterties, E. L. and Wexler, R. M. (1983) *Surv. Prog. Chem.* **10**, 61.
2.   Eischens, R. P. and Pliskin, W. A. (1958) *Adv. Catalysis* 10, **1.**
3.   Blyholder, G. (1964) *J. Phys. Chem.* **68**, 2772.
4.   E.g. Thomas, G. E. and Weinberg, W. H. (1978) *Phys. Rev. Lett.* **41**, 1181.
5.   Bond, G. C. (1966) *Disc. Far. Soc.* **41**, 200.
6.   Morrow, B. A. and Sheppard, N. (1969) *Proc. Roy. Soc.* A **311**, 391.
7.   Ibach, H. and Mills, D. L. (1982) *Electron Energy Loss Spectroscopy and Surface Vibrations*, Academic Press, New York.
8.   Albert, M. R. and Yates, J. T. Jr. (1987) *The Surface Scientist's Guide to Organometallic Chemistry*, American Chemical Society, Washington DC.
9.   Kordesch, M. E., Stenzel, W. and Conrad, H. (1986) *J. Electron Spectrosc. Relat. Phenom.* **39**, 89; *Surface Sci.* **175**, L687.
10. Somers, J., Kordesch, M. E., Lindner, Th., Conrad, H., Bradshaw, A. M. and Williams, G. P. (1987) *Surface Sci.* **188**, 693; Kordesch, M. E., Lindner, Th., Somers, J., Stenzel, W., Conrad, H., Bradshaw, A. M., Williams, G. P. (1987) *Spectrochimica Acta* **43A**, 1561.

18

11. Aminpirooz, S., Schmalz, A. and Haase, J. (1992) *Phys. Rev. B* **45**, 6337; Schindler, K.-M., Hofmann, Ph., Weiß, K.-U., Dippel, R., Gardner, P., Fritzsche, V., Bradshaw, A. M., Woodruff, D. P., Davila, M. E., Asensio, M. C., Conesa, J. C. and González-Elipe, A. R. (1993) *J. Electron Spectrosc. Relat. Phenom.* **65/65**, 75.

12. Wander, A., Hu, P. and King, D. A. (1993) *Chem. Phys. Lett.* **201**, 393. A more recent photoelectron diffraction study (A. Locatelli *et al* (1994), *Phys. Rev. Lett.* **73**, 90) suggests, however, that for the system Pd(110) (2x1)-CO bridge sites are occupied after all.

13. Woodruff, D. P. and Delchar, T. A. (1994) *Modern Techniques of Surface Science*, 2nd edition, CUP, Cambridge.

14. Liang, S. and Trenary, M. (1988) *J. Chem. Phys.* **89**, 3320.

15. Bradshaw, A. M. (1990) in V. Bortolani, N. March and M. Tosi (eds), *Interaction of Atoms and Molecules with Solid Surfaces*, Plenum London, p. 477.

16. Ricken, D. E., Somers, J., Robinson, A. and Bradshaw, A. M. (1990) *Far. Disc. Chem. Soc.* **89**, 291.

17. Stöhr, J. (1992) *NEXAFS Spectroscopy*, Springer-Verlag, Berlin.

18. Aminpirooz, S., Becker, L., Hillert, B. and Haase, J. (1991) *Surface Sci.* **244**, L152.

19. Somers, J., Robinson, A. W., Lindner, Th., Ricken, D. and Bradshaw, A. M. (1989) *Phys. Rev. B* **40**, 2053.

20. Ramsier, R. D. and Yates, J. T, Jr. (1991) *Surface Sci. Rep.* **12**, 243; Madey, T. E. (1987) *Vacuum* **37**, 31.

21. Alvey, M. D. and Yates, J. T. (1988) *J. Am. Chem. Soc.* **110**, 1782.

22. Weiss, P. S. and Eigler, D. M. (1993) *Phys. Rev. Lett.* **71**, 3139.

23. Feidenhans'l, L. L. (1989) *Surface Sci. Rep.* **10**, 105.

24. Woodruff, D. P., Seymour, D. L., McConville, C. F., Riley, C. E., Crapper, M. D., Prince, N. P. and Jones, R. G. (1987) *Phys. Rev. Lett.* **58**, 1460.

25. Heinz, K., Starke, U. and Bothe, F. (1991) *Surface Sci.* **243**, L70.

26. Woodruff, D. P. and Bradshaw, A. M. (1994) *Rep. Progr. Phys.*, in press.

27. Weiss, K.-U., Dippel, R., Schindler, K.-M., Gardner, P., Fritzsche, V., Bradshaw, A. M., Kilcoyne, A. L. D. and Woodruff, D. P. (1992) *Phys. Rev. Lett.* **69**, 3196.

28. Haase, J. and Bradshaw, A. M. (1992) in R. Z. Bachrach (ed), *Synchrotron Radiation Research: Advances in Surface and Interface Science*, Plenum, New York, p. 55.

29. Pangher, N., Schmalz, A. and Haase, J. (1994) *Chem. Phys. Lett.* **221**, 189.

30. Surnev, L., Xu, Z. and Yates, J. T. (1988) *Surface Sci.* **202**, 1, 14.

31. Chen, J. G., Erley, W. and Ibach, H. (1989) *Surface Sci.* **223**, L891.

32. Sheppard, N. and Nguyen, N. T. (1978) *Adv. IR Raman Spect.* **5**, 67.

33. Bradshaw, A. M. and Hoffmann, F. (1978) *Surface Sci.* **72**, 513.

34. E.g. Bradshaw, A. M. and Schweizer, E. (1988) *Adv. Spect.* **23**, 413.

35. *Gmelin Handbuch der Anorganischen Chemie, Bd. 17 Nickel-Organische Verbindungen, Teil 2* (1974) Springer-Verlag, Berlin, S. 381.

36. De la Cruz, C. and Sheppard, N. (1990) *J. Molec. Structure* **224**, 141.

37. Corey, E. R., Dahl, L. F. and Beck, W. (1963) *J. Am. Chem. Soc.* **85**, 1202.

38. Farrugia, L. J., Jeffery, J. C., Marsden, C. and Stone, F. G. A. (1985) *J. Chem. Soc., Dalton Trans.,* 645.

39. Alvey, M. D., Yates, J. T. Jr. and Uram, K. J. (1987) *J. Chem. Phys.* **87**, 7221.

40. Dippel, R., Weiss, K.-U., Schindler, K.-M., Gardner, P., Fritzsche, V., Bradshaw, A. M., Asensio, M. C., Hu, X. M., Woodruff, D. P. and González-Elipe (1992) *Chem. Phys. Lett.* **199**, 625.

41. Kerkar, M., Woodruff, D. P., Avila, J., Asensio, M. C., Fernández-Garcia, M. and Conesa, J. C. (1992) *J. Phys. Cond. Matter* **4**, 6509.

42. Almenningen, A., Andersen, B. A. and Astrup, E. E. (1970) *Acta Chem. Scand.* **24**, 1579.

43. Nitschké, F., Ertl, G. and Küppers, J. (1981) *J. Chem. Phys.* **74**, 5911.

44. Hayden, B. E., Prince, K. C., Woodruff, D. P. and Bradshaw, A. M. (1983) *Surface Sci.* **133**, 589.

45. Sexton, B. A. (1979) *Chem. Phys. Lett.* **65**, 469; *Surface Sci.* **88**, 319.

46. Woodruff, D. P., McConville, C. F., Kilcoyne, A. L. D., Lindner, Th., Somers, J., Surman, M., Paolucci, G. and Bradshaw, A. M. (1988) *Surface Sci.* **201**, 228.

47. Weiss, K.-U., Dippel, R., Schindler, K.-M., Gardner, P., Fritzsche, V., Bradshaw, A. M., Woodruff, D. P., Asensio, M. C. and González-Elipe, A. R. (1993) *Phys. Rev. Lett.* **71**, 581.

48. Goodgame, D. M. L., Hill, N. J., Marsham, D. F., Skapski, A. C., Smart, M. L. and Troughton, P. G. H. (1969) *J. Chem. Soc. Chem. Comm.* 629.

49. Sheppard, N. (1988) *Ann. Rev. Phys. Chem.* **39**, 589.

50. Motyl, K. M., Norton, J. R., Schauer, C. K. and Anderson, O. P. (1982) *J. Am. Chem. Soc.* **104**, 7325.

51. Anson, C. E., Johnson, B. F. G., Lewis, J., Powell, D. B., Sheppard, N., Bhattacharyya, A. K., Bender, B. R., Bullock, R. M., Hembre, R. T. and Norton, J. R. (1989) *J. Chem. Soc., Chem. Commun.* 703.

52. Anderson, O. P., Bender, B. R., Norton, J. R., Larson, A. C. and Vergamini, P. J. (1991) *Organometallics* **10**, 3145.

53. Love, R., Koetzle, T. F., Williams, G. J. B., Andrews, L. C. and Bau, R. (1975) *Inorg. Chem.* **14**, 2653.

54. Anson, C. E., Sheppard, N., Powell, D. B., Bender, B. R. and Norton, J. R. (1994) *J. Chem. Soc. Faraday Trans.* **90**, 1449.

55. Bao, S., Hofmann, Ph., Schindler, K.-M., Fritzsche, V., Bradshaw, A. M., Woodruff, D. P., Casado, C. and Asensio, M. C. (1994) *J. Phys.: Cond. Matter* **6**, L93.

56. Schaff, O., Hofmann, Ph., Schindler, K.-M., Bradshaw, A. M., Davis, R. and Woodruff, D. P., to be published.

57. Netzer, F. P. , Graen, H. H., Kuhlenbeck, H. and Neumann, M. (1987) *Chem. Phys. Lett.* **133**, 49, and references therein.

58. Huber, W., Zebisch, P., Bornemann, T. and Steinrück, H.-P. (1991) *Surface Sci.* **258**, 16.

59. Lin, R. F., Blackman, G. S., Van Hove, M. A. and Somorjai, G. A. (1987) *Acta Crystallogr. B* **43**, 368.

60. Wander, A., Held, G., Hwang, R. Q., Blackman, G. S., Lu, M. L., de Andres, P., Hove, M. A. and Somorjai, G. A. (1991) *Surface Sci.* **249**, 21.

61. Wadepohl, H. (1992) *Angew. Chem.* **104**, 253.

62. Müller, J., Escarpa Gaede, P. and Qiao, K. (1993) *Angew. Chem.* **105**, 1809.

63. Blake. A. J., Dyson, P. J., Johnson, B. F. G. and Matin, C. M. (1994) *J. Chem. Soc., Chem. Commun.* 1471.

64. Ogletree, D. F., Van Hove, M. A. and Somorjai, G. A. (1987) *Surface Sci.* **183**, 1.

AN ATOMIC VIEW OF SURFACE DIFFUSION ON METAL SURFACES

G. L. Kellogg
Sandia National Laboratories
Albuquerque, NM 87185-0344

1. Introduction

The dynamical behavior of individual atoms on single-crystal surfaces is a subject of considerable current interest. A fundamental understanding of how atoms migrate across surfaces and interact with each other is essential to the development of realistic models of thin-film growth, surface alloying, and heterogeneous catalysis. In general, there are two experimental approaches to the study of surface diffusion. Details of the different methods are given in a recent review article by Gomer [1].The first involves observations of changes in a pre-established concentration gradient of the diffusing species as the system approaches equilibrium [2]. One measures the net transfer of material over a surface in a direction that decreases the free energy of the system. The diffusion coefficient is determined from an analysis based on Fick's second law. A wide range of surface analytical techniques may be employed for the measurements, which makes the method applicable to investigations of self-diffusion or adsorbate diffusion on surfaces of essentially any solid material. However, the extraction of quantitative diffusion parameters is difficult because the concentration of diffusing species changes as a function of time, and the effects of adatom-adatom interactions are generally unknown. There is also the possibility of surface defects that can have a significant effect on the transport rate. These problems can be avoided to a certain extent by keeping the net concentration fixed and measuring the time correlation of concentration fluctuations [3] , but adatom-adatom interactions may still be present.

The second approach involves direct measurements of the movements of an isolated atom on a single-crystal substrate [4]. Data analysis is simple in the extreme as the diffusion constant is determined directly from the rate of the atom's displacements. Adatom-adatom interactions are completely eliminated (unless one wishes to study them), and the substrate may be prepared to atomic perfection. The main limitation of this approach is that it requires an experimental probe that can resolve individual atoms on a surface. One such instrument is the scanning tunneling microscope (STM) [5], and recent investigations show considerable promise for obtaining quantitative information on single-atom surface diffusion [6]. However, the primary experimental tool used in single-atom surface diffusion studies is an instrument that pre-dates the STM by about 30 years — the field ion microscope

L. J. Farrugia (ed.), The Synergy Between Dynamics and Reactivity at Clusters and Surfaces, 21–35.
© 1995 Kluwer Academic Publishers.

(FIM) [7]. In fact, much of our current understanding of how individual atoms move and interact on single-crystal surfaces come from studies with the FIM [8].

It is interesting that the first FIM surface diffusion study was reported nearly 30 years ago in a landmark paper by Ehrlich and Hudda [4]. In the intervening years researchers have applied the experimental methods developed in these early studies to a variety of adatom-substrate combinations — mostly metals on metals. The investigations have expanded from studies of single-atom surface diffusion to more complicated processes involving various atom-atom and atom-defect interactions (cluster nucleation is an example). Often, these studies have reinforced our intuitive notions about atomic-level processes at surfaces, i.e., that atom motion takes place by a series of hops between adjacent binding sites on a surface and that cluster nucleation can be explained by simple pair interactions between neighboring atoms. More recently, however, FIM studies have yielded results counter to these intuitive pictures. The discovery of surface diffusion by exchange-mediated mechanisms [9-12] and the observation of atomic-chain configurations during the growth of small clusters [13-16], for example, have forced us to take a closer look at the physics underlying surface diffusion and cluster nucleation. In some cases it makes more sense to view surface diffusion and cluster nucleation in terms of concepts derived from chemistry — concerted reactions, bond making and bond breaking, etc. [17]. It is upon these more recent, unexpected observations that I will concentrate in this review of surface diffusion on metal surfaces.

2. Experimental Procedures

2.1 OBSERVATIONAL METHODS

The experimental procedure used in FIM surface diffusion studies is illustrated by the series of field ion micrographs shown in Fig. 1. The photographs show direct, real-space images of a Rh surface in the region of a (311) plane. The nearly circular ring of spots is an image of the atoms at the plane's edge. Due to the nature of the image formation process in the FIM, only the edge atom of the low-index planes are imaged, the interior of the plane appears uniformly dark. Just beyond the plane edge there is a step down in all directions. The high-contrast spot within the circular region shown in Fig. 1(a) is an Ir adatom that has been vapor deposited on to the top of the plane. The (311) surface of an fcc metal is a corrugated surface consisting of close-packed rows of atoms with intervening surface channels. Analysis of the position of the adatom with respect to the edge atoms indicate that the Ir adatom lies within one of the surface channels.

The FIM images shown in Fig. 1 are recorded with the sample at 77 K. At this low temperature the Ir adatom and all of the Rh surface atoms are immobile. To study diffusion of the adatom, the applied electric field used to image the surface is turned off and the sample temperature is increased for a fixed interval of time. At the end of the heating interval, the sample is re-cooled to the base temperature of 77 K and another image is recorded. Between each photograph shown in Fig. 1(a)-(c), the sample is warmed to a temperature of 180 K for 30 seconds The Rh substrate atoms are still immobile at this temperature, but the Ir adatom moves from one side of the plane to the other. Continued observations like those in Fig. 1(a)-(c) indicate that the adatom performs a one-dimensional random walk with the edge of the plane acting as a reflecting boundary. The presence of a reflection boundary, which is present for many (but not all) adatom-substrate combinations, allows one to make tens to hundreds of observations of the adatom as it diffuses back and forth across the surface [8] .

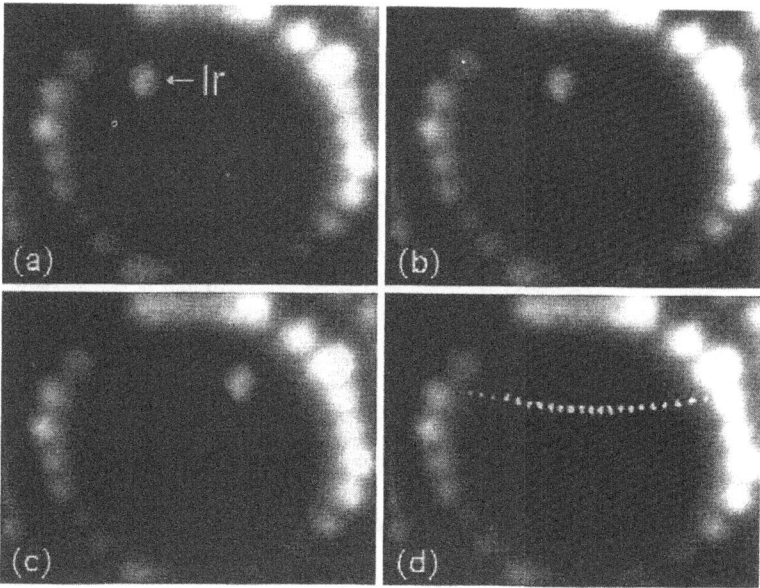

Figure 1 Field ion micrographs illustrating the experimental procedures used in FIM studies of single-atom surface diffusion

The one-dimensional nature of the diffusion is made clear by the photograph shown in Fig. 1(d). This image is from the same Rh(311) plane, but with the Ir adatom removed. Superimposed on the image are small white dots corresponding to the coordinates of the adatom after each of 300 diffusion intervals. The discrete clusters of dots seen in Fig. 1(d) correspond to the individual binding sites that the atom visits as it moves across the surface. The sites clearly fall on the same line indicating that the diffusing adatom remains within a given surface channel of the (311) plane. The deviation from a perfect straight line is due to the non-uniform magnification of the FIM [18] . Binding-site maps such as the one shown in Fig. 1(d) are very useful in that they allow one to calibrate the distance scale on the FIM image exactly at the position where the atom is moving. They are also used to identify the mechanism of surface diffusion on certain surfaces as discussed in section 3.2.

2.2 DATA ANALYSIS

From observations such as those shown in Fig. 1, one can measure the length of the adatom's displacements during each heating interval. The actual number of jumps that the adatom makes in a given interval is not known — only its net displacement. Thus, in FIM surface diffusion studies data analysis in not based on actual displacement rates, but rather on the mean-square displacement of the adatom during the diffusion interval. These values are typically determined at a series of temperatures, limited by the onset of adatom motion (lower limit) and the frequent loss of adatoms off the plane (upper limit). The equation that relates the measurable parameters (i.e., the mean-square displacement and the temperature) to the quantities of interest (i.e., the activation energy of surface diffusion and atomic vibrational frequency) is

$$<x^2>/2\tau = D_0 \exp(-E_d/kT), \tag{1}$$

where $<x^2>$ is the mean-square displacement in one dimension, τ is the diffusion time interval, E_d is the activation energy for surface diffusion, k is Boltzmann's constant, and T is the temperature. D_0 is a prefactor term that is roughly equal to the atomic vibrational frequency times the jump length. Derivations of equation (1) along with detailed discussions of the contribution of entropy effects to D_0, corrections due to the reflecting boundaries, the presence of driving forces, etc. may be found in a number of more extensive review articles [8] .

To obtain the diffusion parameters, one simply makes a standard Arrhenius plot of ln $<x^2>/2\tau$ vs. 1/T and determines the activation energy and prefactor from the slope and intercept, respectively. On surfaces where diffusion is not restricted to one-dimension, the analysis is essentially the same as above except that $<x^2>$ in equation (1) is replaced by $<r^2>=<x^2>+<y^2>$. In either case the accuracy of the activation energy measurement is about ±5%. The Arrhenius prefactor is reliable only to within an order of magnitude. From measurements taken over a large variety of adatom-substrate combinations, it is now generally agreed that the prefactor is essentially invariant from system to system and is given by $\sim 10^{-3}$ cm^2/sec [19] . It is therefore possible to calculate the activation energy for surface diffusion from measurements of the mean-square displacement at a single value of the temperature. For some adatom-substrate combinations this is necessary because the temperature range over which the experiments can be performed is too small to carry out a full Arrhenius analysis.

3. Selected Experimental Results

3.1 CONVENTIONAL HOPPING DIFFUSION

Most of the early FIM investigations of single-atom surface diffusion were concerned with metal adatoms migrating on the (110) plane of W, primarily because the cleaning and imaging procedures for W were well established. Over the years the data base for diffusion on W(110) has been extended to other adatoms — even silicon [20] . Although a detailed discussion of these results is beyond the scope of this article, the measured activation energies of surface diffusion for various adatoms on the W(110) plane provide a nice summary of the general trends associated with single-atom diffusion on metal surfaces. In Fig. 2 the activation energies of surface diffusion for nine elements are plotted against the bulk sublimation energies of the adatoms. A better comparison would be to plot the activation energies of surface diffusion against the binding energies of the adatoms on W(110), but these values have not all been measured. Where they have, the binding energies are found to be of the same order of magnitude as the sublimation energies (several eV) and follow the same trend of increasing energy with increasing melting point [21] . The plot in Fig. 2 thus demonstrates two of the "rules of thumb" associated with diffusion on metal surfaces: (1) surface diffusion activation energies are about 10% of adatom binding energies and (2) on a given substrate, the activation energy for surface diffusion is higher for the higher melting point elements.

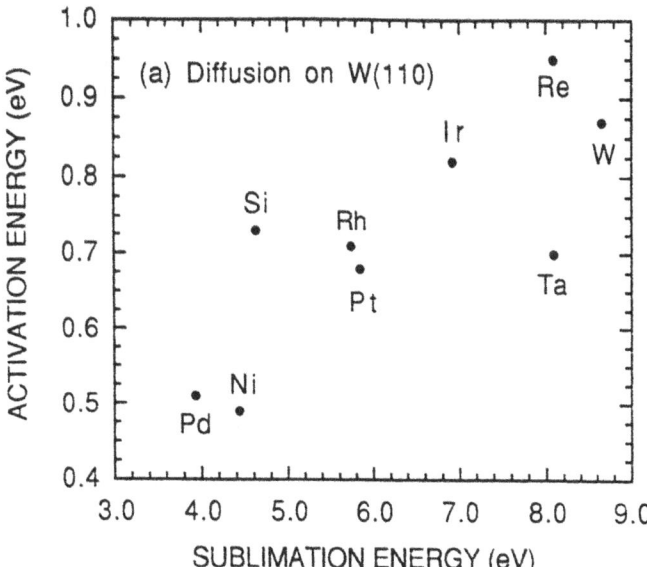

Figure 2 Activation energies of surface diffusion for single atoms on the W(110) plane plotted as a function of the bulk sublimation energy of the diffusing species. References to the original measurements can be found in recent review articles [8]

The scaling of the activation energies of surface diffusion with binding energies fits well with the intuitive notion that an adatom migrates across a surface by a series of hops between adjacent binding sites. An adatom more tightly bound to the surface resides in a deeper potential well and must overcome a higher potential barrier to diffuse. This simple picture of adatom diffusion is further reinforced by measurements of diffusion barriers as a function of surface atomic structure. For both Rh and Pt adatoms migrating on various planes of Rh, the activation energy is quite low on the smooth (111) plane and much higher on the more open (100) plane [22,23]. On the channeled surfaces such as (110), (311), and (331) planes, the motion is confined to one dimension, i.e., along the direction of minimum corrugation. The physical picture of a marble (adatom) rolling over a solid array of marbles (surface atoms) thus appears to adequately describe the results of FIM studies of adatom diffusion on surfaces of W and Rh.

3.2 EXCHANGE-MEDIATED DIFFUSION

Probably the biggest surprise to result from FIM studies of surface diffusion was the discovery that adatoms may migrate across a surface by a mechanism other than conventional hopping. FIM experiments show conclusively that for certain adatom-substrate combinations atom motion may proceed by an "exchange" or "replacement" process. Instead of hopping across the surface, the adatom finds it energetically favorable to push a neighboring atom out of the substrate plane and take its place in the top layer of surface atoms. In the case of self-diffusion (adatom and substrate atoms are the same chemical species) the displaced atom becomes a new adatom and continues the migration process. If the adatom and substrate atoms are different chemical species, then the initial exchange displacement produces an adatom of the substrate element, and diffusion of the original adatom cannot be studied. In the latter case the exchange process leads to surface alloying rather than surface diffusion. As shown below, FIM experiments have confirmed the existence of exchange displacements for both surface self-diffusion and surface alloying.

The first atomic-level, experimental evidence for exchange displacements was reported in 1978 by Bassett and Webber [9]. In FIM studies of self-diffusion on Pt(110), they observed displacements of the Pt adatom both along the surface channels and across the channel walls. In fact, the activation barriers for motion in the two directions were found to be the same within the experimental errors. The observation of cross-channel displacements is totally inconsistent with the conventional picture of metal-on-metal surface diffusion. To explain this unexpected result, they suggested that cross-channel displacements on Pt(110) are due to an exchange process between the adatom and a channel-wall atom. That such an exchange processes actually take place on fcc(110) surfaces was subsequently confirmed in an investigation of W adatoms on the Ir(110) surface using an atom-probe (i.e., a time-of-flight mass spectrometer attached to an FIM). Identification of the adatom appearing after a cross-channel movement provided convincing evidence that an exchange process was involved in the displacement process [10] . More recently, a similar confirmation of exchange processes for Pt atoms on the Ni(110) surface has been reported [24]. These studies have led to the reasonable assumption that all cross-channel displacements on fcc(110) surfaces (including self-diffusion) are due to exchange processes.

For about ten years following the initial work of Bassett and Webber, it was assumed that exchange displacements occurred only on the channeled (110) surfaces of a few fcc metals. The relatively low coordination of atoms in the channel walls was believed to make the exchange process energetically accessible. As a result, the discovery was considered to be fundamentally interesting, but not of much general importance. A considerable amount of renewed interest in exchange diffusion was generated in 1990 when it was shown that a similar exchange process is also energetically favorable on the non-corrugated (100) surfaces of fcc metals. The detailed mechanism involved in exchange displacements on the fcc(100) plane was first proposed by Feibelman [17] based on first-principles calculations for self-diffusion on Al(100). Experimental confirmation of the exchange mechanism came shortly thereafter from FIM studies of self-diffusion on the (100) surfaces of Pt [11] and Ir [12] .

The exchange mechanism for adatoms on fcc(100) surfaces is shown schematically in Fig. 3. An adatom in a four-fold hollow begins the displacement by moving down towards a surface atom. At the same time, the surface atom begins to move up out of its lattice position. At the saddle point of the transition, the adatom and the displaced surface atom have identical bonding configurations above the surface with a vacancy below. The displacement ends when the adatom fills the vacancy and the surface adatom becomes a new adatom at a next-nearest-neighbor position with respect to the original adatom. The calculations predict that the activation barrier for the exchange process is about a factor of three smaller than that for hopping over the two-fold bridge site [17] . The reason for the lower barrier is that the adatom can maintain three "bonds" with neighboring atoms in the concerted exchange process. This bonding configuration is particularly attractive for Al which is trivalent. To hop over a bridge site, the adatom must break one of its bonds with the surface atoms.

Experimental verification of the proposed exchange mechanism relies on the fact that the directions of individual displacements are different for the exchange vs. hopping processes. As shown in Fig. 3, exchange displacements involve the concerted motion of an adatom and a surface atom along the [100] and [010] directions, i.e., displacements to next-nearest neighbor positions. Hopping displacements involve motion along the [110] and equivalent directions, i.e., displacements to nearest neighbor positions. If diffusion proceeds exclusively by exchange displacements, then the adatom can visit only half of the available surface sites. The map of sites visited by the adatom is a square pattern with sides parallel to the [100] and [010] directions. In the notation of surface overlayers, the map has c(2x2) periodicity. The situation is analogous to playing the game of checkers in which the game piece is permitted to move only in diagonal directions and is therefore confined to squares of the same color. If diffusion proceeds by ordinary hopping, then all of the surface sites may be visited. The map of visited sites for hopping displacements is a (1x1) square pattern with sides parallel to the [110]-type directions.

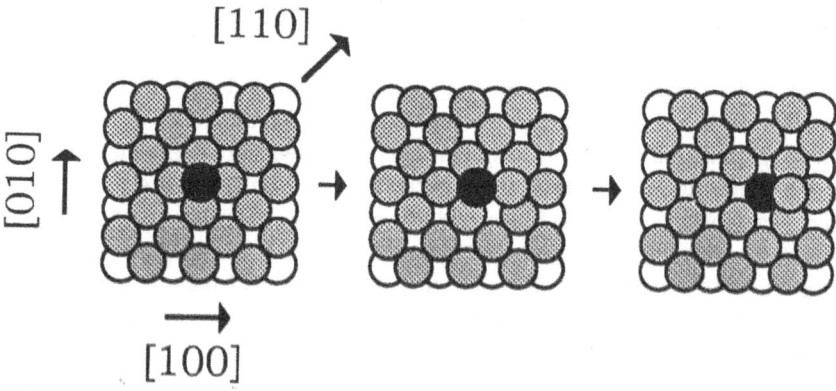

Figure 3 A schematic illustration of the exchange mechanism on fcc(100) surfaces. The shaded circles represent surface atoms of top the (100) layer. The filled circle represents an adatom which becomes incorporated into the surface layer as a result of the exchange process.

Field ion microscope images showing the diffusion of a Pt adatom on a Pt(100) plane along with the resulting site-visitation map are presented in Fig. 4. The first three photographs show the position of the adatom after 30-sec. diffusion intervals at a temperature of 175 K. A small white dot is superimposed on the images to indicate the adatom's position in the current and previous images. Motion of the adatom is evident from the location of the dots in Fig. 4(b)-(c). With continued heating cycles the site-visitation map fills in and eventually produces the pattern shown in Fig. 4(d). This photograph was taken after 300 diffusion intervals. The orientation of the crystallographic directions is obtained from the symmetry of the overall field ion image. The map of sites is clearly a square pattern with sides parallel to the [100]-type directions. This result [11] and similar observations on Ir(100) [12] provides compelling (albeit indirect) evidence that self-diffusion on the (100) surfaces of Pt and Ir takes place by exchange displacements.

A more direct confirmation of the exchange mechanism on fcc(100) surfaces is obtained from FIM studies of Ir adatoms on the (100) plane of Rh [25] . Examples of FIM images from this experiment are shown in Fig. 5. The bright spot near the center of the (100) plane shown in Fig. 5(a) is an Ir atom, identified by its high desorption field (i.e., the electric field required to remove the atom from the surface). Whereas Ir adatoms resist field desorption at electric fields up to the substrate evaporation field (see below), Rh adatoms desorb at about 75% of this value. The same Rh(100) surface is shown in Fig. 5(b) following a 30-sec. heating interval at 330 K. An adatom is present at about the same position as the one in Fig. 5(a), but a careful analysis of the two images indicates that the adatom has made a small displacement. More importantly, the adatom in Fig. 5(b) desorbs from the surface at an electric field strength corresponding to Rh, not Ir. This indicates that an exchange process has occurred producing a Rh adatom and embedding the Ir adatom Rh surface layer.

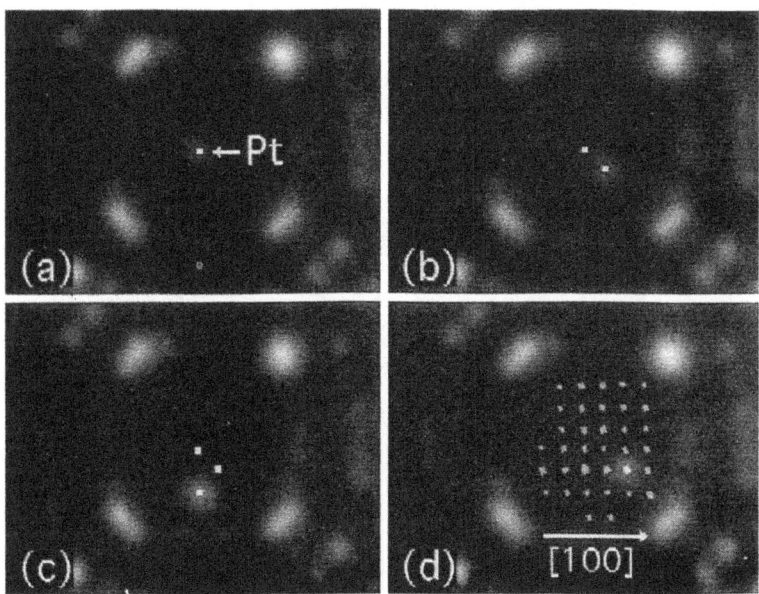

Figure 4   A series of field ion micrographs illustrating the confirmation of the exchange mechanism for self-diffusion on Pt(100) as discussed in the text.

The Rh(100) plane with the Rh adatom removed is shown in Fig. 5(c). The Ir atom, now embedded in the top layer of surface atoms, is not visible because it does not protrude out from the surface. However, the Ir-Rh exchange process may be further confirmed by field evaporation of the top layer of Rh(100) surface atoms. Field evaporation is the removal of the substrate surface atoms by a high electric field. In the region of the low-index crystal planes, this process removes atoms from the edge of the plane inward in a controlled, layer-by-layer fashion. Figs. 5(d)-(f) show the Rh(100) surface at successive stages of field evaporation. It is clear in Fig. 5(f) that when the Rh(100) top-layer atoms are removed, an individual adatom remains on the surface. From its resistance to field desorption, this adatom is identified as Ir — the same Ir adatom that embedded itself into the surface in the exchange process. The sequence shown in Fig. 5 can be repeated over and over. Each time the surface is heated, a Rh adatom is generated and each time the Rh(100) surface is field evaporated, the Ir adatom reappears.

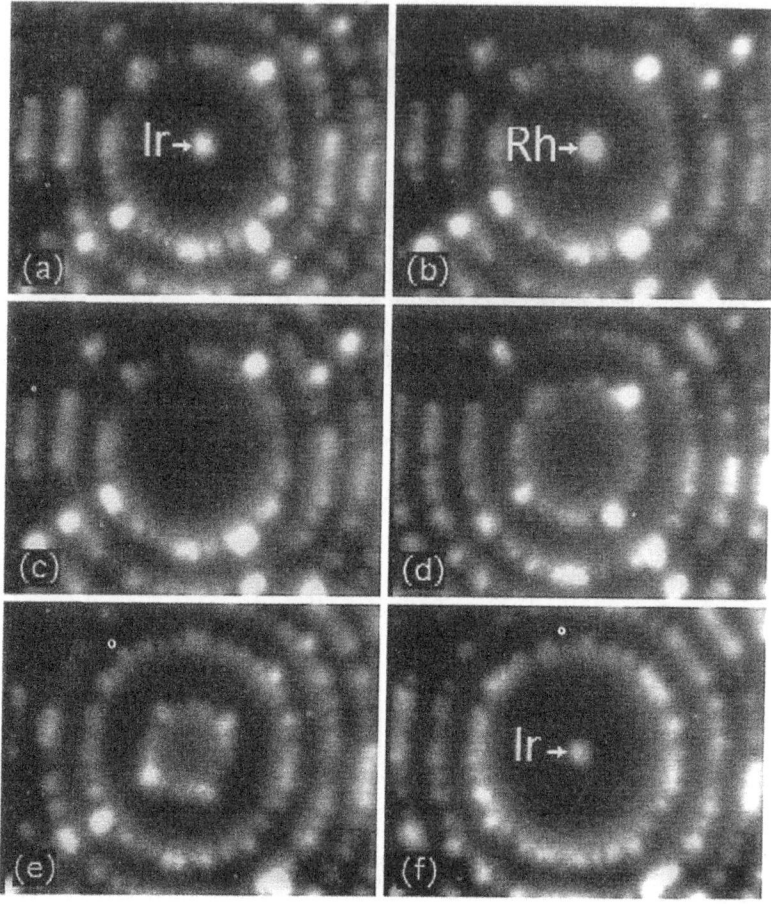

Figure 5 Field ion micrographs illustrating the exchange mechanism for an Ir
adatom on a Rh(100) plane. From ref. [23] .

The general agreement between the theoretical predictions of a lower activatic
barrier for exchange processes [17] and the above experiments provide compellir
evidence that exchange displacements are energetically favorable over hoppir
displacements on certain fcc(100) surfaces. In addition to is fundamental significanc
as a new means of atom transport and surface alloying, the discovery of the exchang
mechanism has interesting practical implications. For example, it has been suggeste
that one may be able to produce patterned, buried interfaces with atomic-lay
dimensions using the exchange process [26]. On the negative side, exchang
displacements may have a deleterious effect in the fabrication of layered materia
such as x-ray optics where atomically sharp interfaces are desired.   In rece

investigations of epitaxial growth on the (100) surfaces of Au, Ag, and Cu (some related to thin-film magnetism) exchange displacements are suspected to be important in defining the nature of the growth process and controlling the resulting morphology of the film [27-31] .

In view of these fundamental and practical implications, it is important to establish how commonly exchange displacements occur. Additional FIM studies indicate that exchange displacements are preferred for some adatom-substrate combinations and hopping displacements for others. Exchange displacements also occur in the migration of small clusters of Pt on Pt(100) [32] . However, the systems for which exchange displacements occur do not seem to be correlated with macroscopic bulk or surface properties of the materials [23] . There is some evidence that the degree of surface relaxations induced by the adatom is correlated with the propensity for exchange [33] . This possibility is consistent with the effect of an external electric field on the rate of exchange displacements [34] , but more systems need to be examined. The current challenge is to define the laws that govern whether adatoms on a given substrate will move by exchange or hopping displacements. Further experimentation and more extensive theoretical modeling are currently underway to address this challenge.

3.3 ADATOM-ADATOM INTERACTIONS AND CLUSTER NUCLEATION

The single-atom surface diffusion studies discussed above represent only a small fraction of the contributions made by the FIM to our understanding of atomic processes on surfaces. Similar methods have been used to investigate the diffusion behavior of two or more atoms on a crystal plane. Detailed measurements of the pair distribution functions of interacting adatoms and dissociation lifetimes of small clusters have produced quantitative values for the interatomic potentials between adatoms on surfaces. The interaction of diffusing atoms with steps and other surface defects has also been studied extensively with the FIM. Often, the interactions energies are found to be non-monotonic or oscillatory in nature, suggesting the presence long-range electronic or elastic interactions on surfaces. The interested reader is referred to several review articles for detailed discussion of these investigations [8] .

In this brief review I discuss a somewhat different approach used in FIM investigations of adatom-adatom interactions on surfaces. As adatoms nucleate into small clusters, FIM observations show that the stable configuration of atoms is not always the expected two-dimensional island. For some systems the clusters are found to be more stable as one-dimensional chains of atoms [13,14,35] . A particularly interesting example is the nucleation of Ir clusters on the Ir(100) plane [15] . FIM studies show conclusively that clusters consisting of five atoms or less are more stable as one-dimensional chains than two dimensional islands. For six atoms or more, the situation is reversed. Pt clusters on the Rh(100) surface exhibit a similar behavior. A simple lattice gas model indicates that in order to explain these configurational stabilities, interactions extending beyond second-nearest neighbors, many body interactions, or both must be invoked [36]. This is an important observation as most simulations of growth still assume that adatom-adatom interactions can be modeled by pair interactions between neighboring atoms.

Even more interesting is the system of Pt on Pt(100). Here, the stable configuration oscillates between a chain and an island as the cluster increases in size from three to six atoms [16]. This behavior is observed in FIM experiments and total energy calculations using embedded atom method (EAM) potentials. Interestingly, the calculations give the correct result only when the substrate surface atoms are allowed to relax in the presence of the cluster atoms. Thus, the response of the substrate atoms to the presence of the adatoms plays an important role in cluster nucleation processes — the surface atoms are not simply a static template on which the adatoms move.

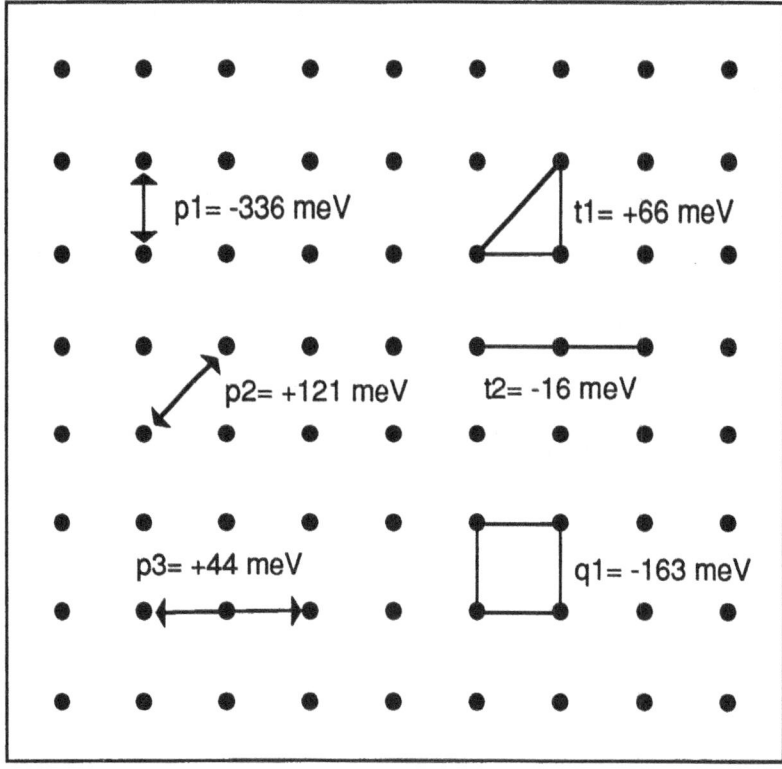

Figure 6 Adatom-adatom interactions energies for Pt on Pt(100) as determined from embedded atom method calculations [16]. p1, 2, etc. are pair energies between atoms at the distances indicated; t1 and t2 are triplet interactions as indicated; and q1 is the quadruplet interaction for four atoms in a square.

The agreement between the total energy calculations and the experimental observations in predicting the configurational stabilities also gives one confidence that the adatom-adatom interaction energies, which can be extracted from the calculations, are valid. The actual values are shown in Fig. 6. Note that pair interactions between third-nearest neighbors and many-body interactions (up to four-body) contribute to the total energy of the cluster. The oscillatory nature of the configurational stability can be rationalized by these interaction energies. Chain structures for three and five atoms result from a rather large repulsive interaction between next-nearest neighbors and a large repulsive three-body interaction for atoms in the triangular configuration. The island structure is stabilized for clusters of four, six, and more atoms by the surprisingly large four-body interaction for atoms in a square.

It should be noted that one may obtain adatom-adatom interaction energies from direct measurements of the diffusion behavior of two or more atoms on a surface, as mentioned above. However, the number of observations required to obtain statistically significant results for even two atoms on a surface already taxes current data gathering methods. To probe three-body or higher-order interactions is beyond present day capabilities. Examination of configurational stabilities combined with theoretical calculations thus provides an alternative approach to investigate the details of adatom-adatom interactions during cluster nucleation and allows one to identify the important physical phenomena (lattice relaxations, many-body interactions, etc.) involved in the nucleation process.

4. Summary

Investigations of surface diffusion and cluster nucleation by field ion microscopy have provided a considerable amount of physical insight concerning the fundamental interactions that control dynamical processes on surfaces. The investigations rely not only on the FIM's ability to resolve and track individual atoms on a surface, but also its ability to manipulate the number of adatoms and the size of clusters by the process of field desorption. The results of the investigations are often quite surprising. Whereas metal atom diffusion was once thought to be a simple, and perhaps uninteresting hopping process, FIM experiments have revealed new mechanisms for atom transport. Whereas cluster nucleation was once thought to be an aggregation process dependent only upon pairwise interactions between atoms, FIM investigations have shown that long-range and many body interactions can make non-negligible contributions to the overall process. By providing a brief overview of the experimental methods used in FIM surface diffusion studies and discussing a few selected applications, I hope to have conveyed some of the rich history as well as the current excitement associated with FIM investigations of dynamical processes on surfaces.

34

5. Acknowledgment

Work performed at Sandia National Laboratories was supported by the U. S. Department of Energy under contract no. DE-AC04-94AL85000.

6. References

1. Gomer, R. (1990), *Rep. Prog. Phys.* **53**, 917-1002.
2. Blakely, J. M. (1963), *Progress in Materials Science* **10**, 395-437.
3. Gomer, R. (1973), *Surf. Sci.* **38**, 373-393.
4. Ehrlich, G. and Hudda, F. G. (1966) *J. Chem. Phys.* **44**, 1039-1049.
5. Binnig, G., Rohrer, H., Gerber, C., and Weibel, E. (1982), *Phys. Rev. Lett.* **49**, 57-61.
6. Lagally, M. G. (1993) *Physics Today* **46**, 24-31 and references therein.
7. Müller, E. W. (1951) Das Feldionenmikroskop, *Z. Phys.* **131**, 136-142.
8. For recent reviews, see:(a) Bassett, D. W. (1983) I. Single Atoms. in V. T. Binh (ed.), *Surface Mobilities on Solid Materials* , Plenum, New York, pp. 63-82. (b) Tsong, T.T. (1990) *Atom-Probe Field Ion Microscopy* , Cambridge University Press, Cambridge, pp. 202-265. (c) Ehrlich, G. (1990), *Scanning Microscopy* **4**, 829-842. (d) Kellogg G.L. (1994), *Surf. Sci. Rep.* (in press).
9. Bassett, D. W. and Webber, P. R. (1978), *Surf. Sci.* **70**, 520-531.
10. Wrigley, J. D. and Ehrlich, G. (1980) *Phys. Rev. Lett.* **44**, 661-663.
11. Kellogg, G. L. and Feibelman, P. J. (1990) *Phys. Rev. Lett.* **64**, 3143-3146.
12. Chen, C. L. and Tsong, T. T. (1990), *Phys. Rev. Lett.* **64**, 3147-3150.
13. Bassett, D. W. (1978), *Thin Solid Films* **48**, 237-246.
14. Fink, H.-W. and Ehrlich, G. (1981) *Surf. Sci.* **110**, L611-L614.
15. Schwoebel, P. R. and Kellogg, G. L. (1988) *Phys. Rev. Lett.* **61**, 578-580.
16. Schwoebel, P. R., Foiles, S. M., Bisson, C. L., and Kellogg, G. L. (1989) *Phys. Rev. B* **40**, 10639-10642.
17. Feibelman, P. J. (1990) *Phys. Rev. Lett.* **65**, 729-732.
18. Müller, E. W. and Tsong, T. T. (1969) *Field Ion Microscopy, Principles and Applications* American Elsevier, New York.
19. Wang, S. C. and Ehrlich, G. (1988) *Surf. Sci.* **206**, 451-474.
20. Casanova, R. and Tsong, T. T. (1981) *Surf. Sci.* **109**, L497-L503.
21. (a) Berlowitz, P. J. and Goodman, D. W. (1987) *Surf. Sci.* **187**, 463-480. (b) Berlowitz,P.J. and Goodman,D.W. (1988) *Langmuir* **4**, 1091-1095. (c) Schlenk,W. and Bauer, E. (1980) *Surf. Sci.* **93**, 9-32.
22. Ayrault, G. and Ehrlich, G. (1974) *J. Chem. Phys.* **60**, 281-294.
23. Kellogg, G. L. (1993) *Phys. Rev. B* **48**, 11305-11312.
24. Kellogg, G. L. (1991) *Phys. Rev. Lett.* **67**, 216-219.
25. Kellogg, G. L. (1994) *Phys. Rev. Lett.* **72**, 1662-1665.
26. Tsong, T. T. (1991) *Phys. Rev. B* **44**, 13703-13710.

27. Schmitz, P. J., Leung, W.-Y., Graham, G. W., and Thiel, P. A. (1989) *Phys. Rev. B* **40**, 11477-11487.

28. Egelhoff Jr., W. F. (1991) *Proc. Mat. Res. Soc.* **229**, 27.

29. Rousset, S., Chiang, S., Fowler, D. E., and Chambliss, D. D. (1992) *Phys. Rev. Lett.* **69**, 3200-3203.

30. He, Y.-L. and Wang, G.-C. (1993) *Phys. Rev. Lett.* **71**, 3834-3837.

31. Johnson, K. E., Chambliss, D. D., Wilson, R. J., and Chiang, S. (1993) *J. Vac. Sci. Technol. A* **11**, 1654-1660.

32. Kellogg, G. L. and Voter, A. F. (1991) *Phys. Rev. Lett.* **67**, 622-625.

33. Kellogg, G. L., Wright, A. F., and Daw, M. S. (1991) *J. Vac. Sci. Technol. A* **9**, 1757-1760.

34. Kellogg, G. L. (1993) *Phys. Rev. Lett.* **70**, 1631-1634.

35. Schwoebel, P. R. and Kellogg, G. L. (1988) *Phys. Rev. B* **38**, 5326-5331.

36. Schwoebel, P. R. and Feibelman, P. J. (1989) *Surf. Sci.* **216**, 263-269.

DYNAMICS OF THE DESORPTION OF CARBON MONOXIDE FROM SIZE-SELECTED SUPPORTED PLATINUM CLUSTERS.

U. Heiz[1,2], R. Sherwood[2], D.M. Cox[2], A. Kaldor[2] and J.T. Yates, Jr[1]
Surface Science Center[1]
Department of Chemistry,
University of Pittsburgh,
Pittsburgh, PA 15260.

and

Exxon Research and Engineering Co.,[2]
Route 22E, Annandale,
New Jersey 08801.

Abstract.

CO chemisorption on monodispersed Pt clusters was investigated. Size-selected Pt_1, Pt_2 and Pt_3 clusters were deposited on an atomically clean SiO_2 film. Surface coverages of these monodispersed metal clusters were less than 1% of a monolayer. CO chemisorption/desorption experiments at 300 K exhibit a distinct 340 K - CO desorption state with an activation energy of ~16 kcal/mol. Comparisons with CO desorption from Pt(111), Pt(112) and bulk Pt films on SiO_2 show that the 340 K - CO desorption process is a unique feature of highly dispersed Pt atoms on SiO_2. CO desorption for the Pt_2 and Pt_3 samples was also detected, and exhibited additional adsorption sites with higher desorption activation energies, but the nature of these sites cannot be ascertained in these experiments with certainty.

1. Introduction.

In the last years much work has been done to investigate the chemical and catalytic behaviour of bare metal clusters from both scientific [1,2] and more practical points of view [3]. The question is whether metal particles with diminishing radii begin to exhibit reactive properties distinctly different from bulk behaviour. Studies with supported Pt-particles on nonmetallic surfaces showed a 1/r (r = particle radius) dependence of a

37

L. J. Farrugia (ed.), The Synergy Between Dynamics and Reactivity at Clusters and Surfaces, 37–47.
© 1995 *Kluwer Academic Publishers.*

specific catalytic activity which levelled off when atomic-scale dispersion was reached. This is to be expected if no size dependence exists beyond the availability of the catalyst atoms for the reaction. On the other hand experiments with much smaller clusters in the gas phase show dramatically selective effects [4] and it is now well established that small unsupported, i.e. gas phase, transition metal clusters exhibit pronounced variations in their physical, electronic, magnetic and chemical properties as the precise number of the metal atoms in the cluster is varied [5,6,7]. Similar effects have not yet been observed with supported catalysts, but these studies have suffered from broad, ill defined size distributions which would mask the effects expected in the smallest size ranges. But the above mentioned cluster results suggest that size-selected supported clusters, i.e. clusters of one size deposited onto surfaces, may also exhibit novel behaviour which is strongly dependent on the cluster size, and in fact size dependent variations in the electronic structure of deposited transition metal clusters has been reported [8,9]. Such considerations lead to speculations that one may be able to build 'designer' catalysts by the judicious choice of metal, metal cluster size, and substrate or support material.

The experiments reported here are a first attempt to deposit size-selected metal particles on a carefully prepared, clean SiO_2 surface and to then study the chemisorption properties of these clusters using a simple model chemical reaction. The key to being able to perform such studies is the provision of sufficient cluster deposition intensity (combined with maintenance of ultrahigh vacuum to maintain clean surface conditions) and to then use highly sensitive detection of the chemisorption and binding of a test molecule, carbon monoxide.

2. Experimental Procedures.

The CO chemisorption/desorption experiments on size-selected supported Pt clusters were performed in an ultrahigh vacuum system consisting of a first chamber with the cluster source and the mass-separation unit (base pressure: 2×10^{-8} Torr), and the analyzing chamber (base pressure: 2×10^{-10} Torr) equipped for Auger electron spectroscopy (AES), high sensitivity temperature programmed desorption (TPD) and ion bombardment. A conventional Pt evaporation source with a moveable shutter was also used in this chamber for comparison experiments. The two units are connected with a gate valve.

The cluster source and mass-separation unit were utilized previously for photoemission experiments on size-selected clusters and have been described elsewhere [8]. Here it will only be discussed briefly. A cold reflex discharge ion source (CORDIS) provided the Xe^+ beam at a typical current density of $5mA/cm^2$ at 25keV kinetic energy. The CORDIS was mounted perpendicular to the cluster beam direction, so that the primary Xe^+ beam bombarded the Pt target at an incidence angle of 45°. Ion extraction lenses, placed close to the target, focused the ionized particles into an energy analyzer (Bessel Box:Extranuclear 616-1), operated in the low resolution broad pass mode. The cluster ions were size-selected with a quadrupole mass spectrometer (Extranuclear 7-162-8) and then transported to an atomically clean SiO_2 surface on a Si(100) crystal in the main chamber through a RF-mode only quadrupole and two Einzel lenses. The beam current and beam profile were measured during deposition by a moveable detector consisting of four copper segments used as Faraday plates.

The size-selected Pt clusters were then supported on an atomically clean Si/SiO_2 surface. For each experiment a Si(100) crystal (13mm x 13mm x 1.5mm, 10Ω/cm, p-type, B-doped) was sputtered with Ar^+(1.5kV) for 45 minutes with the ion gun. The crystal was annealed to 1200 K and then cooled slowly to 300 K. It was found that this procedure produces a clean, well-ordered (2x1) surface [10]. The SiO_2 film was formed at 300 K by bombarding the crystal with Ar^+(0.8kV) in an oxygen background of 2×10^{-5} Torr for 20 minutes. The clean SiO_2 film was then annealed to 900 K. From the ratio of the O (KLL) and Si (LVV) Auger peak-to-peak signals a film thickness of about 40 Å was estimated [11]. High resolution electron energy loss spectroscopy (HREELS) studies have shown that such films grown on a Si(100) single crystal reveal almost identical phonon frequencies as amorphous SiO_2 materials [12]. The Auger spectrum of this substrate was free of carbon and other impurities at the sensitivity limit of Auger spectroscopy (~1 at. %). The cleanliness of the SiO_2 film was also checked by measuring CO desorption after film formation, which reveals no CO desorption feature in the measured temperature range (300 K - 700 K).

Pt coverages for atoms, dimers and trimers were typically 1% of a monolayer, as measured from the cluster bean current and beam profile. These low coverages were used to minimize interactions between Pt_x clusters. Typical Pt_x^+ beam currents were in the range of a few nanoamperes. At a Pt target bias of +10V, the actual beam energy was found to be 10±5eV. The background pressure during deposition was 1×10^{-9} Torr; the increase by a factor of 5 of the background pressure in the analysis chamber was mainly due to the Xe sputter gas.

The Pt_x/SiO_2 surfaces were then dosed with a collimated and calibrated effusive beam doser achievbing accurate CO fluences on the sample [13]. The CO desorption from the Pt_x particles on the SiO_2 surface was measured with a differentially pumped, apertured and multiplexed line-of-sight quadrupole mass spectrometer (UTI 100C equipped with a Cu/Be-multiplier). A skimmer with a diameter of 5mm was used as the aperture, and was located close to the surface for desorption experiments. The aperture was electrically isolated and kept at a voltage of -90V during CO desorption experiments in order to avoid electron stimulated desorption and charging effects on the SiO_2 by electrons emitted from the filament of the mass spectrometer [14]. The oxidized Si crystal could be resistively heated by a home built linear-program temperature controller at a rate of 1.8 K/s [15]. Temperature programmed desorption (TPD) measurements were done with CO exposure of 2×10^{13} to 5.6×10^{15} molecules/cm^2. Isotopically labeled ^{13}CO was used to improve the signal-to-noise ratio, avoiding the 28 amu background signal.

40

3. Results.
3.1 Temperature Programmed Desorption of Pt_x/SiO_2 Samples.

CO adsorption/desorption experiments were achieved on four different samples, a SiO_2/Si substrate with 0.009 ML of Pt atoms, a SiO_2/Si substrate with 0.009 ML of Pt dimers, a SiO_2/Si substrate with 0.006 ML of Pt trimers and a clean SiO_2/Si substrate. Figure 1 shows results of the CO desorption from each of the four samples after an exposure of 5.6×10^{15} CO molecules/cm^2. For the three samples containing Pt_1, Pt_2 and Pt_3 cluster deposits, CO is readily chemisorbed at 300 K. For Pt_1 three desorption peaks are observed one at 340 K and two lower intensity peaks at 400 K and 440 K. For Pt_2 the relative intensity of the two higher temperature peaks is enhanced compared to the 340 K peak. Pt_3 shows three high temperature CO desorption peaks superimposed on the 340 K desorption peak. The relative yield of the higher-temperature CO desorption product is further enhanced on Pt_3 compared to Pt_1 or Pt_2 deposits. On the other hand no evidence for CO desorption is observed for the SiO_2/Si substrate devoid of any platinum. Subsequent Auger analysis shows no evidence of carbon buildup attributable to CO decomposition on the SiO_2 and thus we conclude, as have many others, that CO does not adsorb onto SiO_2 at room temperature [16,17].

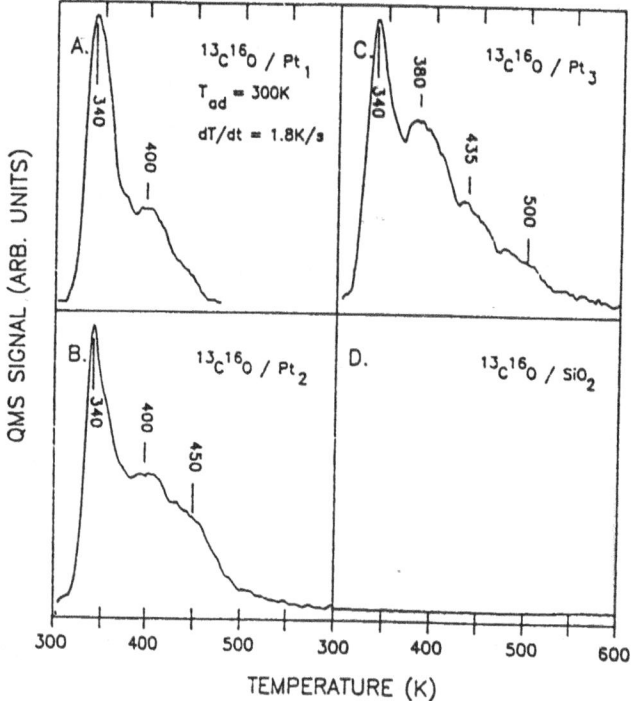

Figure 1. TPD spectra for four different samples: (A) SiO_2/Si substrate with 0.009 ML of Pt_1; (B) SiO_2/Si substrate with 0.009 ML of Pt_2; (C) SiO_2/Si substrate with 0.006 ML of Pt_3; (D) Clean SiO_2/Si substrate

In Figure 2, we show a series of desorption spectra for different CO exposures on deposits made from Pt_2. These experiments were done on the same Pt_2 deposited layer, interrupting the temperature program at 550 K in each case, followed by different exposure to CO for the next experiment. The results of the experiments shown in Figure 2 indicate that the distribution between CO species in the 340 K-desorption state and the higher temperature states does not change markedly as the CO coverage is increased. The inset in Fig. 2 shows that the quantity of observed CO increases linearly with CO exposure, which also indicates that the cluster samples are stable in the measured temperature range for the duration of the experiment.

CO Desorption from Pt_2 Deposits on SiO_2

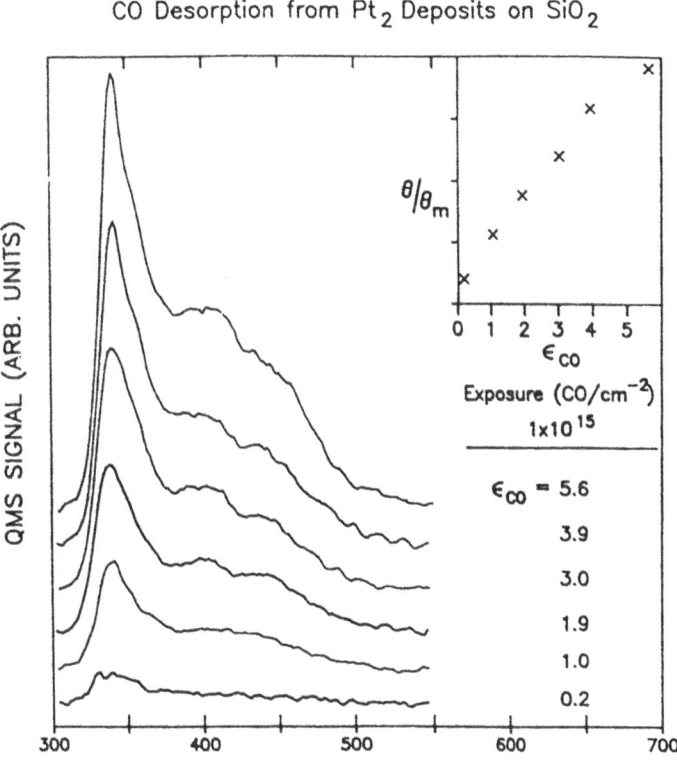

Figure 2. TPD spectra of deposits from Pt_2 at different CO exposures. The inset shows the relative quantity θ/θ_m of CO as a function of exposure of CO, ϵ_{co}

In order to better understand the nature of the desorption behaviour we performed isotopic labeling experiments as follows: the sample was prepared and then dosed at 300 K with 5.6×10^{15} $^{13}CO/cm^2$. The sample was then heated to just above a specific desorption peak, cooled to 300 K and dosed again with ^{12}CO. The TPD experiment was then executed, and both ^{12}CO and ^{13}CO desorption signals were measured. Due to a poor signal-to-noise ratio for ^{12}CO, the experiment was repeated by dosing first with ^{12}CO, followed by ^{13}CO after the heating and cooling cycle. The results from two of these experiments are shown in Figure 3. The higher temperature curve (3A.2) show the results for first dosing with ^{13}CO, heating to 353 K, cooling to 300 K, dosing with ^{12}CO and then monitoring ^{13}CO during TPD. Note that almost no ^{13}CO is observed for T<350 K. In addition the amount of ^{13}CO desorbed is identical (within our experimental uncertainties) to that which would have been obtained in this high temperature range if ^{12}CO had not been adsorbed. When the Pt$_2$ deposit is initially dosed with ^{12}CO, heated to 353 K, cooled and finally dosed with ^{13}CO, the ^{13}CO desorption is as shown in A.1 of Figure 3. For Figure 3B the same procedure is followed except that the Pt$_2$/SiO$_2$ sample was heated to 398 K. Similar results were obtained for CO desorption from Pt$_1$. These results show that no exchange of CO occurs between the different binding sites and suggests that the different binding sites are independent.

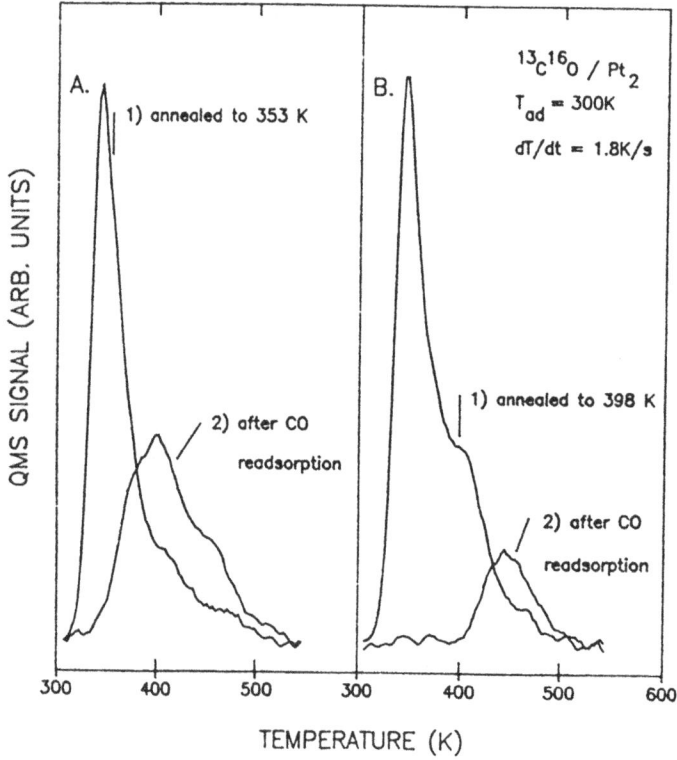

Figure 3. TPD spectra of ^{13}CO from deposits produced from Pt$_2$. (A1) TPD after dosing with ^{12}CO, annealing to 353 K and re-adsorption of ^{13}CO at 300 K. (A2) TPD after dosing with ^{13}CO, annealing to 353 K, and re-adsorption of ^{12}CO at 300K. (B1) TPD after dosing with ^{12}CO, annealing to 398 K and re-adsorption of ^{13}CO at 300 K. (B2) TPD after dosing with ^{13}CO, annealing to 398 K, and re-adsorption of ^{12}CO at 300 K.

3.2. Temperature-Programmed Desorption of Pt Bulk Samples.

Figure 4 shows results of CO desorption for three different Pt deposits on a SiO_2 film. The Pt is deposited from a conventional Pt evaporation source: CO desorption is shown from a thick Pt-film, a medium (~0.3 ML) and a low (~0.01 ML) Pt coverage. The desorption kinetics from the thick film are quite similar to that from Pt(111). As expected for highly dispersed Pt deposits on SiO_2, the CO TPD from the 0.01 ML Pt/SiO_2 preparation is nearly identical to that for the Pt layer prepared by deposition of Pt_1^+ ions.

CO Desorption from Evaporated Pt Films/SiO_2

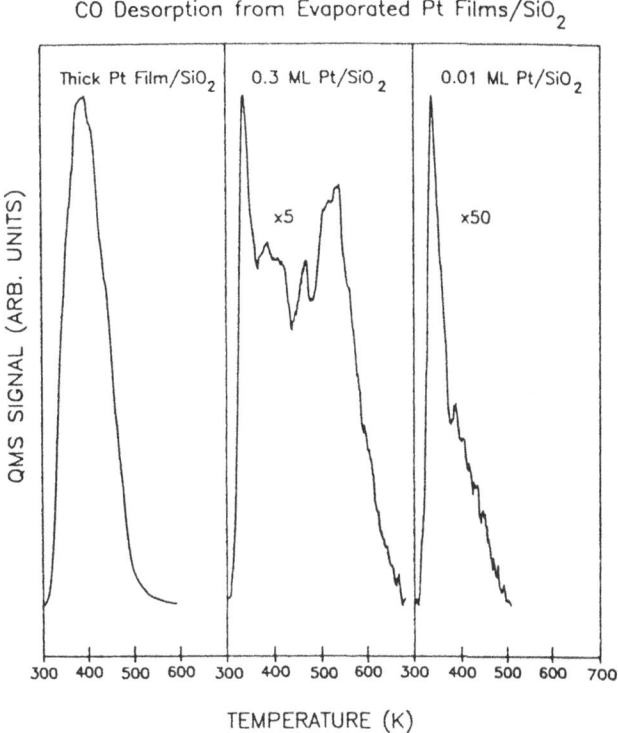

Figure 4. CO desorption from three different Pt films on Si/SiO_2 prepared using conventional evaporation source

As can be seen by comparing the results of Figures 4 with those of Figures 1 and 2, the CO desorption kinetics behaviour for Pt_1 , Pt_2 and Pt_3 deposits on SiO_2 are different from CO desorption from bulk films of deposited Pt on SiO_2.

4. Discussion.

To our knowledge these are the first adsorbate desorption measurements from metal particles deposited from mass-selected metal cluster beams. In the last 10 years much effort has been made to investigate small supported metal particles by evaporating metal from hot filaments. All these studies [18,19,20] on supported Pt particles produced adsorbate desorption profiles similar to the bulk metal even for particles as small as 2nm in diameter (about 30-40 atoms). The desorption behaviour of these particles was compared to flat and stepped Pt surfaces; changes as a function of particle size were interpreted as a change in the population of bulk-like adsorption sites of flat and stepped single crystals. For Pt single crystals several measurements of the initial (zero coverage) desorption activation energy (E_d°) of CO were measured [21], obtaining E_d° values of 36.5 and 39.5 kcal/mol for Pt(557) and Pt(112) and 33 kcal/mol for Pt(111).

Ab initio generalized valence band and correlation-consistent configuration interaction studies of CO interacting with Pt atoms [22] leads to adiabatic Pt-CO dissociation energies of 15.4 kcal/mol; a corrected estimate of 18.5 kcal/mol was obtained for E_d° (CO). The authors of this work report that CO bonds to a single Pt atom by a-donor/-backbonding mechanism yielding a linear geometry with sigma repulsive effects due to unpaired 6s electrons.

Pt-Pt bonding in the case of the dimer and trimer involves primarily 6s electrons, which diminishes these sigma repulsive effects. CO primarily interacts with the attractive d^9 configuration for bigger clusters and in the bulk limit, which leads to higher CO adsorption energies in comparison to the adsorption energy on single Pt atoms.

The basic experiments reported here may be summarized as follows:

1. All highly dispersed Pt_x/SiO_2 surfaces exhibit a distinct ~340 K CO desorption state which is not observed for CO on Pt single crystals or for bulk Pt films on SiO_2. The relative yield of CO in this 340 K state is highest for depositions made from Pt_1 and decreases systematically for deposits made from Pt_2 and Pt_3 clusters (Figure 2). Using the Chan, Aris and Weinberg method, a desorption activation energy of 16 kcal/mol is estimated for this CO desorption state on deposits made from Pt_1 clusters.

2. Al Pt_x/SiO_2 made from clusters (Pt_1 , Pt_2 and Pt_3) and from bulk sublimation exhibit one or more CO desorption states above ~ 400 K (Figures 2 and 4).

3. Isotopic exchange of CO does not occur between the 340 K CO state and the CO desorption states desorbing above ~ 400 K, indicating that the states do not interconvert upon heating and therefore the species responsible for these states do not coexist on the same Pt_x particle. (Figure 3).

4.   The CO desorption states fill at a constant rate independent of CO coverage, with little variation in the relative population of the 340 K-CO and the higher-temperature CO states over the entire CO coverage range (Figure 2). This indicates that CO...CO interaction between CO molecules in the different types of bonding sites does not occur, again suggesting that the desorption state does not originate as a result of CO...CO interaction on a single Pt_x particle.

5.   Thick Pt films (bulk like Pt) deposited on SiO_2 closely resemble Pt(111) in CO desorption kinetics (Figure 4).

One explanation for these results is that the 340 K-CO desorption state originates from single Pt atoms adsorbed on SiO_2. This is based on the observation of this 340 K-CO desorption state for Pt_1 , Pt_2 , Pt_3 and low coverage Pt sublimation-deposition experiments on SiO_2. Furthermore, the measured activation energy of CO desorption (16 kcal/mol) agrees well with *ab initio* calculations for a Pt-CO species [22]. The observation of the 340 K-CO state for Pt_2 and Pt_3 suggests therefore that the cluster deposition onto SiO_2 is dissociative in some cases.

The higher temperature CO-desorption states may be interpreted in differemt ways, as listed below:

1.   The high temperature CO-desorption states could originate from larger clusters of Pt_x which behave more like bulk-like Pt_x clusters in their CO adsorption/desorption properties. These clusters could be formed from Pt migration/nucleation effects which might be induced by 5-10eV Pt_x^+(for x>1) collisions with SiO_2 at 300 K.

2.   The even higher temperature desorption states could originate from preserved Pt_2 and Pt_3 clusters, due to the possibility of bridged- and 3-fold sites being available. The only calculations found in the literature are density functional studies of CO chemisorbed on Rh_4 and Pd_4 , which compare these different bonding sites [23]. If one assigns the additional desorption peak of the dimer at 445 K to bridge bound CO and the specific two peaks of the trimer at 377 K and 474 K to bridge-bound and 3-fold-bound CO respectively, one can observe the same trend as for the Pd_4 CO-complex, whose desorption energies were calculated as follows: 30.4 kcal/mol (top), 41.2 kcal/mol (bridge) and 54.2 kcal/mol (3-fold). For Rh_4CO the authors of ref. [23] postulate little differences in the desorption energies of these three sites: 53.3 kcal/mol (top), 54.4 kcal/mol (bridge) and 49.5 kcal/mol (3-fold).

A second interpretation is that the clusters land and remain intact upon deposition. In this case the different CO desorption states of each cluster size are proposed to originate from monodispersed Pt_x species which interact with different binding sites on the SiO_2 substrate to produce various Pt coordinations. These different binding sites on the SiO_2 substrate may be characteristic of our SiO_2 film or could be introduced by the impact of the Pt_x clusters upon collision with the substrate. It is well established that in a silicon dioxide film all oxidation states of silicon may be present in the near SiO_2/Si interface region (up to 10 Å) [24,25] In a second layer, up to 50 Å in thickness, distorted amorphous SiO_2 was described with four-, six-, seven-, eight-, and nine-membered rings of SiO_2 tetrahedra joined by bridging oxygen [26]. It is possible that differently coordinated Pt species are present on the surface, which leads to different CO adsorption energies. Similar effects were observed for small Pt-particles supported on alumina [27]. In this work, the authors observed a weak metal-substrate interaction for the smallest particles, and a growth mechanism on defects of alumina was suggested.

In this case the multiple CO desorption states observed here for the Pt clusters may be due to differently bound Pt particles, with the 340 K-CO desorption state originating from the most gas-phase Pt atom like CO-adsorption site. This is in accordance with the good agreement between the 340 K-CO desorption energy and the theoretical Pt-CO dissociation energy.

Our experiments cannot completely discriminate between these two interpretations.

Acknowledgement.

We acknowledge with thanks the support of the Army Office of Research for the purchase of the Auger spectrometer used in this work under Contract No. DAAL03-91-G-0323. U.H. acknowledges support of an A.W. Mellon postdoctoral fellowship from the University of Pittsburgh and a Swiss National Science Foundation postdoctoral fellowship. We also acknowledge the support of the Exxon Research and Engineering Corporation in hosting U.H. and in supplying the major apparatus used in this work.

References

1.   F. Blatter, M. Frey, U. Heiz, U. Roethiisberger, M. Schaer, A. Vayloyan, C. Yeretzian and E. Schumacher (1988) *Chimia*, **42**, 357-376.
2.   P. Fayet, A. Kaldor and D.M. Cox (1990) *J. Chem. Phys.* **92**, 254-260.
3.   S. Ichikawa, H. Poppa and M. Boudard (1984) *The Effect of Particle Size on the Reactivity of Supported Palladium*, ACS Symposium Series **248**, 439-451.
4.   S.C. Richtmeier, E.K. Parks, K. Liu, L.G. Pobo and S.J. Riley (1985) , *J. Chem. Phys.* **82**, 3659-3665.
5.   A. Kaldor, D.M. Cox, M.R. Zakin, *Adv. Chem. Phys.* 70, 211 (1988); D.M. Cox, A. Kaldor, P. Fayet, W. Eberhardt, R. Brickman, R. Sherwood, Z. Fu and

D. Sondericher (1990) *Effect of Cluster Size on Chemical and Electronic Properties*, ACS Sym. Series, **437**, pp. 172.

6. R.S. Berry, T.L. Beck, H.L. Davis, J. Jellinek, *Adv. Chem. Phys.*, 70, 75 (1988).

7. U. Heiz, U. Roethlisberger, A. Vayloyan and E. Schumacher (1990) *Israel J. of Chem.*, **30**, 147-155.

8. W. Eberhardt, P. Fayet, D.M. Cox, Z. Fu, A. Kaldor, R. Sherwood and D. Sondericher (1990) *Phys. Rev. Lett.*, **64**, 780-783.

9. H.-V. Roy, P. Fayet, F. Patthey, B. Delley, C. Massobrio and W.-D. Schneider (1994*Phys. Rev. B.*, **49**, 5611-5620.

10. M.L. Colaianni, P.J. Chen, H. Gutleben and J.T. Yates Jr., (1992) , *Chem. Phys. Let.* **191**, 561-567.

11. C.F. Yu, S.S. Todorov and E.R. Fossum (1987) *J. Vac. Sci. Technol.*, **A5(4)**, 1569-1571.

12. P.J. Chen, M.L. Colaianni, M. Arbab and J.T. Yates, Jr., (1993) *J. of Non-Crystalline Solids*, **155**, 131-140.

13. A. Winkler and J.T. Yates, Jr., (1988) *J. Vac. Sci. Technol.* **A6**, 2929-2932; M.J. Bozack, L. Muehlhoff, J.N. Russell, Jr., W.J. Choyke and J.T. Yates, Jr., (1987) *J. Vac. Sci. Technol.*, **A5**, 1-8.

14. V.S. Smentkowski and J.T. Yates, Jr., (1989) *J. Vac. Sci. Technol.*, **A7**, 3325-3331.

15. R.J. Muha, S.M. Gates, P. Basu and J.T. Yates, Jr., (1985) *Rev. Sci. Inst.* **56**, 613-616.

16. H. Miura, R.D. Gonzalez (1982) *J. Phys. Chem.*, **86**, 1577-1582.

17. M.A. Vannice, L.C. Hasselbring and B. Sen (1986) *J. of Catalysis*, **97**, 66-74.

18. E.I. Altman and R.J. Gorte (1988) *Surf. Sci.* **195**, 392-402; E.I. Altman and R.J. Gorte (1986) *Surf. Sci.*, **172**, 71-80.

19. D.L. Doering, H. Poppa and J.T. Dickinson (1982) *J. Vac. Sci. Technol.*, **20**, 827-830.

20. Y. Zhu and L.D. Schmidt (1983) *Surf. Sci.*, **129**, 107-122.

21. H.R. Siddiqui, P.J. Chen, X. Guo, I. Chorkendorff and J.T. Yates, Jr., unpublished results.

22. G.W. Smith and E.A. Carter (1991) *J. Phys. Chem.*, **95**, 2327-2339.

23. A. Goursot, I. Papai and D.R. Salahub (1992) , *J. Am. Chem. Soc.* **114**, 7452-7458.

24. W. Braun, H. Kuhlenbeck (1987) *Surface Science*, **180**, 279-288.

25. F.J. Himpsel, F.R. McFeely, A. Taleb-Ibrahimi and J.A. Yarnoff (1988) *Phys. Rev. B.* **38**, 6084-6096.

26. F.J. Grunthaner, P.J. Grunthaner, R.P. Vasquez, B.F. Lewis, J. Maserjian and A. Madhukar (1979) ,*Phys. Rev. Let.* **43**, 1683-1686.

27. A. Masson, B. Bellamy, Y. Hadji Romdhane, M. Che, H. Roulet and G. Dufour (1986) *Surf. Sci.* **173**, 479-497.

NMR INVESTIGATION OF BINDING OF AROMATICS AT CATALYTIC SURFACES

A Deuterium NMR Study

Cecil Dybowski and Mark A. Hepp[1]
Department of Chemistry and Biochemistry
University of Delaware
Newark, Delaware 19716

1. Introduction

Dynamic nuclear magnetic resonance allows the study of kinetic processes in systems at chemical equilibrium. For NMR parameters to be sensitive to dynamic processes, the chemical exchange rate constants must be approximately the same magnitude as the spread of frequencies in the NMR spectrum. Under these conditions, dynamic effects in the chemical system profoundly influence the NMR response. Line-shape analysis was used very early in the evolution of NMR spectroscopy to address chemical problems.[1,2,3,4] Over the years, the manner in which chemical exchange affects the line shape has become well understood and analyses based on the effects of exchange have been applied to systems undergoing internal rearrangements like rotations,[5] pseudorotations,[6] ring inversions,[7] and configurational inversions,[8] and to a range of biomolecular exchange problems.[9] The activation energies of chemical reactions to which dynamic NMR methods have been applied range from a few kJ/mol to the borderline where intermediates become too unstable for the reaction to proceed.[10]

In some cases, not only rate constants and activation energies are determined, but the mechanism of rearrangement may also be unambiguously assigned.[11] With the development of more powerful computers and advances in computational methods, line-shape analysis is being applied to increasingly more sophisticated problems.[12]

The exchange of molecules between surface sites and free fluids above a surface is an important process in catalytic chemistry and adsorption in general. In this paper we examine the deuterium NMR spectrum of phenanthrene-d_{10} adsorbed in a Rb-exchanged X-zeolite. We observe the effects of chemical exchange, from which we extract

---

[1]   Current address: Code 6120, Naval Research Laboratory, Washington, DC

L. J. Farrugia (ed.), The Synergy Between Dynamics and Reactivity at Clusters and Surfaces, 49–61.
© 1995 *Kluwer Academic Publishers.*

information on the dynamic state of phenanthrene in these systems through comparison of the NMR spectra to line-shape simulations. The results suggest the formation of an ion-molecule complex whose lifetime is not long on the NMR time scale.

2. Theory

The deuterium NMR spectrum of a powdered sample is often inhomogeneously broadened by contributions from spins whose principal axes of the quadrupolar coupling are at various orientations with respect to the static magnetic field. In general, any line shape may be thought of as the Fourier partner of a time-dependent function, G(t), called the relaxation function:

$$G(t) = \int_{-\infty}^{+\infty} e^{i\omega t} P(\omega) d\omega \tag{1}$$

Here $P(\omega)$ is the probability distribution function governing the set of frequencies in the spectrum. For a sample which is stationary on the NMR time scale, $P(\omega)$ is a map of the distribution of orientations of the nuclear electric-field gradients (efg) relative to the Zeeman field. For example, the spectrum of a powder with a random distribution of orientations of the electric-field-gradient (efg) principal axes leads is the familiar powder pattern.

Molecular motion introduces an additional effect -- a change of the precessional frequency of each unique nuclear species with time. In this case, G(t) incorporates the effects of the exchange and $P(\omega)$ is the spectrum incorporating the distribution of frequencies and the frequency modulation caused by exchange among sites of different quadrupolar coupling.

Chemical exchange among several discrete sites involves random change among discrete sets of frequencies. By assuming that the exchange may be modeled as a stationary Markov process, the effects of chemical exchange can be expressed by a constant probability of occupancy at site i, W_i, and a conditional probability that a molecule will leave site i for site j within a time Δt. This conditional probability can be written as:

$$W(\omega_i | \omega_p, \Delta t) = \delta_{\omega_i, \omega_j} + \Pi(\omega_p, \omega_j) \Delta t \tag{2}$$

where Π is a matrix of transition probabilities. The relaxation function governed by exchange among a finite set of frequencies is then given as a sum over all possible paths:

$$G(t) = \sum W_1 W(\omega_1 | \omega_2, \Delta t) ..$$
$$.. W(\omega_{n-1} | \omega_n, \Delta t) e^{i(\omega_1 + \omega_2 + ... + \omega_n) \Delta t} \tag{3}$$

where W_1 is the initial equilibrium probability distribution and the sum is over all possible paths the molecule could take from each initial site to the final site during the time $n\Delta t$. Defining $G_\alpha(t)$ as the sum over all paths which lead ultimately to site α, the relaxation function may be expressed as a sum of contributions from each site:

$$G(t) = \sum_\alpha G_\alpha(t) \tag{4}$$

After a subsequent time increment, the molecule may move to another site, ß, and the sum over all paths leading to that new site is given as:

$$G_\beta(t + \Delta t) = \sum_{\omega_\alpha} G_\alpha(t) W(\omega_\alpha | \omega_\beta, \Delta t) e^{i\omega_\beta t} \tag{5}$$

If Δt is very small, the exponential term may be replaced by $(1 + i\omega \Delta t)$. Then, in the limit that Δt goes to 0, one obtains the master equation for chemical exchange:[13]

$$\frac{\partial G}{\partial t} = G \cdot (i\omega + \pi) \tag{6}$$

Here, G is a vector representing the response of the components of the magnetization at each site, ω is a diagonal matrix of the site frequencies and π is a matrix containing the probabilities that the molecule jumps from one site to another.

This master equation may be solved by direct integration to yield the matrix G in terms of W, the equilibrium probability vector, and Λ, the exchange matrix. The line shape, $I(\omega)$, is obtained by Fourier transformation of this function.[14]

$$I(\omega) = Re \int_{-\infty}^{+\infty} 1 \cdot e^{\Lambda t} \cdot e^{-i\omega_1 t'} \cdot W dt' \tag{7}$$

This equation has been used, for example, to calculate two-site exchange in a liquid.[15] Moreover, the technique can be straightforwardly extended to the case of deuterated molecules exchanging between two sites in a solid, powdered sample.

An additional complication is manifested in deuterium NMR experiments on solids. The rapid decay of the FID makes it difficult to detect the early-time behavior of the system. To circumvent this problem, spectroscopists use the quadrupole-echo pulse sequence (Figure 1). However, using this experiment introduces other distortions of the spectrum which must be considered in a line-shape simulation. Spiess and Sillescu [16] have derived a master equation which accounts for the intensity losses due to distributions of homogeneous broadening in the quadrupole-echo experiment. The exact

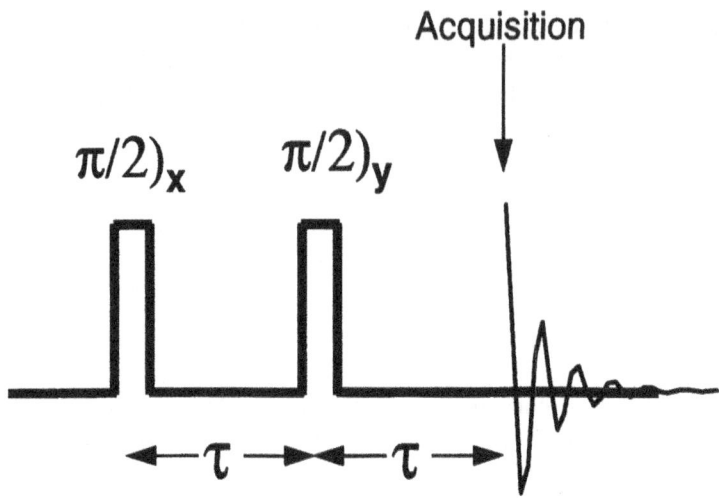

Figure 1 The quadrupole-echo sequence for obtaining spectra of deuterated materials in the solid state.

solution has been calculated only for the simple case of two-site exchange.[16] In general, more complicated problems must be solved numerically.

The distortions in the deuterium NMR line shape are due to chemical exchange during the time between the two pulses of the quadrupole-echo experiment. They are readily observed by comparing calculated spectra of the free-induction decay with the calculated response to the quadrupole-echo experiment. An example is shown in Figure 2, from a calculation we made for a deuterated water molecule performing π flips about the C_2 axis with varying jump frequencies. In the left column are the undistorted spectra obtained if one could detect the full free induction decay; on the right are the spectra determined by the quadrupole-echo sequence with a delay τ of 15 μs between the pulses. One can clearly see the distortions introduced by the quadrupole-echo sequence; however the major features of the resonances can be discerned from the quadrupole-echo spectra. If τ is increased, the intensity losses become more severe. Above $\tau \approx 40\mu$s, the line shape is roughly independent of τ.[16]

The foregoing analysis works well for describing problems in which instantaneous jumps among a set of well-defined sites are occurring, *e.g.* π flips of water or of phenyl rings [17,18] or $2\pi/5$ jumps about the C_5 axis of ferrocene.[19] A more general formalism has been developed by Greenfield *et al.* which is based on recursion relations among first-rank tensors. This method allows for the possibility of simultaneous rotations about as many as four coincident axes.[20] Thus, they have calculated line shapes due to the simultaneous π flips of phenyl rings about an axis that is slightly librating. These spectra have a rather different appearance from those calculated without such librations, showing that the details of motion are exhibited in the spectral lineshape.

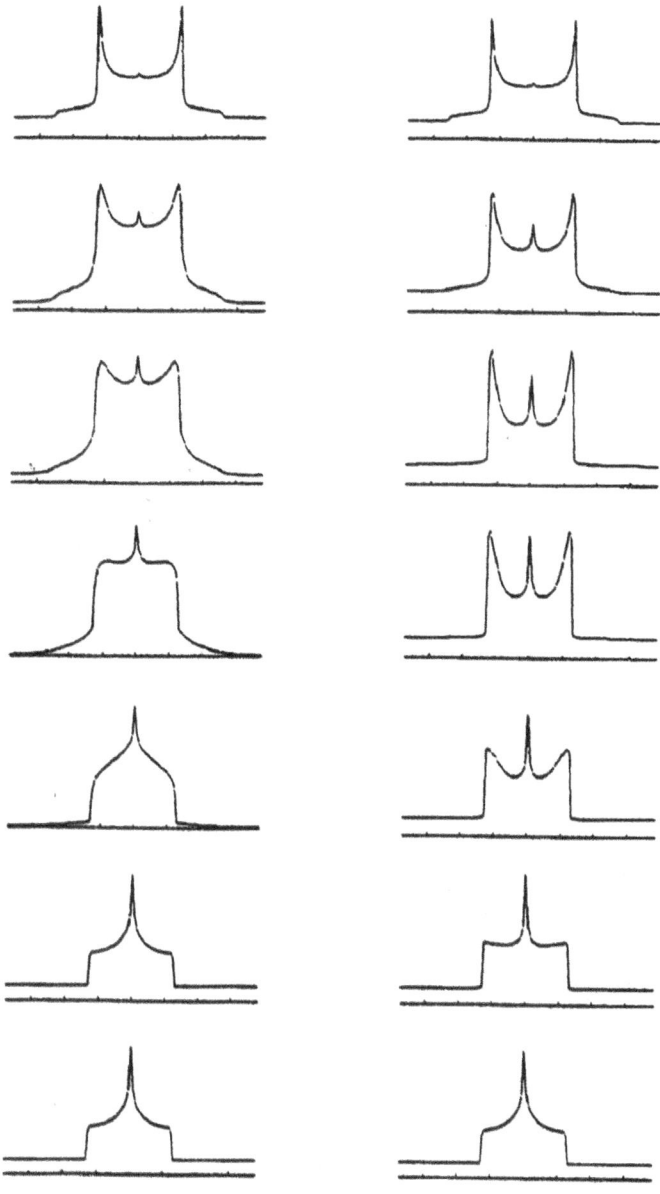

Figure 2 Calculated spectra for the one-pulse (left) and the quadrupole echo (right) experiments illustrate the effects of a two site π jump for a deuterated water molecule.

In our work, we simulate the effects of motion on line shapes using a Monte Carlo approach. In these calculations, the evolution of large numbers of magnetic moments is simulated by a random walk through sites with various local frequencies. The exchange sites are defined and labelled with frequencies corresponding to the orientation of the deuterium electric-field gradient (efg) at each site. Each site is characterized by a steady-state occupancy and a lifetime that describes the probability that the molecule will leave that site at the end of a single time increment, Δt. The frequency of the nucleus at any site is assumed to remain constant as long as the molecule remains at that site, but changes instantaneously if the molecule changes site. The calculation begins with a fictitious magnetic moment placed randomly in an initial site. The precession angle during the subsequent time intervals is calculated as

$$\phi(t) = \phi(j\Delta t) = \sum_{J=0}^{n} \omega_j \Delta t \qquad (8)$$

where the sum is over all intermediate frequencies the molecule may have during the interval, $n\Delta t$. To mimic the random hopping, at the end of each subperiod, a uniform random variable is generated to determine if the molecule jumps to a new site for the subsequent period. Spin evolution is simulated by incrementing the precession angle while the random walk among all possible sites is occurring. The relaxation function for the quadrupole-echo experiment with a single exchange path, K_α, is calculated at a time t after the second pulse as

$$K_\alpha(t,\tau) = K_\alpha(j\Delta t, m\Delta t) = 2\cos\phi_\alpha(j\Delta t) \qquad (9)$$

This evolution is then averaged over a statistically relevant number of exchange paths. The powder average is obtained by performing this operation over a random distribution of orientations of the reference frame. Figure 2 shows the results of the application of this random-walk calculation to the two-site exchange problem of a water molecule undergoing π flips about its C_2 axis. The results of this Monte Carlo calculation are similar to the numerical solution of the master equation which has been demonstrated in the past,[16] giving confidence in the validity of the Monte Carlo technique.

This Monte Carlo approach is computationally intensive; however, it allows considerable flexibility in the definition of sites and is considerably easier to program because it avoids matrix diagonalization and other manipulations used in numerical solutions. Multiple-axis exchange problems can be easily solved with the method, but the time and expense of the calculation will increase much more rapidly than with the numerical solutions of the master equations. We shall use this method to simulate deuterium NMR spectra of phenanthrene-d_{10} included in Rb X-zeolite.

3. Experimental Conditions

Samples of the rubidium-exchanged X-zeolite were obtained from Dr. V. J. Ramamurthy of the DuPont Company. They were prepared as follows: Zeolite 13X (Na-X) was obtained from the Linde division of Union Carbide. Sodium ions were exchanged by contacting the material a number of times with rubidium nitrate solution (10%) at 363K in aliquots of 10 mL. The samples were then thoroughly washed with water and dried. Prior to use, these zeolites were heated in air at 1K/min to 773K and maintained at 773K for 12h. After cooling to 373K, the samples were removed and stored under anhydrous conditions at room temperature. Phenanthrene-d_{10} was introduced into the zeolite by spreading approximately 5 mg of the organic powder in a mortar and adding approximately 200 mg of the activated zeolite. The contents of the mortar were mixed well under a nitrogen atmosphere and sealed in a 5-mm NMR tube. Such a mixing should produce a loading of approximately 0.2 molecule of phenanthrene-d_{10} per supercage. The samples were then annealed for three days at 333K to ensure an equilibrium distribution of guest molecules in the zeolite.

NMR experiments were performed on a Bruker MSL300, operating at a frequency of 46.073 MHz. The quadrupole-echo pulse sequence21 was used to acquire spectra, with a $\pi/2$ pulse width of 2.5 µs, and a delay between pulses of 15 µs. The echo was collected prior to the echo maximum and shifted left to ensure that the signal used in the Fourier transformation began precisely at the echo maximum. The temperature was controlled by a Bruker B-VT1000 temperature controller to ±0.1K, with an accuracy of ±0.5K.

4. Application of the Random-Walk Model to Spectra of Phenanthrene-d_{10} Included in Rb X-Zeolite

Previously, we showed that in Na, K and Cs X-zeolites, phenanthrene-d_{10} exhibits a two-component spectrum, the shape of which depends on the temperature at which the spectrum is obtained. We interpreted these data as indicating that phenanthrene exists in two distinct states in the zeolite. The temperature-dependent spectra result from the equilibrium partitioning of phenanthrene between the two states. Deconvolution of these spectra gave equilibrium occupancies of the two states as a function of temperature, from which we extracted concentration equilibrium constants, and enthalpies and entropies of exchange between the two states. We concluded that the two states were an ion-bound phenanthrene molecule and a "free" phenanthrene molecule.

The analysis of the data for Na, K and Cs X-zeolites impregnated with phenanthrene-d_{10} assumed the phenanthrene molecules were in slow exchange, thereby allowing the NMR spectrum to give a "snapshot" of the sample. The deuterium NMR spectra of phenanthrene-d_{10} included in Rb X-zeolite, shown in Figure 3, can not be analyzed so easily because the phenanthrene-d_{10} is exchanging among sites on a time scale comparable to that on which the NMR spectrum is acquired. As a result, the spectra are homogeneously broadened. To analyze these spectra, one must extract information from a simulation of the exchange and its effects on the NMR spectrum.

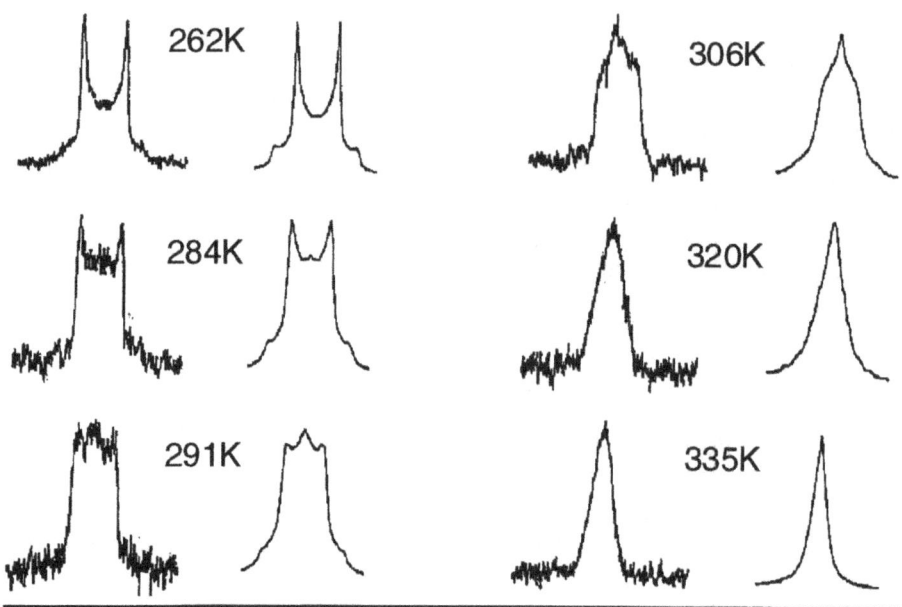

Figure 3 Comparison of experimental and simulated spectra for phenanthrene-d_{10} in Rb X-zeolite at various temperatures.

The faujasite structure is composed of sodalite cages linked in a tetrahedral arrangement by rings of six oxygen atoms, to form hexagonal prisms and supercages. The negative charge on the aluminate group requires the presence of charge-balancing cations to preserve electroneutrality. There are three positions for these ions in the faujasite structure, S_I, S_{II} and S_{III}. Because of the inaccessibility of S_I and the steric hindrance to contact S_{III}, we consider interaction only with the ions at S_{II}, the ion at the hexagonal face of the sodalite cage adjacent to the supercage. These S_{II} sites are arranged at the apices of a regular tetrahedron which can be drawn inside the supercage.

As a model, we consider five-site exchange. The five sites are the ion-bound phenanthrene at the four S_{II} sites (each described by a specific quadrupolar coupling of the deuterons) and a free phenanthrene molecule not strongly bound to the structure and having an average quadrupole coupling that is zero. The energetics of binding at each of the S_{II} sites is considered to be the same as at the others. The NMR parameters will be different from site to site due to the different orientation of the principal axes of the efg at each site. Finally, we insist that exchange between bound sites be disallowed; thus, a phenanthrene-d_{10} molecule may pass from one bound site to another only by passing through the unbound state. There are then two parameters that specify the exchange. There are three related parameters: the dissociation rate constant, k_d, the adsorption rate constant, k_a, and the equilibrium constant, K_{eq}. We shall specify k_d and K_{eq} in what follows. k_a can be calculated from these two. In addition, one must specify the frequencies at each site, which are determined by the quadrupole coupling constant and the orientation of the supercage in the field. In the averaging process, each initial frequency at a site is randomly chosen, which then fixes the others.

The simulated spectrum depends on only two parameters, the dissociation rate constant, k_d and the equilibrium constant K_{eq}. Thus, we calculated a series of spectra for various values of these two parameters. Comparison to the experimental spectra (Figure 3) gives sets of these two parameters at various temperatures. The agreement between the experimental and calculated line shapes is quite good. While it is possible that other models may also reproduce the observed spectra, the agreement over a range of conditions that gives easily interpretable results indicates that the model is reasonably accurate.

Figure 4 gives the temperature dependence of the equilibrium constant and Figure 5 gives similar data for the dissociation rate constant, k_d. From the slope of the van't Hoff plot in Figure 4, we estimate that the enthalpy for the process of dissociation of the complex to form the free phenanthrene and the bare ion is 33.4±5.0 kJ/mole. The entropy change for this process is estimated to be 99.5±17.2 J/mole-K. This is a reasonable value for a process which involves the gain of three degrees of translational freedom.

From the Arrhenius plot of Figure 5, one may estimate the activation energy for dissociation of the ion-molecule complex to be $E_{a,d} = 23.4±3.8$ kJ/mole. This activation barrier is less than ΔH, implying that the reverse reaction has a negative activation energy, $E_{a,a} = -10.0±5.0$ kJ/mole or that the rate constant for the reaction

$$Rb^+ + Ph \rightarrow Rb-Ph \tag{10}$$

has a negative temperature dependence. The negative temperature dependence of a rate constant for a reaction between an ion and a neutral was first observed for the association of He$^+$ with He.[22] Subsequently, systematic studies of the temperature dependence of rate constants by Bohme et al.[23] and Durden et al.[24] have shown this to be a common feature of ion-molecule association reactions in the gas phase. The commonly accepted mechanism for this feature of ion-molecule reactions is interaction with a third body, as first proposed by Rabinowitz in 1937. [25] Activation energies for recombinations

58

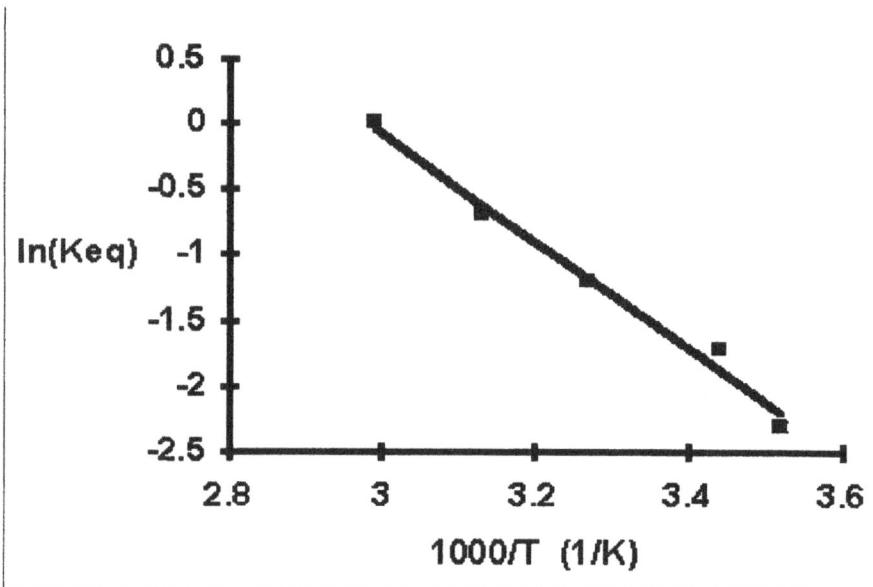

Figure 4 The plot of ln(Keq) versus 1/T (1/K), from the temperature dependence of the calculated NMR spectra based on the five site exchange model.

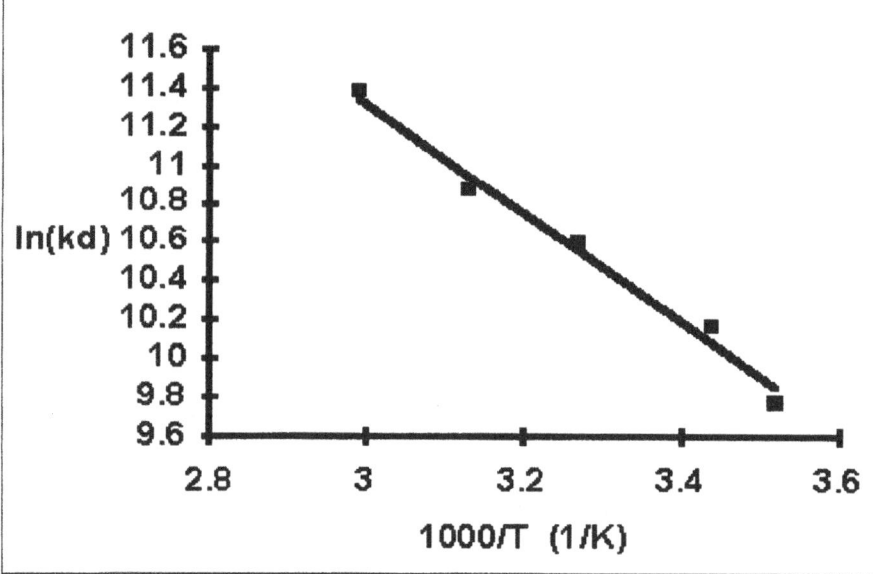

Figure 5 The Arrhenius plot of the dissociation rate constants versus 1/T for determining the energy of dissociation of the ion-molecule complex.

of this kind are typically in the range from -4 to -16 kJ/mole.[26]

For the reaction of equation 10, the interaction is between a polarizable charge and an induced dipole. *Ab initio* SCF-MO calculations have shown that such bonds are purely electrostatic when group 1 ions are involved. [27,28,29,30] When phenanthrene-d_{10} binds to the rubidium ion, 33 kJ/mole of stabilization energy are deposited in the vibrational mode of the ion-molecule complex. This low-energy bond between two relatively large masses will have a rather low vibrational frequency and will not couple strongly with the high-frequency fundamental modes of phenanthrene-d_{10}. This kind of nonstatistical behavior (non-RRKM) is also observed in the formation and vibrational predissociation of van der Waals complexes of rare-gas atoms with large aromatics. [31,32,33]. It is possible that dinitrogen, added to the sample to prevent contamination is the collision partner that carries away the excess energy. In addition, it may be possible that the low-frequency vibrational modes of the zeolite structure may absorb this energy.

5. Conclusions

The deuterium NMR spectroscopy of phenanthrene-d_{10} adsorbed in Rb X-zeolite shows a temperature-dependent line shape. The line shape can be quantitatively analyzed by a simple exchange model, with phenanthrene exchanging between four energetically equivalent sites at rubidium ions and a fifth "free" site. The thermodynamic and kinetic parameters are consistent with those previously determined by NMR spectroscopy and with analogous gas-phase reactions of aromatics. Thus, we conclude that the formation of ion-molecule complexes in X-zeolites is similar to the formation of gas-phase ion-molecule complexes. These complexes are the origin of the temperature-dependent shifts in the absorption spectroscopy of these systems and may be important in the catalytic activity of these materials.

6. Acknowledgments

We acknowledge the generosity of Dr. V. J. Ramamurthy of the DuPont Company, who provided the materials and the impetus to study the materials with his investigations of their optical properties. Dr. D. R. Corbin, also of the DuPont Company, first suggested NMR spectroscopy might address the nature of the optical activity and he provided many suggestions about the zeolite and its interactions with substrates; his advice and counsel are acknowledged. We acknowledge the support of the National Science Foundation under grant CHEM-9013926.

60

7. References

1.   Gutowsky, H. S., McCall, D. W. and Slichter, C. P. (1953) *J. Chem. Phys.* **21**, 279.

2.   Gutowsky, H. S. and Holm C. H. (1957) *J. Chem. Phys.* **25**, 1288.

3.   Grunwald, E. and Jumper, C. F. (1963) *J. Amer. Chem. Soc.* **85**, 2051.

4.   Arnold, J. T. (1957) *Phys. Rev.* **102**, 136.

5.   Kessler, H. W. (1970) *Angew. Chem., Intl. Ed. Engl.* **9**, 219.

6.   Westheimer, F. H. (1968) *Acc. Chem. Res.* **1**, 90.

7.   Martin, G. L. and Martin, M. L. (1972) *Prog. Nucl. Magn. Reson. Spectrosc.* **8**, 163.

8.   Bottini, A. T. and Roberts, J. D. (1958) *J. Amer. Chem. Soc.* **80**, 5203.

9.   Saunders, M. (1967) in Ehrenberg, A., Malstrom, B. G. and Vanngard, T. (eds), *Magnetic Resonance in Biological Systems*, Pergamon, Oxford, p. 85.

10.   Binch, G. (1968) in E. L. Eliel and N. L. Allinger (eds), *Topics in Stereochemistry*, Interscience, New York **3**, 97.

11.   Derendyaev, B. D., Mamatyuk, V. I. and Koptyug, V. A. (1971) *Bull. Acad. Sci. USSR, Chem. Sci.* **5**, 972.

12.   Moro, G. and Freed, J. H. (1981) *J. Chem. Phys.* **74**, 3757.

13.   Abragam, A. (1961) *The Principles of Nuclear Magnetism*, Oxford University Press, London.

14.   Mehring, M. (1983) *Principles of High Resolution NMR in Solids*, Springer-Verlag, Berlin.

15.   Hepp, M. A. (1992) Ph.D. dissertation, University of Delaware.

16.   Spiess, H. W. and Sillescu, H. (1981) *J. Magn. Reson.*, *42*, 381.

17.   Spiess, H. W. and Sillescu, H. (1981) *J. Magn. Reson.*, *42*, 381.

18.   Roy, A. K., Jones, A. A. and Inglefield, P. T. (1985) *J. Magn. Reson.*, *64*, *441*.

19.   Wemmer, D. E., Rubin, D. J. and Pines, A. (1981) *J. Amer. Chem. Soc.*, *103*, 28.

20.   Greenfield, M. S., Ronemus, A. D., Vold, R. L., Vold, R. R., Ellis, P. D. and Raidy, T. E. (1987) *J. Magn. Reson.*, *72*, 89.

21.   Metzger, D. S. and Gaines, J. R. (1966), *Phys. Rev.*, *147*, 644.

22.   Niles, F. E. and Robertson, W. W. (1965) *J. Chem. Phys.*, *42*, 3277.

23.   Bohme, D. K., Dunkin, D. B., Fehsenfeld, F. C. and Ferguson, E. E. (1969) *J. Chem. Phys.*, *49*, 5201.

24.   Durden, D. A., Kebarle, P. and Good, A. (1969) *J. Chem. Phys.*, *50*, 805.

25.   Rabinowitz, E. (1937) *Trans. Faraday Soc.*, *33*, 283.

26.   Kebarle, P. (1972) in Franklin, J. L. (ed), *Ion-Molecule Reactions*, Plenum, New York, volume 1, p. 315.

27.   Diercksen, G. H. F. and Kraemer, W. P. (1972) *Theor. Chim. Acta*, *23*, 387.

28.   Clementi, E. E. and Popkie, (1972) *J. Chem. Phys.*, *57*, 1077.

29.   Kistenmacher, H., Popkie, H. and Clementi, E. (1973) *J. Chem. Phys.*, *58*, 1689.

30.   Kistenmacher, H., Popkie, H. and Clementi, E. (1973) *J. Chem. Phys.*, *59*, 5892.

31.   Kenney, J. E., Brumbaugh, D. V. and Levy, D. H. (1979) *J. Chem. Phys.*, *71*, 4757.

32.   Amirav, A., Even, U. and Jortner, J. (1980) *J. Phys. Chem.*, *85*, 309.

33.   Butz, K. W., Catlett, D. L., Ewing, G. E., Krajnovich, D. and Parmenter, C. S. (1986) *J. Phys. Chem.*, *90*, 3533.

FROM MOLECULAR CARBONYL CLUSTERS TO SUPPORTED METAL PARTICLES *Synthesis, Characterization, Catalysis.*

Roberto Giordano[a], Enrico Sappa[a] and Giovanni Predieri[b].
Dipartimento di Chimica Inorganica, Chimica Fisica e Chimica dei Materiali, Università di Torino. Via Pietro Giuria 7, I-10125 Torino, Italy. (b) Dipartimento di Chimica Generale ed Inorganica, Chimica Analitica e Chimica Fisica, Università di Parma. Viale delle Scienze, 43100 Parma, Italy.

1. Introduction.

The development of metal carbonyl cluster chemistry in the last 30 years was stimulated by the perspective of using these derivatives as active and selective homogeneous catalysts; unfortunately this aspect was overestimated. The efficiency of clusters is severely limited by the low stability of the metal frame under the conditions required for some reactions, or by the difficulty of releasing the organic moieties without affecting the metal core. Only a limited number of examples of **intact cluster** catalysis are known [1]. On the other hand, the difficulty of clusters in releasing organic moieties has been exploited for obtaining "models" of labile intermediates which could not be trapped in the reactions of mono- or binuclear catalysts [2].

The clusters have also been considered as soluble models for the chemisorption, activation and reactivity of small substrate molecules on the surface of metal particles; it has been shown that the bonding energies and the metal packing in clusters and in the metal lattices are comparable [3]. An attractive development of this approach, is the use of homo- or hetero-metallic carbonyl clusters as precursors of heterogeneous catalysts. These are likely to afford (i) the presence of low-valent metals (easy activation on oxide surfaces), (ii) the absence of anionic ligands (e.g. chloride) that would act as catalyst modifiers or poisons and (iii) controlled metal composition. The very small **cluster-derived metal particles (CDMP)** supported on inorganic oxides are highly active in a wide range of chemical transformations and may result in more selectivity than the systems formed by the corresponding metal salts with the same metal stoichiometry [4]. There is however an important aspect of the cluster-surface analogy that has been very little explored. Comparisons between the reactivity associated with similar bonding

63

L. J. Farrugia (ed.), The Synergy Between Dynamics and Reactivity at Clusters and Surfaces, 63–73.
© 1995 *Kluwer Academic Publishers.*

modes both on clusters and on surfaces have been rarely attempted. In particular, an evaluation of the behaviour of molecular clusters as homogeneous catalysts with respect to that of the CDMP prepared from the same clusters, vis-à-vis their reactivity with the same substrates is nearly nonexistent. This work represents a preliminary attempt in this direction.

We have focused our attention onto the tetrahedral, hydridic clusters $H_4Ru_4(CO)_{12}$ (**Complex 1**), $H_2Ru_4(CO)_{13}$ (**Complex 2**), $H_2FeRu_3(CO)_{13}$ (**Complex 3**) [5], $(CpNi)M_3(\mu-H)_3(CO)_9$ (M=Ru,Os) (**Complexes 4a, 4b**) [6] and $HRuCo_3(CO)_{12}$ (**Complex 5**) [7] whose structures are shown in **Figure 1**.

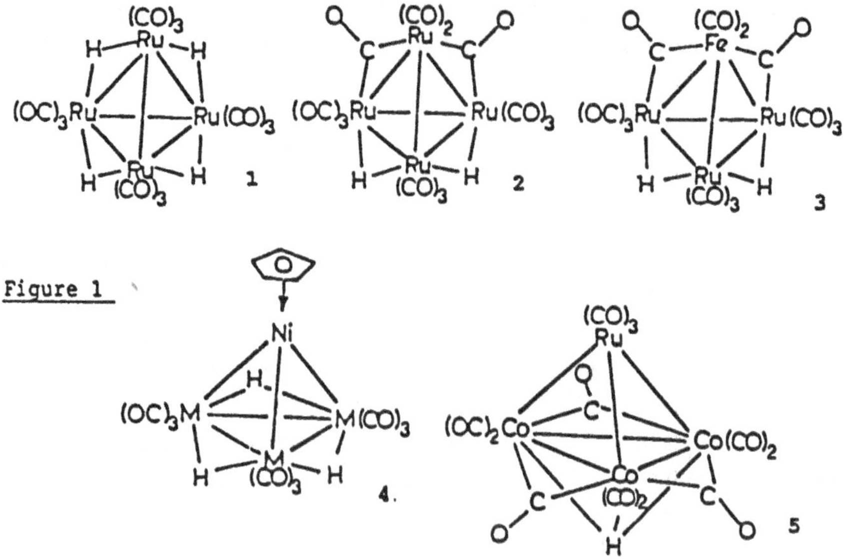

Figure 1

We have considered (ii) their homogeneous catalytic activity with the aim at understanding the role of intermediate organometallic complexes; (iii) their heterogeneous catalytic activity and (iv) their transformation into metal particles on inorganic oxide supports. (v) A comparison of the behaviour of the homogeneous and heterogeneous systems vis-à-vis of the same substrates has also been started.

2. Homogeneous hydrogenation of alkynes and dienes with clusters 1-4.

Clusters **4** are hydrogenation catalysts for linear [8] and cyclic dienes [9]: "cluster catalysis" was evidenced and reaction mechanisms proposed. Only one metal centre is involved and intermediates with the dienes coordinated to one metal in a π-fashion via one double bond were identified It is worth noting is that the capping (CpNi) acts as a stabilizing group. When it is lost, trinuclear derivatives with multi-site bound hydrocarbyls are formed, which are still active in isomerization, but not in hydrogenation.

Clusters **1**, **2** and **3** are active catalysts for the hydrogenation of C_2Ph_2 and the isomerization of cis-stilbene (Table 1) [5]. There is evidence for competition between formation of **catalytically active metal fragments** (induced by hydrogen) and condensation of metal fragments (induced by the alkyne) to **inactive alkyne-substituted organometallics**. Only $H_2Ru_3(CO)_9(C_2Ph_2)$ (**Complex A**, Figure 2) [10] was shown to be a catalytic intermediate. Apparently hydrogen is first added to the cluster metals and then transferred to the coordinated substrate. A number of examples (closely related to **A**) are known of catalytically active triangular ruthenium or osmium systems with bridging hydrogens and alkynes coordinated in a parallel fashion on the edges.[11] Therefore, a tempting hypothesis is that this structural arrangement promotes cluster-catalyzed homogeneous hydrogenation.

Clusters **1**, **2** and **3** are also active in hydrogenation and isomerization of cyclic dienes [12], **Table 1**. Once again we have found that catalysis is due to metal fragments, presumably to $Ru(CO)_3(C_6H_8)$ (**Complex B**). In the reaction solutions we could identify the organometallic complexes **C-F**, some of which already reported [13], **Figure 2**.

Table 1

Homogeneous catalytic activity of clusters 1-3 in the hydrogenation of 1,3- and 1,4-cyclohexadienes.

Complex[a]	maximum TON	organic products
	Hydrogenation of C_2Ph_2[b]	
$H_4Ru_4(CO)_{12}$	58.6	cis-stilbene
$H_2Ru_4(CO)_{13}$	41.6	
$H_2FeRu_3(CO)_{13}$	54.1	
	Isomerization of cis-stilbene[c]	
$H_4Ru_4(CO)_{12}$	32.1	trans-stilbene
$H_2Ru_4(CO)_{13}$	139.7	
$H_2FeRu_3(CO)_{13}$	182.1	
	Hydrogenation of 1,3-CHD	
$H_4Ru_4(CO)_{12}$	6.6	cyclohexene
$H_2Ru_4(CO)_{13}$	53.4	
$H_2FeRu_3(CO)_{13}$	26.8	
	Hydrogenation of 1,4-CHD	
$H_4Ru_4(CO)_{12}$	36.7	1,3-CHD
$H_2Ru_4(CO)_{13}$	86.7	
$H_2FeRu_3(CO)_{13}$	42.1	
	Isomerization of 1,4-CHD	
$H_2Ru_4(CO)_{12}$	6.6	1,3-CHD
$H_2Ru_4(CO)_{13}$	24.4	
$H_2FeRu_3(CO)_{13}$	18.8	

(a) Reaction conditions; t=120°C, H_2=1 atm (except for the isomerization experiments), solvent n-octane. (b) See Ref. 6.

In contrast with the behaviour of **A**, complexes **C-F** are side-products in the catalytic reactions. An explanation could be the following; complex **A** is formed by coordination of alkyne and addition of extra hydrogen, whereas complexes **C-F** are formed upon oxidative addition (dehydrogenation) of the dienes. Therefore, external hydrogen would reform the dienes rather than give hydrogenation products. This process should be compared with the behaviour of the metal particles in the presence of the same dienes, as discussed below.

3. Heterogeneous catalytic activity of CDMP from supported clusters 2,3,5.

The hydrogenation-dehydrogenation experiments on dienes were carried out in a catalytic-analytic gas-chromatographic system [14], while the methanation, ammonia synthesis and Fischer-Tropsch reactions were carried out in a pulse reactor connected to a gas chromatograph [15].

Hydrogenation of dienes, benzene and toluene in the presence of CDMP from clusters 2,3,5. This occurs at low temperatures and under atmospheric pressure of hydrogen (**Table 2**).

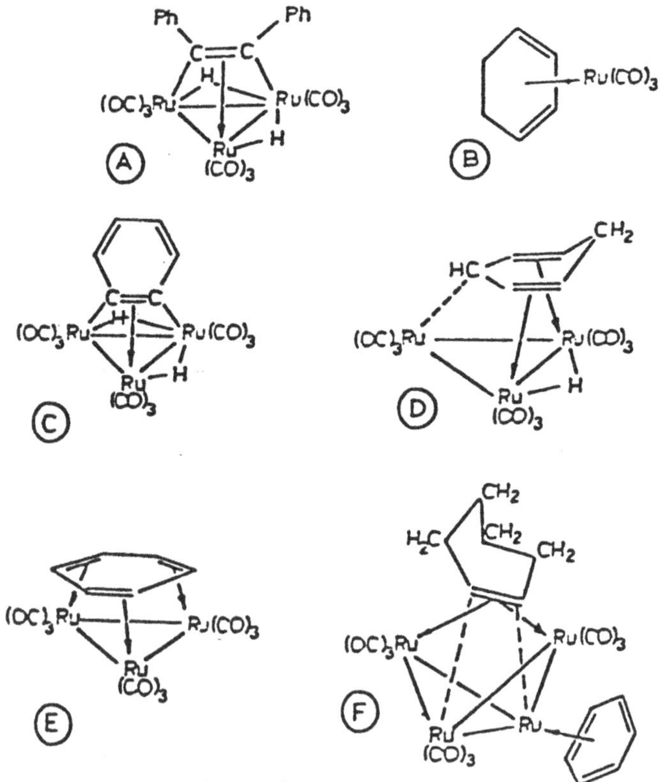

Figure 2

Table 2

Hydrogenation and hydrogenolysis experiments on metal particles from clusters **2,3,5**[a]

Substrate	Cluster	Products and yields
	Reactions at 60°C	
1,3-CHD[c]	$H_2Ru_4(CO)_{13}$	Cyclohexane 99.8%[b]
	$H_2FeRu_3(CO)_{13}$	Cyclohexane 99.9%
	$HRuCo_3(CO)_{12}$	Cyclohexane 100%
Benzene	$H_2Ru_4(CO)_{13}$	Cyclohexane 100%
	$H_2FeRu_3(CO)_{13}$	Cyclohexane 100%
	$HRuCo_3(CO)_{13}$	Cyclohexane 100%
Toluene	$H_2Ru_4(CO)_{13}$	Methyl-cyclohexane 100%
	$H_2FeRu_3(CO)_{13}$	Methyl-cyclohexane 99.8%
	$HRuCo_3(CO)_{12}$	Methyl-cyclohexane 100%
	Reactions at 100°C	
1,3-CHD	$H_2Ru_4(CO)_{13}$	Cyclohexane 99.7%
	$H_2FeRu_3(CO)_{13}$	Cyclohexane 99.3%
Benzene	$H_2Ru_4(CO)_{13}$	Cyclohexane 100%
	$H_2FeRu_3(CO)_{13}$	Cyclohexane 96.3%
Toluene[d]	$H_2Ru_4(CO)_{13}$	Methyl-cyclohexane 99.8%
	$H_2FeRu_3(CO)_{13}$	Methyl-cyclohexane 96.1%

(a) Reaction conditions; H_2 flow 37.5 mL/min; substrate injected 0.2 μL. Metal charge: cluster (2) 0.40%, cluster (3) 0.29%, cluster (5) 0.42%. (b) In all reactions the difference to 100% is given by hydrogenolysis products. (c) Cyclohexane and 1,4-CHD behave as 1,3-CHD. (d) Cyclohexane and methane as byproducts.

More interesting are the dehydrogenation and disproportionation reactions observed when N_2 is used as a carrier (**Table 3**); these occur at higher temperatures (starting from 100°C) and some cracking is observed. Uptake of hydrogen from the substrate, its chemisorption on the metal particles and release at higher temperatures has been already hypothesized for **4a** [16]. It is worth noting that these reactions correspond to the oxidative addition reactions observed in homogeneous conditions.

In our attempts at obtaining kinetic and mechanistic informations, there were some indications that the hydrogenation of 1,3-CHD on $H_2Ru_4(CO)_{13}$ derived CDMP apparently follows the Michaelis-Menten approach (**Figure 3**), with a unimolecular first-order mechanism at the beginning shifting to zero-order and reaching a limiting value for the rate when the catalyst surface is saturated by the chemisorbed reactant. We have also some indications for the catalytic activity being inversely related to the chemisorption strength of the substrate at high surface coverages.

Methanation, ammonia synthesis and Fischer-Tropsch reactions. We have achieved ammonia synthesis on CDMP from $Ru_3(CO)_{12}$ or **4a** on alumina added with potassium hydroxide [15c]. We also published preliminary results on CO and CO_2 methanation in the presence of CDMP derived from **4b** and related derivatives [15a,b]; we had hypothesized that methanation of CO_2 on different CDMP was dependent on the formation of CO in a reverse WGSR reactions, but we had no evidence for this behaviour.

Table 3

Dehydrogenation-disproportionation reactions on metal particles from clusters **2,3,5**,[a,b].

Substrate	Cluster	Products and yields
		Reactions at 100°C
Cyclohexene	(2)	Cyclohexane 0.3%[c], Benzene 99.6%
	(3)	Cyclohexane 0.9%, Benzene 98.3%
	(5)	Cyclohexane 0.7%, Benzene 16.8%
1,3-CHD[d]	(2)	Cyclohexane 0.2%, Benzene 99.6%
	(3)	Cyclohexane 0.6%, Benzene 99.2%
	(5)	Cyclohexane 0.8%, Cyclohexene 14.0% Benzene 85.1%
		Reactions at 150°C
Cyclohexene	(2)	Cyclohexane 16.8%, Benzene 83.1%
	(3)	Cyclohexane 2.5%, Benzene 97.5%
	(5)	Cyclohexane 19.8%, Benzene 64.9%
1,3-CHD	(2)	Cyclohexane 0.2%, Benzene 96.0%
	(3)	Cyclohexane 0.9%. Benzene 99.1%
	(5)	Cyclohexane 13.2%, Benzene 86.8%
		Reactions at 200°C
Cyclohexene	(2)	Cyclohexane 32.9%, Benzene 64.1%[e]
	(3)	Cyclohexane 32.6%, Benzene 66.8%[e]
1,3-CHD	(2)	Cyclohexane 12.2%, Benzene 87.7%[e]
	(3)	Cyclohexane 0.2%, Benzene 98.3%[e]

(a) Reaction conditions; N_2 flow 25.0 mL/min; injected substrate 0.2 μL. Metals charge see Table 2. (b) Cyclohexane gives only traces of cyclohexene and benzene. Methyl-cyclohexane is hydrogenated in small yields to toluene, benzene and cyclohexane. (c) Difference to 100%, unreacted substrate. (d) 1,4-CHD behaves in the same way. (e) Traces of cracking products.

Further studies on methanation and Fischer-Tropsch reactions in the presence of CDMP from **4a** supported onto alumina showed that, during hydrogenation of CO_2 at different temperatures under a CO_2/H_2 flow, chemisorbed CO is formed from reverse WGSR via HCOOH. Formic acid (gc-ms, 46 m/z) is both a reaction product and an intermediate for the production of methane via CO, mostly at high temperatures (**Scheme 4**).

$$CO_2 + 2\ H(M) \Longleftrightarrow HCOOH(ad)$$
$$HCOOH(ad) \Longleftrightarrow CO(M) + H_2O$$
$$CO(M) + 6\ H(M) \longrightarrow CH_4 + H_2O$$

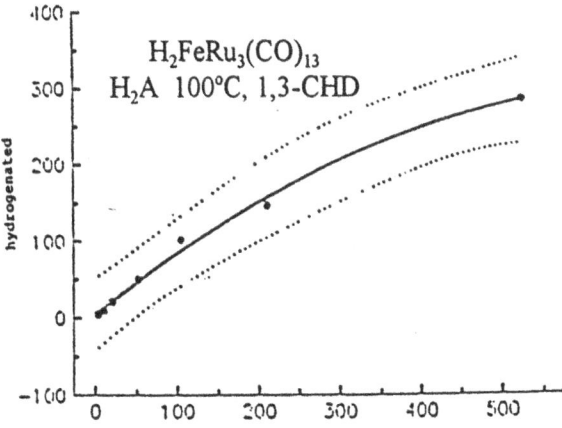

Figure 3: injected

Figure 4 shows two significant regions of a IR spectrum at the intermediate temperature of 315°C when chemisorbed formate is still present (2910 and 1590 cm^{-1}) and gaseous methane (3020 cm^{-1}) is appearing. The maximum concentration of adsorbed formate occures between 200 and 300°C, whereas at 500°C it disappears and methane production reaches its maximum. Interestingly, there is a recent "homogeneous model" for this reaction [17].

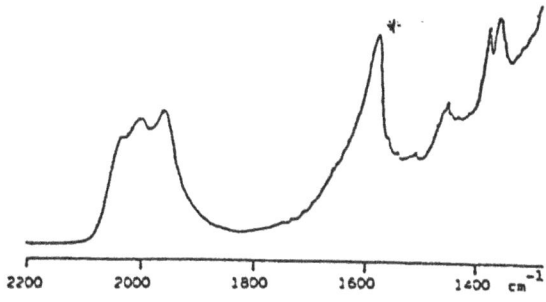

Figure 4

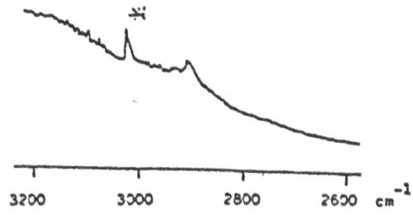

Finally, CDMP from **4a** have been tested in the Fischer-Tropsch reaction, to compare with a corresponding system obtained from a mechanical mixture. The molar selectivity towards C_3-C_7 hydrocarbons at 235 and 250°C [$(H_2/CO)=2$, P=5 bar] is respectively 37%-28% with the **4a** CDMP (molar yields 8-10%) and 19%-37% with the Ni/Ru$_3$ mechanical mixture (molar yields 2-10%).

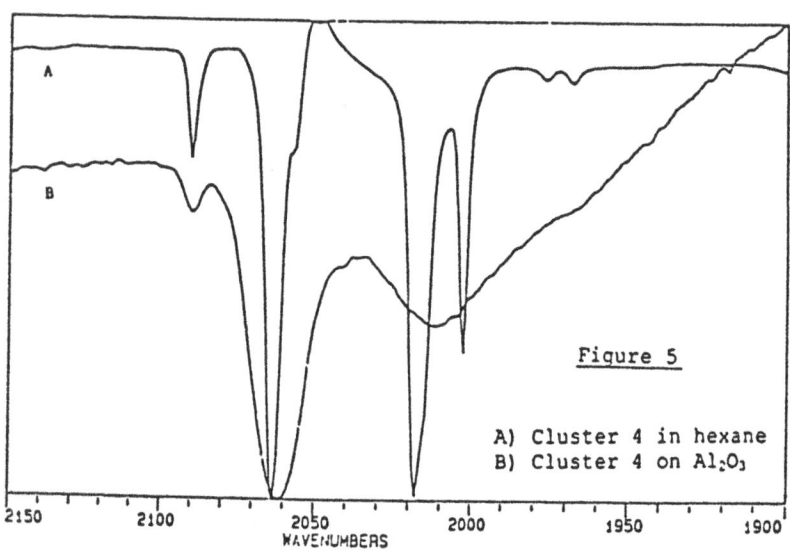

Figure 5

A) Cluster 4 in hexane
B) Cluster 4 on Al$_2$O$_3$

4. Transformation of clusters into supported metal particles.

The preparation of the heterogeneous catalysts for the above hydrogenation-dehydrogenation reactions was carried out as described in refs. 15-18.

The formation of supported metal particles during ammonia synthesis was followed for **4a** [18]: further evidence has been obtained by following the trans-formations of **4a** on alumina (metal loading 2%) during CO and CO$_2$ hydrogenation experiments. Initially **4a** is simply physisorbed on the support (**Figure 5**); on heating, different carbonyl species are produced as shown by diffuse-reflectance FTIR spectroscopy (**Figure 6**). Under a constant helium flow, at 300°C, two pairs of carbonyl bands are observed at 2060, 1990 and 2040, 1960 cm^{-1}, probably due to geminal mono- or poly-nuclear carbonyl species [19]. At 450°C only a single carbonyl band at 1950 cm^{-1} attributable to chemisorbed CO on metal aggregates is detected. It disappears after 20 min at 500°C. Under hydrogen, a comparable behaviour is observed, a significant difference being the appearance of only one pair of bands (2040, 1960 cm⁻¹) at 300°C; these being typical of pure ruthenium carbonyl species. [19] From these data, it is not possible to deduce the fate of the nickel atoms. It is likely that metal segregation occurs and that nickel is partially trapped as a surface aluminate: nevertheless, at the end of the thermal decomposition bimetallic particles of nanometric size have been observed by TEM.

The decarbonylated material, when exposed to a flow of carbon monoxide is able to fix CO, giving rise to a complex carbonyl stretching pattern in which the pair of bands at 2040, 1960 cm⁻¹ are well distinguishable. The catalysts, obtained under

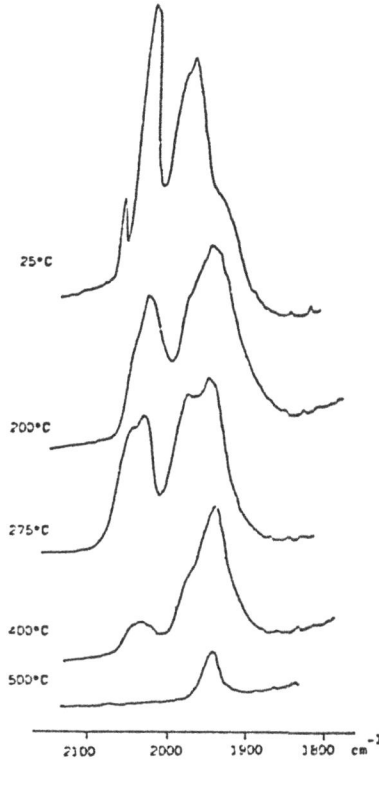

Figure 6

helium or hydrogen at 450°C are active in CO and CO_2 methanation under atmospheric pressure, as described above; CO is completely converted to methane at 300°C over the catalyst treated under hydrogen.

5. Comparison of the catalytic activity of clusters in solution and of supported metal particles, vis-à-vis of the same substrates.

We are beginning a solid-state spectroscopic study of the behaviour of cluster-bound, 1,3-CHD and benzene ligands - as found in the organometallic products C, D and E from the homogeneous experiments - with that of the same ligands on the supported metal particles during heterogeneous catalytic runs. A characterization of the CDMP from clusters 2 and 3 when supported and treated on silica is also under progress.

Acknowledgements. We thank Prof. M. Castiglioni (Università di Torino) for helpful discussions. DRIFT spectra were recorded by Dr. L. Basini (SNAM Progetti, S. Donato Milanese). Fischer-Tropsch tests were performed by Prof. V. Ragaini (Università di Milano). Financial support to this work was insured by MURST and CNR, Programmi Finalizzati Chimica Fine II°.

References and Notes.

1. Süss-Fink, G. and Meister, G. (1993), *Adv. Organomet. Chem.*, **35**, 41-134.
2. Boroni, E., Costa, M., Predieri, G., Sappa, E., Tiripicchio, A. (1992), *J. Chem. Soc. Dalton Trans.*, 2585-2590, and references therein.
3. Muetterties, E.L., Rhodin, T.N., Band, E., Brucker, C.F. and Pretzer, W.R. (1979), *Chem. Rev.*, **79**, 91-137.
4. (a) Gates, B.C., Gucz,i L. and Knözinger, H. (Eds), (1986), *Metal Clusters in Catalysis*, Elsevier, Amsterdam, (b) Whyman, R. in Basset, J. M., et al (Eds), *Surface Organometallic Chemistry: Molecular Approaches to Surface Catalysis*, Kluwer, Dordrecht, 1988, pp. 75-95. (c) Braunstein, P. and Rosé, J. (1989) in I. Bernal (Ed.) *Stereochemistry of Organometallic and Inorganic Compounds*, Vol. 3, Elsevier, Amsterdam. (d) Fackler Jr, J.P. (Ed.) (1990) *Metal-metal Bonds and Clusters in Chemistry and Catalysis*, Plenum, New York. (e) Psaro, R., Dossi, C., Sordelli, L., Ugo, R (1992), *Chim. e. Ind.*, **74**, 154-165.
5. Giordano, R. and Sappa, E. (1993), *J. Organomet. Chem.*, **448**, 157-166, and references therein.
6. Castiglioni, M., Sappa, E., Valle, M., Lanfranchi, M. and Tiripicchio, A (1983), *J. Organomet. Chem.*, **241**, 99-111.
7. Mays, M.J. and Simpson, R.N.F., (1968), *J. Chem. Soc (A)*, 1444-1447.
8. Castiglioni, M., Giordano, R., Sappa, E., Tiripicchio, A., Tiripicchio Camellini, M. (1986), *J. Chem. Soc., Dalton Trans.*, 23-30.
9. Castiglioni, M., Giordano, R. and Sappa, E. (1987), *J. Organomet. Chem.*, **319**, 167-181.

10. Cauzzi, D., Giordano, R., Sappa, E., Tiripicchio, A., Tiripicchio Camellini, M. (1993) *J. Cluster Sci.* **4**, 279-296.

11. (a) Castiglioni, M., Gervasio, G., Sappa, E. (1981), *Inorg. Chim. Acta.* **49**, 217-225. (b) Amadelli, R., Bartocci, C., Carassiti, V., Aime, S., Osella, D., Milone, L. (1985), *Gazz. Chim. Ital.* **115**, 337-342. (c) Aime, S., Gobetto, R., Milone, L., Osella, D., Violano, L., Arce, A.J., and De Sanctis, Y. (1991) *Organometallics,* **10**, 2854-2856. (d) Castiglioni, M., Giordano, R., Sappa, E., (1989), *J. Organomet. Chem.*, **369**, 419-431. (e) Basu, A., Bhaduri, S., Sharma ,K., Jones, P.G., (1987), *J. Chem. Soc., Chem. Commun.*, 1126-1127. (f) Adams, R.D., Li, Z., Swepston, P., Wu,W., Yamamoto, Y., (1992), *J. Am. Chem. Soc.*, **114**, 10657-10658. (g) Bianchi, M., Menchi, G., Matteoli, U., Piacenti, F. (1993), *J. Organomet. Chem.*, **451**, 139-146 and references therein.

12. Castiglioni, M., Giordano, R., Sappa, E., manuscript in preparation.

13.   (a) Domingos, A.J.P., Johnson, B.F.G. and Lewis, J. (1972), *J. Organomet. Chem.*, **36** C43-44. (b) Aime, S., Milone, L., Osella, D., Vaglio, G.A., Valle, M., Tiripicchio, A, and Tiripicchio Camellini ,M. (1979), *Inorg. Chim. Acta.* **34**, 49-55. (c) Goudsmit, R.J., Johnson, B.F.G., Lewis, J., Raithby, P.R. and Rosales, M. (1983), *J. Chem. Soc., Dalton Trans.*, 2257-2261. (d) Johnson, B.F.G., Lewis, J., Martinelli, M., Wright, A.H., Braga, D. and Grepioni, F. (1990), *J. Chem. Soc., Chem. Commun.* 364-366. (e) Braga, D., Grepioni, F., Johnson, B.F.G., Lewis, J., Housecroft, C.E. and Martinelli, M. (1991), *Organometallics*, **10**, 1260-1268.

14. (a) Castiglioni, M., Giordano, R., Sappa, E., Predieri, G., Tiripicchio, A. (1984), *J. Organomet. Chem.* **270**, C7-10. (b) Castiglioni, M., Giordano, R., Sappa, E. (1986), *J. Mol. Cat.* **37**, 287-296. (c) Idem, *ibidem* (1987), **40**, L65-69. (d) Castagno, F., Castiglion,i M., Sappa, E., Tiripicchio, A., Tiripicchio Camellini, M., Braunstein, P., Rosé, J. (1989), *J. Chem. Soc., Dalton Trans.*, 1477-1482.

15. (a) Moggi ,P., Albanesi, G., Predieri, G., Sappa, E. (1983), *J. Organomet. Chem.*, **252**, C 89-92. (b) Albanesi, G., Bernardi, R., Moggi, P., Predieri, G., Sappa, E. (1986), *Gazz. Chim. Ital.* **116**, 385-390. (c) Moggi, P., Predieri, G., Albanesi, G., Papadopoulos, S., Sappa, E. (1989), *Appl. Cat.* **53**, L1-4.

16. Castiglioni, M., Giordano, R., Sappa, E. (1987), *J. Mol. Cat.* **42**, 307-322.

17. Sakamoto, M., Shimizu, I and Yamamoto, A (1994), *Organometallics* **13**, 407-409.

18. (a) Predieri, G., Moggi, P., Papadopoulos, S., Armigliato, A., Bigi, S., Sappa, E. (1990) *J. Chem. Soc., Chem. Commun.*, 1736-1737. (b) Armigliato A., Bigi S., Moggi, P., Papadopoulos, S., Predieri, G., Salviati, G., Sappa, E. (1991), *Materials Chem. and Phys.* **29**, 251-260.

19. (a) Zecchina, A., Guglielminotti, E., Bossi, A. and Camia, M. (1982), *J. Catal.* **74**, 225-239. (b) Guglielminotti, E., Zecchina, A., Bossi, A. and Camia, M. (1982), *ibidem* **74**, 240-251. (c) Idem (1982), *ibidem* **74**, 252-265.

CLUSTER EQUILIBRIA. RELEVANCE TO THE ENERGETICS AND REACTIVITY OF SURFACE BOUND FRAGMENTS

Thomas P. Fehlner
Department of Chemistry and Biochemistry
University of Notre Dame
Notre Dame, IN 46556 USA

1. Introduction

The possibility of connections between the properties of clusters on the one hand and surfaces on the other has stimulated a considerable amount of research in the area of cluster chemistry. [13]   Much has concerned structure (geometric and electronic) and spectroscopic properties and the successes have been detailed elsewhere. [4, 5]   Hence, it is highly appropriate that the other half of cluster chemistry, that concerned with reactivity, be examined in light of possible connections with surface reactivity. What follows is a review of studies of cluster reactivity, mainly drawn from our own work, with an emphasis on its relevance to the reactivity of molecules on surfaces.

1.1 ROLE OF INTERMEDIATES

Most chemical reactions of practical interest proceed from reactants to products via a complex mechanism involving intermediate species of varying stabilities.   As illustrated in the reaction coordinate diagram in Figure 1, the energies and structures of the intermediates resemble those of the crucial activated complexes more closely than those of the reactants or products, eqn (1). This is an essential reason why the characterization of reaction intermediates continues to be an important method for the elucidation of reaction mechanisms. Although the experiments can be difficult, both direct and indirect methods have been successfully employed

$$\text{Reactants} \rightarrow [I_1] \rightarrow [I_2] \rightarrow \text{Products} \qquad (1)$$
$$\uparrow\downarrow$$
$$[I_x]$$

75

L. J. Farrugia (ed.), The Synergy Between Dynamics and Reactivity at Clusters and Surfaces, 75–94.
© 1995 *Kluwer Academic Publishers.*

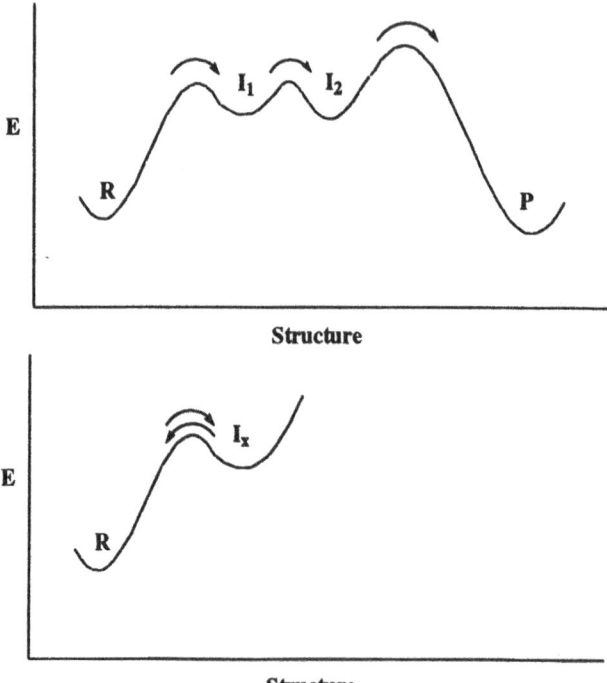

Structure

Figure 1. Reaction profiles for equation (1).

for the detection and characterization of intermediates in homogeneous reactions in solution or in the gas phase. [6]   Because of the mechanistic value of these data, substantial efforts to isolate individual reactions steps as well as interconversions of individual states of reactants and products continue.

It follows that the characterization of intermediates in reactions taking place on the surface of a solid, e.g., in a heterogeneous, catalyzed reaction, will also yield mechanistically important information. Unfortunately, measurement of the structure and energy of a stable surface species is a difficult task in itself. The characterization of unstable intermediate species is even more so and the surface mechanistic problem becomes a formidable one. As with homogeneous systems, the identification of an abundant surface species does not guarantee that it is an actual intermediate on the reaction pathway. Highly reactive species at low abundances can carry a reaction pathway. In spite of these difficulties, the application of modern spectroscopic methods have yielded considerable information on active surface species. [7, 8] The mechanistic insight thereby gained can be impressive and direct connections to important processes adds a practical touch. [9] Still, the price of these experiments in both time and money is sufficiently large that less direct, model-based approaches contribute valuable understanding of the nature of surface reactions.

1.2. CLUSTER-SURFACE HYPOTHESIS

The cluster-surface hypothesis[5, 10] was well received when it was presented in the 1970's. It added impetus to the drive to develop cluster chemistry as a new area of inorganic chemistry. Indeed, it was thought that clusters themselves might serve as catalysts combining the advantages of homogeneous catalysts with the multinuclear metal sites of heterogeneous catalysts. [2] Although this particular application has developed slowly, the hypothesis of a cluster-surface analogy remains of considerable interest. In particular, its success in developing models of organic fragments in unusual bonding arrangements with respect to an array of metal atoms is well known. For example, the characterization of the vibrational behavior of a C=C=O moiety bound to a three metal cluster provides a spectroscopic signature useful in identifying the same species on an analogous metal surface site. [11]

The semi-localized nature of the fragment-metals interaction permits cluster compounds to serve as useful platforms for the study of some aspects of the bonding of fragments to surfaces. However, the cluster-surface model has been criticized in the sense that the metal atoms in a metal cluster are very different from metal atoms in a metal crystallite of some size. [12] Certainly, the ancillary cluster ligands greatly perturb the metal centers and ligand effects on the metal-fragment bonding must be substantial. This is easily verified by partitioning the disruption energy of a cluster into that derived from metal-ligand interactions and that derived from metal-metal bonding. The former is considerably larger reflecting the weaker, albeit important, metal-metal interactions. If surface species are to be modeled with cluster systems, ancillary ligand effects must be taken into account.

Another aspect of the model approach is illustrated in Figure 2. By their nature, model compounds (M) are stable entities and sit at the bottom of potential wells. In general metal cluster rearrangements have low barriers and, thus, the putative surface species constitutes a local

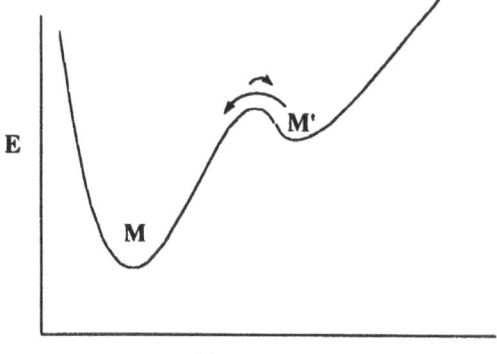

Structure

Figure 2. Relationship of alternate cluster structures to observed form M.

thermodynamic minimum. The probability that such a surface species is a dead-end intermediate is high, i.e., the cluster models $[I_x]$ in Eqn (1), rather than $[I_i]$. The characteristics of the model compound itself give little indication of the relative energy and structure of closely related species (M') that may better mimic $[I_i]$ on the reaction pathway. Further, even when models for reasonable consecutive intermediates are found, the ancillary ligands are often different making any direct comparison of energies impossible. Thus, the relevance of a single stable cluster to the problem of surface reactivity, as opposed to surface structure, is limited.

1.3. RELEVANT CLUSTER TYPES

If all clusters were irrelevant to the problem, this contribution would end here. It does not as there are clusters which can be used to address the question of surface reactivity. I will focus on two types in the following.

1.3.1. *Isoelectronic Clusters*
In one, the structure and properties of a transition metal cluster containing a complex organic ligand are compared with those of an isoelectronic cluster containing a related main group element ligand. We have pointed out previously that the contrast of isoelectronic organometallic and inorganometallic clusters leads to useful ideas of reactivity. [13] Figure 3 illustrates one example. As elaborated below, the structure with M-H-E bonds, E = main group atom, M = metal, is favored for E = B$^-$ rather than C. Thus, changing carbon to another main group element (B vs C) changes the energetics and permits characterization of a structure which, for the other element (carbon), is not the most stable structure. It follows that intermediate species containing carbon can be modeled by isoelectronic compounds. An interesting and useful extension of this idea is the comparison of isoelectronic clusters in which the metals are different but the main group element fragment of interest is the same.

A good analogy comes from biochemistry of all places. In an enzyme in which an intermediate or transition state resembling a tricoordinate borane is the crucial structure on the reaction pathway, a borane mimic of the substrate binds tightly to the active form of the enzyme thereby lowering its energy so much that activity is suppressed. This idea has been effectively extended to the synthesis of borane analogs of biologically important molecules which are found to have significant anticancer activity, i.e., they shut down metabolic pathways. [14] We suggested[15] that the same game can be played with the cluster models of surface species.

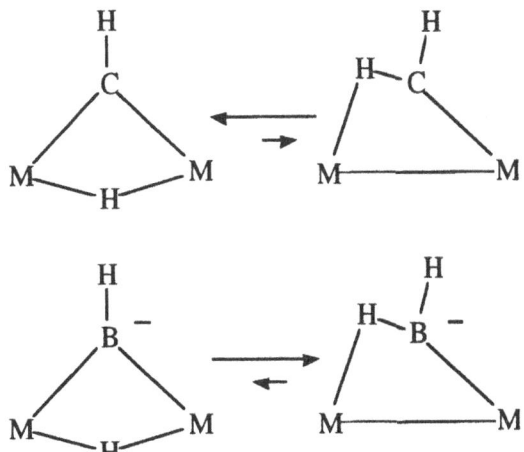

Figure 3. Relative stability of CHM vs BHM interactions.

1.3.2. *Clusters in Equilibrium*

The second relevant type of clusters encompasses systems that exhibit observable equilibria between isomeric cluster forms or different compositions. Equilibrium properties yield information on both reactivities and energetics. Importantly, one gains precise and detailed information on an actual reaction involving a fragment of interest. In addition, although a single cluster may not be a good absolute model for a surface bound species (see above), the differences between two surface fragment model compounds, which are directly related by a chemical reaction, should be connected to the differences between actual surface species, i.e., much of the effect of the ancillary ligands should cancel out in the comparison.

Recall that there is an exponential relationship between equilibrium constant and energy, Eqn (2), and therefore the equilibrium constant, K, is a very sensitive measure of energy. Even crude measurements of K and the temperature variation of K yield detailed energetics with a precision

$$K = e^{-\Delta G/RT} = [\text{reactants}]/[\text{products}] \qquad (2)$$

that can't be approached by present day calculations for these large systems. This permits selected low barrier reactions of relevance to catalytic pathways to be discussed with confidence, e.g., the properties of M that affect the relative stabilities of the structural forms illustrated in Figure 3. The fact that only selected systems can be observed constitutes an intrinsic drawback. However, as the present fixation of the cluster chemist with solid state structural data wanes, more studies of cluster reactivity in solution should reveal additional examples.

In the following these two ideas are applied to (a) the distribution of H atoms between E = main group atom and surface M = transition metal atom, (b) the effect of heterometals on the E-M interaction and (c) the general relevance of cluster reactions to problems in surface reactivity.

2. Distribution of H Atoms on Clusters and Surfaces.

The addition and removal of hydrogen atoms to and from a surface species is an important component of many surface reactions, e.g. C-H bond activation. [16] That is, the interaction of a hydrocarbon with a surface must involve, at least transiently, a C-H-M_n interaction of some type. Likewise the formation of a C-H bond from a surface carbide and a chemisorbed H atom must involve a C-H-M_n interaction. Clearly the factors that affect well depths and barrier heights for the interconversions shown in Figure 4 are fundamental to the understanding of these systems.

In this section three questions are addressed. (a) What do the structures of these intermediates look like? (b) What is the role of the ancillary ligands on cluster structure? (c) What properties of the metals

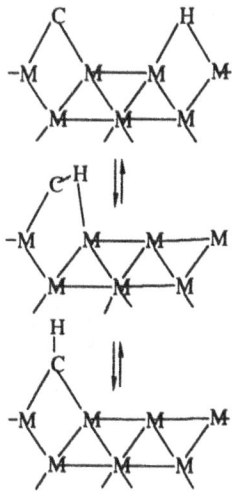

Figure 4. Interconversion of surface carbide to methylidyne fragment.

and main group atom, e.g., carbon, are important in determining the equilibrium distribution of H atoms with respect to E and M?

2.1. STRUCTURES

The trimetal cluster having the molecular formula $Fe_3(CO)_9CH_4$ exists in solution as an equilibrium mixture of three tautomers, $(\mu\text{-H})_3Fe_3(CO)_9(\mu_3\text{-CH})$, $(\mu\text{-H})_2Fe_3(CO)_9(\mu_3\text{-HCH})$, and $(\mu\text{-H})Fe_3(CO)_9(\mu_3\text{-H}_2CH)$. [17] These are models for methyne, **2**, methylene, **3**, and methyl, **4**, fragments bound to a trimetal site (Figure 5). The structures were established by NMR experiments including partial deuteration in the case of fluxional species. A crystal structure of the ethylidyne complex which exhibits a single tautomeric form, establishes the basic framework structure. [18] The most stable isomer has the structure consistent with conventional organometallic wisdom, e.g, that of the ethylidyne cluster. The existence of the equilibria as well as the stability order of the three species was unambiguously determined by deprotonation to yield $HFe_3(CO)_9HCH^-$ which on reprotonation at low temperature gave exclusively the cluster

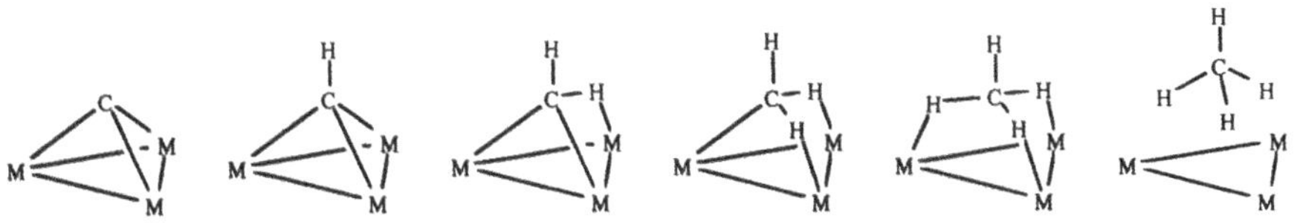

Figure 5. CH_n fragments on a triangular metal surface site.

with a coordinated methyl fragment, **4**, which is the least stable tautomer. This species rearranges as the temperature increases to yield the equilibrium mixture at room temperature. Addition of Lewis base led to the formation of CH_4 thereby establishing this set of structures as realistic models for the hydrogenation of carbon to form CH_4. The ruthenium analog apparently only exhibits the tautomeric form with a methylidyne fragment, **2**, showing that metal identity affects the equilibrium position of this tautomerization. [19]

The isoelectronic boron derivative, $[HFe_3(CO)_9H_2BH]^-$, is fluxional even at low temperature and no tautomeric forms can be observed by NMR in solution. [20] It exhibits the structure of the least stable form of the isoelectronic carbon derivative, i.e., an EH_3 metal bound fragment, **4**. In this case, the neutral $HFe_3(CO)_9H_3BH$ has been crystallographically characterized (Figure 6) [21] and the tautomeric form of the fluxional monoanion was proven by NMR and Mössbauer spectroscopies. Clearly, changing the identity of the main group atom in this pair of isoelectronic species does result in the stabilization of a higher energy structural form of the organometallic cluster. This verifies that such an approach can be used to model structures of selected carbon based intermediates with analogous main group compounds.

The neutral ferraborane (Figure 6) can be viewed as a BH_4^- ligand (isoelectronic with CH_4) coordinated to a three metal site and a model for structure **5** in Figure 5. [15] This may be viewed as a model for an intermediate or transition state for the reaction of $Fe_3(CO)_9CH_4$ with a Lewis base resulting in the formation of methane.

The ruthenium analog of $HFe_3(CO)_9BH_4$ exists in solution as a 1 : 1 mixture of tautomeric forms having main group fragment structures **4** and **5**. [22] Changing the metal from Fe to Ru is sufficient to shift the equilibrium from **5** to **4** in the direction of dehydrogenation of the main group atom. Note that changing from Fe to Ru has the same relative effect on tautomer stability for the isoelectronic carbon and boron systems.

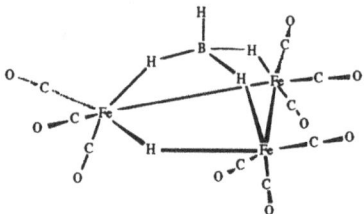

Figure 6. X-ray crystal structure of (μ-H)Fe$_3$(CO)$_9$(μ_3-H$_3$BH).

The trimetal system is the smallest system that has important cluster properties, e.g., a triangular metal array. But a metal surface presents many other surface site geometries to an admolecule. Thus, it is important to note that a four-metal system exhibits the same type of behavior, albeit with different tautomeric distributions. The similarities and differences shed light on the role of metal cluster size and metal site geometry which is presented to the main group species of interest by the surface.

For the cluster HFe$_4$(CO)$_{12}$CH structure **8**, Figure 7, is observed. [23] On deprotonation, **9** and then **10** are produced. No tautomeric equilibria are observed. However, for ruthenium, **8** and **8'** are observed in equilibrium in solution. [24] Again changing from Fe to Ru promotes M-H-M interactions over M-H-E interactions. For boron, HFe$_4$(CO)$_{12}$BH$_2$ exhibits structure **7** and deprotonates via **8**, **9'** and **10**. [25] However, for ruthenaborane analog the sequence is **7**, **8**, **9** and **10**. [26]   No tautomeric equilibria been reported for the ferra- or ruthenaboranes but the comparison reveals the same effect of metal as seen in the isoelectronic carbon system.

The trimetal cluster is representative of one type of site on a flat metal surface whereas the four metal cluster models a site at a step on a metal surface (Figure 8). [27]   The cluster results suggest that the distribution of surface bridging hydrogens between M-H-E and M-H-M

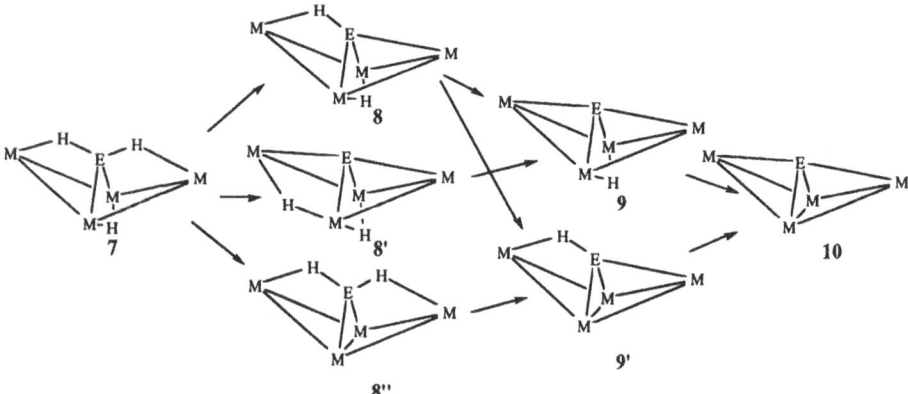

Figure 7. Tautomeric forms of the M$_4$E "butterfly" clusters.

types depends not only on atom properties but also site characteristics including geometry, i.e., in going from a M_3 triangular site to a M_4 "butterfly" site an equilibrium like that in Figure 4 shifts towards the carbide, i.e.., in the same direction as changing cluster atoms from Fe to Ru.

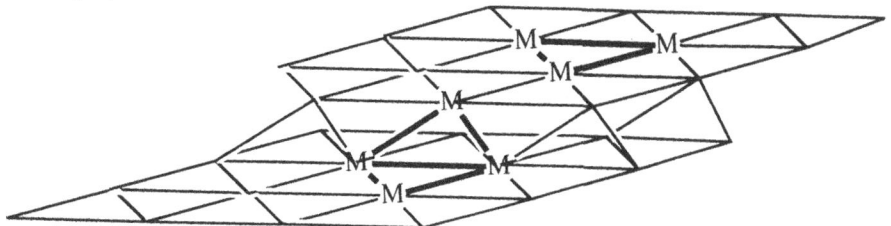

Figure 8. Representation of M_3 and M_4 sites on a stepped metal surface.

2.2. ROLE OF ANCILLARY LIGANDS

The observations described above appear relevant to the question of surface fragment reactivity but the effects of the ancillary ligands must be more fully understood if the cluster to surface extrapolation is to be meaningful. Clearly, the numbers and positions of the hydrogens on a cluster framework do affect cluster structure. Some of the consequences are evidenced in the geometric structures. Although bridging hydrogens are not considered to occupy vertices of the ideal polyhedron defining cluster geometry they do have steric demands. For example, there is a commonly observed spatial relationship with respect to main group or transition metal atom in a cluster environment. For the main group atom near tetrahedral and for the transition metal atom near octahedral coordination is optimal.

Thus, in going from $Co_3(CO)_9CMe$ to $H_3Fe_3(CO)_3CMe$ the dispositions of the CO ligands are considerably affected. We have reported the systematics of these changes for differing numbers of M-H-M and M-H-E bridging hydrogens in a M_3E cluster in terms of a "tilt angle" defined by the angle the pseudo c_3 axis of the $M(CO)_3$ group makes with the M_3 plane. [28] Because the positions of the three CO ligands affect the orientation and properties of the three frontier orbitals of the $M(CO)_3$ fragment, the number and positions of the bridging hydrogens will also affect the M-M and M-E bonding interactions.

Like H, there are preferential positions for the CO ligands as well. However, as reflected by the differing CO positions in $M_4(CO)_{12}$, M = Co, Rh, Ir, as well as the facile fluxionality of many metal carbonyl clusters, a number of ligand positions of equal energy are accessible for metal carbonyl clusters. [2] The inter CO repulsive interactions can determine, or strongly affect both the core structure as well as the structure adopted by the ligand envelope. [29]

The placement of the hydrogen and CO ligands and the resulting geometric differences between cluster structures reflect electronic differences within the cluster bonding network. These constitute perturbations on the cluster bonding type rather than qualitative differences. Note that there is an essential difference between C-H-M interactions in clusters and mononuclear metal complexes that has not been generally recognized in the various treatises on "agostic" hydrogen interactions. [30, 31]   That is, in mononuclear complexes the agostic hydrogen occupies a coordination position. In its absence, the compound would be unsaturated and such unsaturation has been taken as a necessary condition for the existence of an agostic hydrogen. However, in a cluster the placement of the H atom has no such restriction. Structures 2, 3, and 4 differ in the number of C-H-M interactions but none can be considered unsaturated. Indeed, viewed as M₃E clusters rather than M₃ clusters, all have the same electron count.

The differing possible positions of H (and CO) ligands do affect the electronic structure of the putative surface fragment. The experimental confirmation of this point comes from the examination of the nature of the E-H bond in a M_3EH cluster. The E-H coupling constant is a measure of the s character of the hybrid E orbital forming the bonding interaction with the hydrogen atom. In turn, this reflects the bonding of the EH fragment to the three metal atoms. In fact, the magnitude of J_{EH} decreases monotonically as the number of E-H-M interactions, E = B, C, increase. [20]   That is, as the number of E-H-M interactions increases, the hybrid orbital of the E atom forming a bond with the terminal hydrogen gains p character and approaches that of a saturated species. Molecular orbital parameters also reflect the differences although the interpretation of quantities such as Mulliken charges and overlap populations is fraught with difficulties. But this does not invalidate the equilibrium comparison as one expects the E atoms to reflect the number of E-H interactions.

2.3. IMPORTANT ATOMIC PROPERTIES

Neither the empirical correlations nor the calculations give definitive information on the ultimate origin of relative stabilities of these various structures. The problem is that the balance between the various factors governing ultimate stability is an extremely delicate one and one that is beyond the precision the approximate models available for these problems. Thus the measured equilibrium constants, which precisely define the relative energetics, become extremely important buoys in the sea of clusters pointing us in the correct direction. Consonant with this limping analogy, only a limited number of equilibria exist to keep us on course.

By focusing attention on these defined points we have been able to develop a relationship between the properties of E and M and the disposition of hydrogen atoms between M-M and E-M atom pairs in a variety of clusters. [32]   It is a relationship that is based on the atomic properties of E and M and, thus, is directly transferable to main group species on the surface of a metal. Specifically, the difference in the electronegativities of E and M is found to be directly proportional to the tendency of H to bridge an M-E edge vs a M-M edge. For constant M (or E) the greater the electronegativity of E (or M) the lower the tendency to form a hydrogen bridge. In essence the bridging hydrogen acts as proton chasing negative charge. This is demonstrated quite nicely by the comparison of the cluster systems $M_3(CO)_9CH_4$ and $M_3(CO)_9BH_5$ for M = Fe and Ru discussed above. For M = Fe, the observed tautomeric equilibria and most stable structures demonstrate unambiguously that the Fe-H-Fe bridge is energetically favored over the Fe-H-E bridge for E = C whereas the reverse is true for E = B. Comparing M = Fe and Ru for E = B, the observed equilibrium demonstrates a greater tendency for Ru-H-Ru compared to Ru-H-B relative to iron. Although cluster size and geometry (see section 2.1) as well as other factors play a role, for any given structure the difference in electronegativities is the fundamental factor controlling H atom position.

It is important to note that the barriers for both CO and H ligand migration on clusters are low enough such that site exchange is often observed on the NMR time scale. Presumably the same ligands are mobile on a metal surface. Hence, the thermodynamic tendencies defined by the cluster equilibria discussed should have equal import for surface processes. Thus, electronegativity differences will be important in determining main group atom hydrogenation vs dehydrogenation. So, for example, we have pointed out that in going from Fe to Co to Ni the C-H-M interaction become thermodynamically more favorable than the M-H-M interaction, i.e., the equilibria in Figure 6 are driven to the right. [32] Provided that thermodynamics is the controlling factor, this correlates nicely with the fact that nickel is a methanation catalyst while iron yields both methane as well as higher hydrocarbons under Fischer-Tropsch conditions. Obviously this observation may well be fortuitous but perhaps not. Fundamental atom properties are the same, albeit modified, whether found in discrete molecules or solid state species. The essential point is that as more cluster equilibria are discovered and characterized, even more information on the factors that control the reactivity of surface species will become available. At minimum, these data will place limitations on the reactions of surface species.

3. Heterometal Effects

It is the sweep of transition metal properties that permits the seemingly endless variation that makes them such important chemical entities particularly as catalysts. That is the variation in properties as one moves across a row of the d block elements is much less pronounced than for the p block elements and the fine tuning necessary to avoid dead-end intermediates is possible. However, the question of exactly how different metals subtly perturb a surface species and change its reactivity is certainly not completely understood. A comparison of cluster equilibria between isoelectronic species is one way of defining the effects of changing metals even if only in a relative sense. Again, it is the behavior of the hydrogen atom that is used to explore this question. That is, the geometric location, Brönsted acidity, and CO vs H site preferences are important properties of the hydrogen atom that reflect processes that are directly relevant to surface reactions. Each of these is illustrated with a cluster example in the following.

3.1. MOST STABLE LIGAND POSITIONS

The cluster system $M_3(CO)_9BH_3(CO)$ is a very interesting one. For M = Os the cluster structure is well defined as $(\mu-H)_3Os_3(CO)_9(\mu_3-BCO)$ (**11**, Figure 9) [33, 34] whereas for M = Fe the cluster adopts the structure

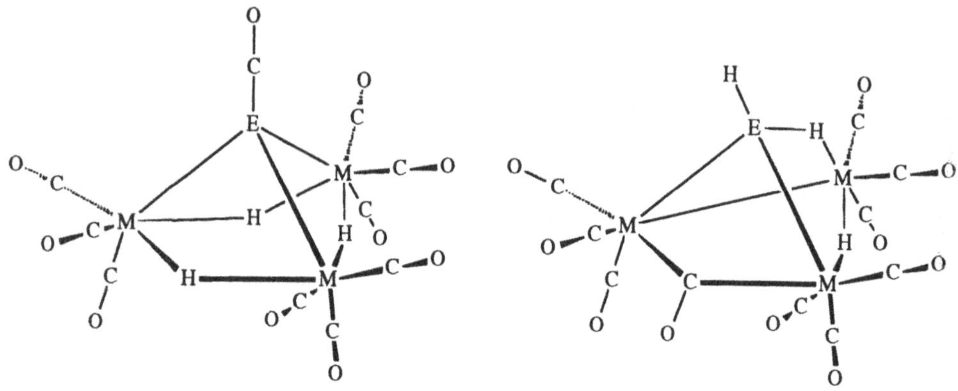

11   **12**
Figure 9. Known structural forms of the $M_3(CO)_{10}EH_3$ cluster.

$(\mu-H)Fe_3(CO)_9(\mu-CO)(HBH)$ based on spectroscopic data[20] as well as the crystal structure of $[Fe_3(CO)_9(\mu-CO)(HBCl)]^-$ (**12**, Figure 9). [35] Because one must deal with the preferences of CO and H for Os/Fe vs B, the analysis of this situation is not a simple one. However, it serves as a dramatic example of the fact that, subtle or not, a change in transition metal can qualitatively change the relative ligand preferences of main group vs transition metal atom. On a surface then, the distribution of EH + MCO vs ECO + MH will also be affected by the identity of the metal.

3.2. HYDROGEN ACIDITY

The effect of a heterometal is explicitly reflected in the differences in the Brönsted acidity of B-H-M hydrogens in the isoelectronic $B_2H_5FeCo(CO)_6$, **14**, and $B_2H_6Fe_2(CO)_6$, **13** molecules (Figure 10).

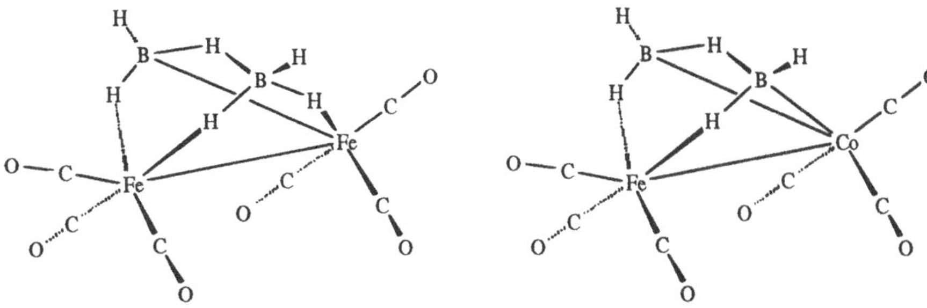

13   **14**

Figure 10. Isoelectronic metallaboranes.

Comparison of the reactions of $[B_2H_4FeCo(CO)_6]^-$ with $B_2H_6Fe_2(CO)_6$ vs $[B_2H_5Fe_2(CO)_6]^-$ with $B_2H_5FeCo(CO)_6$ shows that $B_2H_6Fe_2(CO)_6$ is the better acid. [36]   That is, despite the slightly greater electronegativity of Co than Fe substitution of Co for FeH results in a lower Brönsted acidity. The only reasonable explanation is that the unshielded proton in $B_2H_6Fe_2(CO)_6$ (a framework bridging proton) vs the shielded proton in $B_2H_5FeCo(CO)_6$ (a proton in the metal nucleus) makes the difference.   Clearly the effects of the heterometal are completely overwhelmed by the ligand (H) effects in spite of the fact that the clusters compared are strictly isoelectronic. Of course this requires that the effect of the bridging hydrogens not be localized but rather affect the properties of the cluster as a whole. Although the dependence of Brönsted acidity on ancillary ligands as well as metal identity has been noted previously, it is important to reemphasize the fact that the ligand-metal interaction is a large factor in determining metal fragment behavior and heterometal effects in non-isoelectronic compounds with very different ligands are liable to be overwhelmed by ligand effects. That being said, these results show that the acidity of a given surface bound H atom will be affected by the presence or absence of nearby H atoms. This creates a connection between surface coverage by hydrogen and effective acidity.

3.3. METAL VS SUBSTITUENT EFFECTS

Again it is in an equilibrium system that one can precisely assess the effects of both a heterometal as well as substituents on the position of cluster bridging hydrogens. The equilibrium constant of the tautomerization $(\mu\text{-}H)FeCo_2(CO)_9(\mu_3\text{-}CR)$ $\Leftrightarrow$ $FeCo(CO)_9(\mu_3\text{-}HCR)$ has been evaluated as a function of R and compared with the related tautomerization observed for $Fe_3(CO)_9CH_4$ discussed above. [37]   It was demonstrated that the effect of R is mainly a steric one. Curiously, electronic effects of the R group are small, i.e., a series of para-substituted phenyl derivatives exhibited little change in the equilibrium constant despite substantial variation of the Hammett s constants for the substituents. On the other hand, change of the metal to Fe (with the additional two hydrogens) leads to a significant change in the equilibrium position.

Fe3(CO)9CH3Me exists as a single tautomer namely (μ-H)3Fe3(CO)9CMe whereas FeCo2(CO)9CHMe exists as an equilibrium mixture of two tautomers in solution. We have estimated that the difference in free energy amounts to 4 kcal/mol which amounts to a discrimination factor of $\approx 10^3$ at 300 K. We ascribed this to the difference in metals although it is not known how to take precisely into account the effect of the differing numbers of skeletal hydrogens which we know to be important (see section 3.2). In contrast to the situation with respect to acidity, we suspect in this case that the difference in metals is the dominant factor.

4. Relevance to Surface Transformations

The two preceding sections are, in a sense, a foundation for this section. Here the focus is on the energetic and mechanistic aspects of cluster reactions of relevance to understanding selected surface processes.

4.1. ENERGETICS

Because of the essentially delocalized aspect of the core bonding in a cluster, the energetics of clusters are not easily treated with a simple bond energy term value analysis. On the other hand, the comparison of two closely related species connected by an established equilibrium process leads to relative energetics that are unambiguous. As demonstrated by the system discussed below, the interpretation of the energetics permits comment on significant aspects of surface reactivity.

The equilibrium process, Eqn (3), shown in Figure 11 has an

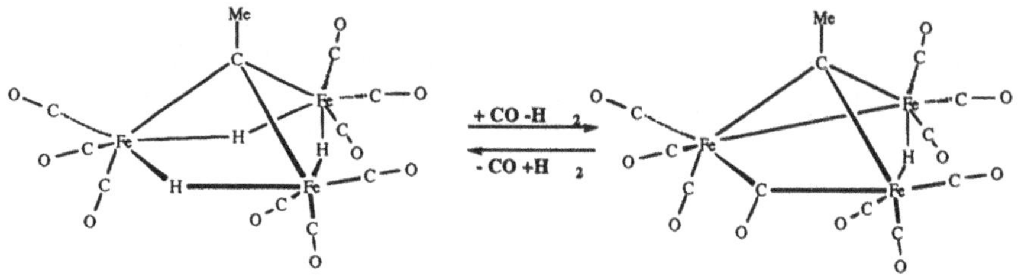

Figure 11. Cluster equilibrium involving CO and H_2.

$$(\mu\text{-H})_3Fe_3(CO)_9(\mu_3\text{-CCH}_3) + CO \Leftrightarrow (\mu\text{-H})Fe_3(CO)_9(\mu\text{-CO})(\mu_3\text{-CCH}_3)$$

$$\mathbf{2'} \qquad\qquad\qquad\qquad\qquad \mathbf{15}$$

$$+ H_2 \quad (3)$$

equilibrium constant of 1.5 at 60°C. [38]   Cluster degradation of **2'** by CO is a competitive side reaction at high CO concentrations.   An analysis of the energetics allow a value for the Fe-H-Fe energy to be obtained. It is significantly larger than the energy of Fe-Fe and Fe-H. The important point is that despite the large H-H bond energy the overall reaction has a small enthalpy because the Fe-H-Fe energy is large enough to compensate. It is the multicenter interactions then, found both in clusters and surface structure, that are important in allowing complex displacements such as these to take place. For example, although the process of b-hydrogen elimination of $FeCH_2CH_3$ from a single metal site to yield FeH and C_2H_4 is significantly endothermic, it becomes energetically favorable if the final product contains Fe-H-Fe rather than a terminally bound hydrogen.

4.2. MECHANISM

The equilibrium in Eqn (3) by itself does not provide much information on the mode of displacement of H_2 by CO and vice versa. Further, the reaction is a rather slow one in that it takes ca 10h at 60°C to achieve equilibrium. However, we discovered that a more facile interconversion could be achieved via deprotonation/protonation of the iron clusters. [39]

The Brönsted acidity of cluster hydrides and basicity of metal-metal bonds with respect to a proton is well known. [3]   What is not as generally recognized is that Brönsted acid/conjugate base equilibria are easily established in these clusters. Indeed in THF sometimes an equilibrium mixture can be observed in solution. For example, $B_2H_6Fe_2(CO)_6$ in THF is present as the neutral and the monoanion. [36]   Removal of the THF and addition of hexane results in total recovery of the $B_2H_6Fe_2(CO)_6$. In the case of these metallaboranes, protonation/protonation equilibria can also be observed taking place on a silicagel column. For example, $HFe_4(CO)_{12}BH_2$ has a much smaller R_f than expected and one that depends on silicagel properties and column temperature. [36]   Finally, some of these clusters irreversibly deprotonate on typical silicagel columns, e.g., $HFe_3(CO)_{10}BH_2$,   whereas others are unaffected, e.g., $HFe_3(CO)_9H_3BH$.   Considering the nature of common supports used in catalytic practice, e.g., silicagel, one can envision similar proton exchanges taking place on a metal supported catalyst. Hence, a cluster reaction in solution driven by addition and loss of protons is not at all ruled out as a model for a surface reaction.   As shown in the following, an indirect, complex pathway constitutes a perfectly reasonable model for a surface reaction.

The entire complex process whereby Eqn (3) can be carried out under mild conditions is shown in Figure 12. The detailed justification for this mechanism has been presented earlier[39]   and only the conclusions relevant to these discussions are presented here. Removal of a proton from the methyl group leads to an unstable anion that exists in tautomeric equilibrium between M-H-M and M-H-C forms. Above 0°C H_2 is lost and the formal C=C double bond of the vinylidene moiety coordinates to the metals replacing the two electrons of the departing hydrogen thereby forming **16**. Addition of CO followed by reprotonation leads to **15**. Alternatively, addition of H_2 in the presence of H^+ leads to **2'**. An unsaturated ethylidyne intermediate is proposed that either adds H_2 or rearranges to the neutral vinylidene cluster **17**. The site of protonation, deprotonation was shown by labeling experiments to take place at the b-carbon atom. As the methyl hydrogens are not the most acidic protons on the cluster, the reaction must be kinetically controlled. Ancillary ligand steric effects were shown to be important. Other labeling experiments demonstrate that the H_2 eliminated and added is derived from or added to the hydrogens bridging the metal-metal edges. In this process steric effects appear less important.

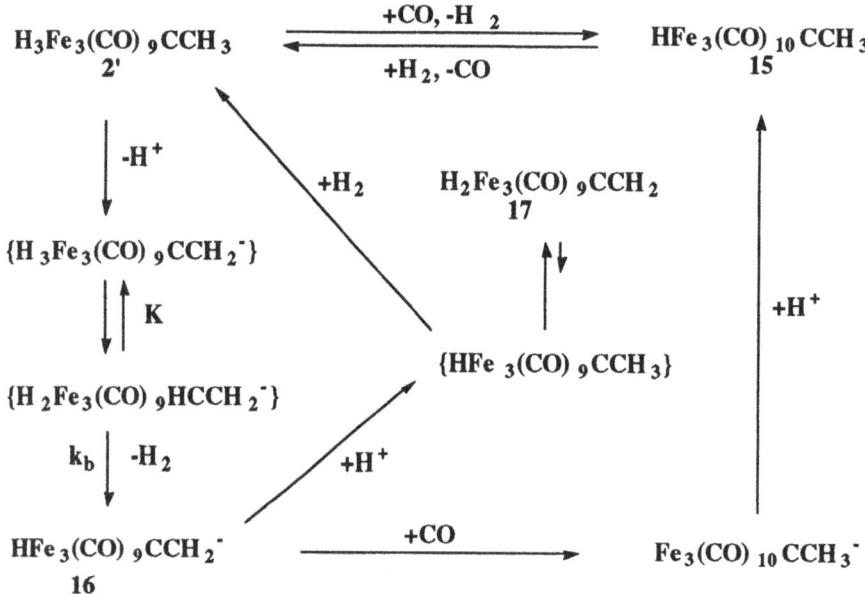

Figure 12. Alternate route for the reaction in eqn (3) via intermediates in { }.

In effect, removal of a proton permits facile tautomerization to yield a species that is unstable with respect to hydrogen elimination. Addition of a proton generates a species that is formally unsaturated and can undergo facile ligand addition. Considering the fact that an efficient and facile reaction pathway can be supported by low concentrations of very active intermediates the idea of either H^+ or H atom rearrangement generating active intermediates on a surface is an attractive one.

4.3. SURFACE VS SOLID SITES

There are other cluster systems that have interesting, if less obvious, relationships to surface-substrate interactions. The build-up of surface carbides, for example, constitutes a poisoning reaction for catalysts involved in hydrocarbon formation and transformation processes. Clearly, the formation, loss and diffusion of such species into the interior of the catalyst are important to surface chemistry. Cluster reactions and equilibria can provide some worthwhile insights here as well.

The boride cluster 1,6-[Rh$_2$Fe$_4$(CO)$_{16}$B]$^-$ (Figure 13) has carbide relatives and serves as a model for an interstitial boron site. We have studied the mechanism of the rearrangement of 1,2-[Rh$_2$Fe$_4$(CO)$_{16}$B]$^-$ to 1,6-[Rh$_2$Fe$_4$(CO)$_{16}$B]$^-$ [40, 41] and the results provide provocative insight into a way in which other ligands may facilitate the reconstruction of the immediate metal environment of a main group atom. This rearrangement is facilitated by the presence of soft ligands, i.e., motion of the metals relative to the boron atom is enhanced by the coordination of a soft base to the metals. It is important to note that overall cluster substitution by external ligands is an order of magnitude slower than the base promoted rearrangement, i.e., weak coordination is sufficient. This suggests that the reconstruction of a ligand saturated surface can be enhanced by weakly interacting ligands and that such reorganization may well be fast with respect to overall ligand replacement.

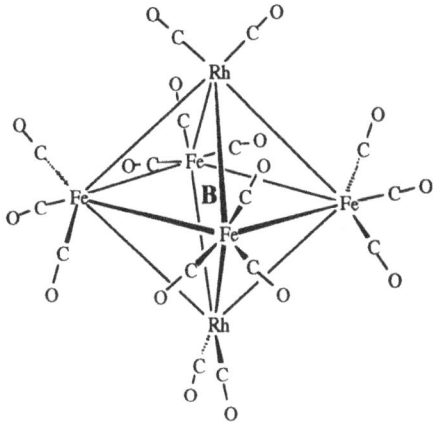

Figure 13. Structure of 1,6-[Rh$_2$Fe$_4$(CO)$_{16}$B]$^-$.

5. Summary

The cluster-surface hypothesis remains a provocative one. But it can be more than that. Element variation of either main group, e.g., in the form of C vs B$^-$, or transition metal, e.g., Co vs Fe$^-$, element is an effective method of obtaining qualitative information on the behavior of less stable carbon species in a multinuclear transition metal environment. Cluster reactions, particularly equilibria, can provide quantitative insight into surface processes. Measured equilibrium properties of cluster species provide an unambiguous way of defining selected reactions that also take place on metal surfaces.

92

6. Acknowledgments

My coworkers listed in the pertinent references carried out the experimental studies upon which this contribution is based. Their excellent work is greatly appreciated. The financial support of the National Science Foundation for the research carried out at the University of Notre Dame is also gratefully appreciated.

7. References

1. Gonzalez-Moraga, G. (1993) *Cluster Chemistry*. Introduction to Transition Metal and Main Group Element Clusters, Springer Verlag, New York.
2. Shriver, D. F., Kaesz, H. D. and Adams, R. D. (1990) *The Chemistry of Metal Cluster Complexes*, VCH, New York.
3. Johnson, B. F. G. and Lewis, J. (1981) Transition metal clusters, *Adv. Inorg. Chem. Radiochem.* **24**, 225.
4. Gates, B., Guczi., L. and Knözinger, H., (1986) *Studies in Surface Science and Catalysis*, Elsevier, New York.
5. Muetterties, E. L., Rhodin, T. N., Band, E., Brucker, C. F. and Pretzer, W. R. (1979) Clusters and surfaces, *Chem. Rev.* **79**, 91.
6. Fontijn, A. and Clyne, M. A. A. (1983) *Reactions of Small Transient Species*, Academic Press, New York.
7. Rhodin, T. N. and Ertl, G. (1979) *The Nature of the Surface Chemical Bond*, North-Holland, New York.
8. Somorjai, G. A. (1981) *Chemistry in Two Dimensions: Surfaces*, Cornell Univ. Press, Ithaca.
9. Bent, B. E., Nuzzo, R. G. and Dubois, L. H. (1989) Surface organometallic chemistry in the chemical vapor deposition of aluminum films using triisobutylaluminun. β-Hydride and β-alkyl elimimination reactions of surface alkyl intermediates, *J. Am. Chem. Soc.* **111**, 1634.
10. Mason, R. (1976/77) The Evolution of a coordination and organometallic chemistry of surfaces, *Israel J. Chem.* **15**, 174.
11. Sailor, M. J., Went, M. J. and Shriver, D. F. (1988) Characteristic vibrational frequencies and normal modes of the CCO ligand in trinuclear ketenylidene clusters, *Inorg. Chem.* **27**, 2666.
12. Moskovits, M. (1979) metal cluster complexes and heterogeneous catalysis-A heterodox view, *Accts. Chem. Research* **12**, 229.
13. Fehlner, T. P. (1992) *Inorganometallic Chemistry*, Plenum, New York.
14. Spielvogel, B. F. (1980) Synthesis and biological activity of boron ànalogues of the a-amino acids and related compounds, Pergamon Press, Oxford.
15. Fehlner, T. P. (1988) Metal-rich ferra- and cobaltaboranes. Mimics of organometallic clusters, *New J. Chem.* **12**, 307.
16. Muetterties, E. L. (1982) Hydrocarbon reactions at metal centres, *Chem. Soc. Rev.* **11**, 283.

17. Dutta, T. K., Vites, J. V., Jacobsen, G. B. and Fehlner, T. P. (1987) The making and breaking of C-H bonds on a metal cluster framework. analysis of the tautomerization and deprotonation of $Fe_3(CO)_9CH_4$ leading to the interconversion of FeHFe and CHFe interactions, *Organometallics* **6**, 842.

18. Wong, S. W. and Fehlner, T. P. (1981) Preparation of $H_3Fe_3(CO)_9CCH_3$ from $Fe(CO)_5$, *J. Am. Chem. Soc.* **103**, 966.

19. Carty, A. J., Johnson, B. F. G., Lewis, J. and Norton, J. R. (1972) The synthesis of $H_3Ru_3(CO)_9CCH_3$, a ruthenium ethylidyne complex, *J. Chem. Soc. Dalton Trans.* 477.

20. Vites, J. C., Housecroft, C. E., Eigenbrot, C., Buhl, M. L., Long, G. J. and Fehlner, T. P. (1986) The structure and properties of $HFe_3(CO)_9BH_3R$ and the conjugate bases $[HFe_3(CO)_9BH_2R]^-$ (R=H and CH_3), *J. Am. Chem. Soc.* **108**, 3304.

21. Vites, J. C., Eigenbrot, C. and Fehlner, T. P. (1984) The preparation and characterization of $HFe_3(CO)_9H_3BH$, *J. Am. Chem. Soc.* **106**, 4633.

22. Chipperfield, A. K. and Housecroft, C. E. (1988) The preparation of the metal-rich ruthenaborane, $Ru_3(CO)_9BH_5$: a case of isomerisation involving *endo*-hydrogen migration, *J. Organomet. Chem.* **349**, C17.

23. Beno, M. A., Williams, J. M., Tachikawa, M. and Muetterties, E. L. (1981) The closed three-center carbon-hydrogen-metal interaction. A neutron diffraction study of $HFe_4(\eta^2\text{-}CH)(CO)_{12}$, *J. Am. Chem. Soc.* **103**, 1485.

24. Cowie, A. G., Johnson, B. F. G., Lewis, J. and Raithby, P. R. (1986) Structural isomers of $H_2Ru_4C(CO)_{12}$; X-ray crystal structure of $Ru_4C(CO)_{13}$, *J. Organomet. Chem.* **306**, C 63.

25. Rath, N. P. and Fehlner, T. P. (1987) Multiple deprotonation of a ferraborane. Evidence for the formation of a discrete transition-metal boride, *J. Am. Chem. Soc.* **109**, 5273.

26. Hong, F. E., McCarthy, D. A., White III, J. P., Cottrell, C. E. and Shore, S. G. (1990) Reaction. of $THF \cdot BH_3$ with $H_2Ru_4(CO)_{13}$ and $H_4Ru_4(CO)_{12}$; Preparation and structure of $HRu(CO)_{12}BH_2$, *Inorg. Chem.* **29**, 2874.

27. Housecroft, C. E. (1992) Transition metal main group cluster compounds, in T. P. Fehlner (ed), *Inorganometallic Chemistry*, Plenum Press, New York.

28. Lynam, M. M., Chipman, D. M., Barreto, R. D. and Fehlner, T. P. (1987) Endo hydrogens on main group-transition metal clusters. Theoretical analysis of the interconversion of FeHFe and EHFe interactions and deprotonation of $Fe_3(CO)_9EH_x$ (E = B, X = 5; E = C, X = 4), *Organometallics* **6**, 2405.

29. Mingos, D. M. P. and Wales, D. J. (1990) *Introduction to Cluster Chemistry*, Prentice Hall, New York.

30. Brookhart, M. and Green, M. L. H. (1983) C -H - transition metal bonds, *J. Organomet. Chem.* **250**, 395.

31. Brookhart, M., Green, M. L. H. and Wong, L.-L. (1988) Carbon hydrogen transition metal bonds, *Prog. Inorg. Chem.* **36**, 1.

94

32. Fehlner, T. P. (1990) On the energetics of the formation of E-H-M for M-H-M interactions on transition metal clusters, *Polyhedron* **9**, 1955.

33. Shore, S. G., Jan, D.-Y., Hsu, L.-Y. and Hsu, W.-L. (1983) Insertion of boron into an osmium-carbonyl bond. Preparation and crystal structure of the carbonyl borylidyne $H_3(CO)_9Os_3BCO$, *J. Am. Chem. Soc.* **105**, 5923.

34. Jan, D.-Y., Workman, D. P., Hsu, L.-Y., Krause, J. A. and Shore, S. G. (1992) Clusters derived from the hydroboration of $(\mu-H)_2Os_3(CO)_{10}$ and their derivatives, *Inorg. Chem.* **31**, 5123.

35. Crascall, L. E., Thimmappa, B. H. S., Rheingold, A. L., Ostrander, R. and Fehlner, T. P. (1994) Synthesis of $[AsPh_4][Fe_3(CO)_9(CO)(HBCl)]$ by oxidative chloride substitution of $[Fe_3(CO)_9(HBCO)]^{2-}$, *Organometallics* **13**, 0000.

36. Jun, C.-S. (1994) Synthesis of New Metallaboranes, Thesis, University of Notre Dame,

37. Barreto, R. D., Puga, J. and Fehlner, T. P. (1990) Electronic and steric control of the formation of CHM from MHM interactions in carbon atom capped transition metal carbonyl clusters. Effect of the metal fragment, *Organometallics* **9**, 662.

38. Vites, J. and Fehlner, T. P. (1984) A facile equilibrium involving CO and H_2. The Fe-H-Fe bond energy in $H_3Fe_3(CO)_9CMe$, *Organometallics* **3**, 491.

39. Dutta, T. K., Meng, X., Vites, J. C. and Fehlner, T. P. (1987) Reactions of $H_3Fe_3(CO)_9CMe$. H_2 displacement by CO and H_2 elimination following deprotonation, *Organometallics* **6**, 2191.

40. Bandyopadhyay, A. K., Khattar, R. and Fehlner, T. P. (1989) Skeletal rearrangement of $Rh_3Fe_4(CO)_{16}B^-$. An example of an associative cluster isomerization, *Inorg. Chem.* **28**, 4434.

41. Bandyopadhyay, A. K., Khattar, R., Puga, J., Fehlner, T. P. and Rheingold, A. L. (1992) Boron in new environments. Structure and reaction properties of discrete, mixed-metal, hexanuclear borides, *Inorg. Chem.* **31**, 465.

REACTIONS AND DYNAMICS OF RUTHENIUM CLUSTERS

K. Vrieze and C.J. Elsevier
Inorganic Chemistry Laboratory,
J.H. van't Hoff Research Institute,
University of Amsterdam,
Nieuwe Achtergracht 166,
1018 WV Amsterdam, The Netherlands.

Abstract.

Reactions of monoazadienes with ruthenium carbonyl complexes lead to a large variety of bis, tris and tetranuclear ruthenium compounds. In this article it is shown that understanding has been gained on how clusters may build up and break down. In particular interest is focussed on the reactivity and dynamics of the linear tetranuclear complex $Ru_4(CO)_{10}[R^1C=C(H)(H)C=N-R^2]_2$. Our insight into the build-up of this compound has allowed the rational synthesis of isostructural iron-ruthenium clusters $Fe_2Ru_2(CO)_{10}[R^1C=C(H)(H)C=N-R^2]_2$.

1. Introduction.

A long standing interest of our group has been the purposeful design of metal cluster complexes and the study of their dynamics and reactivity behaviour [1,2,3,4]. In the late seventies the fascinating versatile coordination behaviour of diazadienes (R-DAB = RN=C(H)(H)C=NR) became apparent. They were shown to behave as a 2e σ-N donor ligand and as a 4e σ-N,σ-N' bidentate ligand bonded to one metal atom or to two metal atoms. Very interesting was the discovery by us of the 6e donor σ-N, σ-N' , η^2-C=N' coordination mode and of the 8e σ-N, σ-N' , η^2-C=N, η^2-C=N' coordination mode, where the R-DAB ligand bridges in both cases two metal atoms [1,2,3,4] (*Figure 1*). An important feature of the coordinated R-DAB ligands is that the 2e, but in particular the 4e-, 6e and 8e-donor modes may interconvert easily in response to the electronic and coordination requirements of the metal complex moiety to which the R-DAB is bonded.

L. J. Farrugia (ed.), The Synergy Between Dynamics and Reactivity at Clusters and Surfaces, 95–111.
© *1995 Kluwer Academic Publishers.*

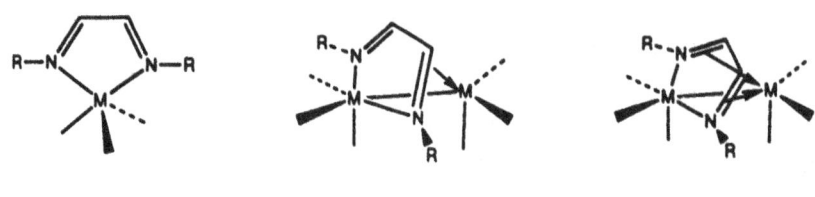

Figure 1

It was further demonstrated that the metal coordinated R-DAB ligands may not only behave as flexible spectator ligands, but may themselves also undergo reversible C-C, C-N, C-H and N-H bond forming and bond breaking reactions [4,5,6,7,8,9].

In view of these fascinating properties it was rather logical to turn our attention to the related monoazadienes (R^1,R^2-MAD = R^1-C(H)(H)C=NR^2). It was anticipated that these enimines would show coordination behaviour rather similar to that of R-DAB, but would also by abstraction of an H-atom be amenable to the formation of metallacyclic compounds, as to some extent had already been demonstrated by us in very early work [10].

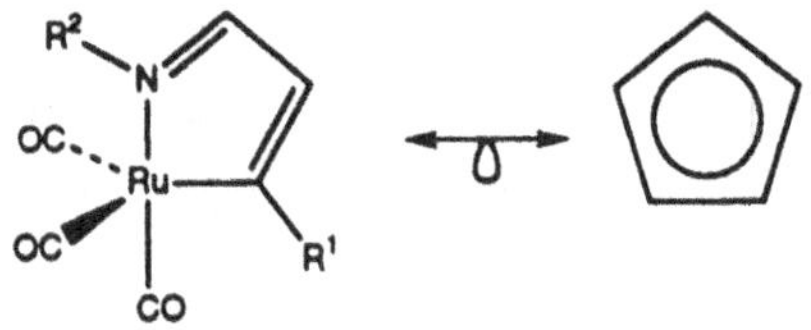

Figure 2

At an early stage we realized that the Ru(CO)$_3$(MAD-yl) unit, which contains a five-membered ring with a metallated MAD ligand is isolobal with c-C$_5$H$_5$, since Ru(CO)$_3$ is isolobal with CH$^+$ and NR with CH$^-$ (*Figure 2*). Not known were the donor modes shown in Figure 3, which depict the cyclometallated 5e σ-N, σ-C, η^2-C=C, 5e σ-N, σ-C, η^2-C=N and the 7e σ-N, σ-C, η^2-C=N, η^2-C=C moieties which are directly analogous to the 6e and 8e bridging modes of R-DAB.

Figure 3

The first real opening in this field came from the serendipitous discovery of the novel linear 66e tetranuclear cluster $Ru_4(CO)_{10}(MAD-yl)_2$ [11], of which the structures of the two diastereomers are shown in Figure 4. This cluster $Ru_4(CO)_{10}[CH_3C=C-(H)(H)C=N-i-Pr]_2$ contains two ß-metallated MAD-yl ligands in the 7e coordination mode (*Figure 4*). This compound plays a central role in this chemistry and therefore the discussion will be mainly focussed on its formation, dynamic properties and the reactivity behaviour.

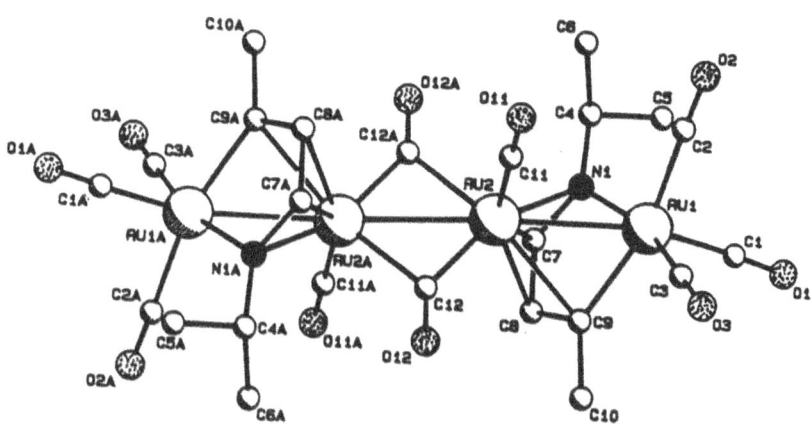

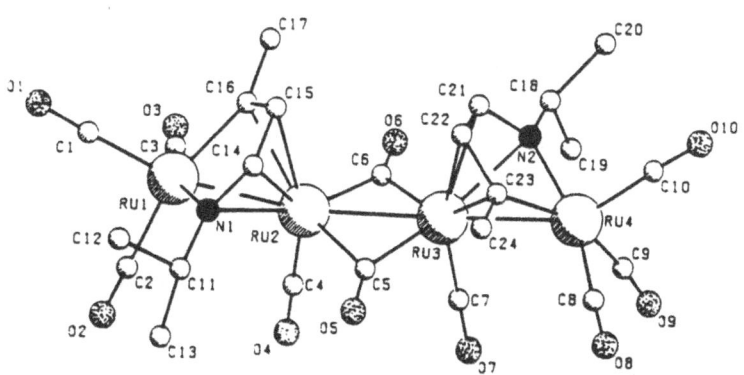

Figure 4
X-ray molecular structures of *trans*-(AC) (top) and of
cis-(AA) (bottom) $Ru_4(CO)_{10}(MAD-yl)_2$.

2. Relevant reactions of the $Ru_3(CO)_{12}$/MAD system.

At an early stage we realized that $Ru_4(CO)_{10}(MAD-yl)_2$ is the isolobal analogue of $Cp_2Ru_2(CO)_4$ and we wondered, not in hindsight we might add, whether the isolobal analogies could not be stretched further to predict the precursors leading to this tetranuclear complex and even to its dynamic and reactivity properties. So just as $Ru_4(CO)_{10}(MAD-yl)_2$ is isolobally related to $Cp_2Ru_2(CO)_4$, one might consider $Ru_2(CO)_6(R^2NCH_2C(H)=CR^1$ as an isolobal analogue of $(\eta^4-C_5H_6)Ru(CO)_3$, while $HRu_2(CO)_5(R^2N=C(H)(H)C=CR^1)$ is isolobal to $CpRuH(CO)_2$. It has been well established that $Cp_2Ru_2(CO)_4$ is formed from $(\eta^4-C_5H_6)Ru(CO)_3$ via loss of CO to form $CpRuH(CO)_2$ and subsequent loss of H_2 [12]. So the question to which we addressed ourselves was whether we could isolate or observe the precursors leading to the tetranuclear complex according to the route shown in Figure 5.

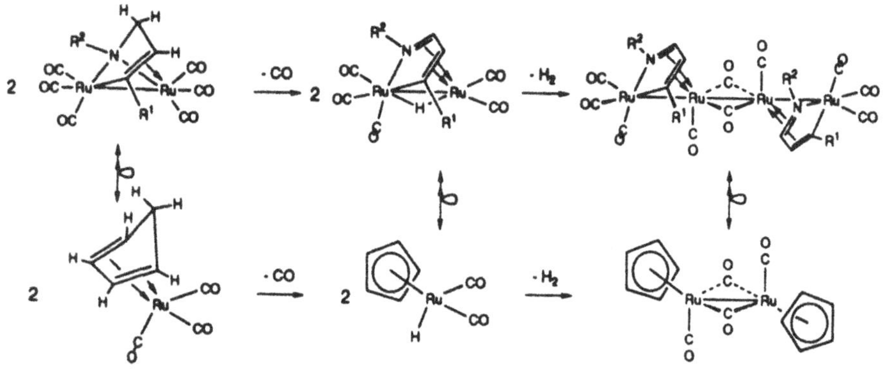

Figure 5
Relevant isolobal relations.

Therefore we carried out an extensive investigation as to the product formation and the reactivity of the complexes formed in the reaction of $Ru_3(CO)_{12}$ with R^1,R^2-MAD as a function of R^1 and R^2 [13,14].

In Scheme 1 a broad overview of the results of our research is presented, but we will only discuss the reactions relevant for this article.

From this Scheme it is evident that the resulting picture is much more complicated than we thought, but it is very interesting indeed that we found the key intermediates in the formation of the tetranuclear cluster i.e. $Ru_2(CO)_6[R^2NCH_2-C(H)=CR^1]$ and $HRu_2(CO)_5[R^2N=C(H)(H)C=CR^1]$.

The first compound shows the presence of a ß-metallated formally dianionic

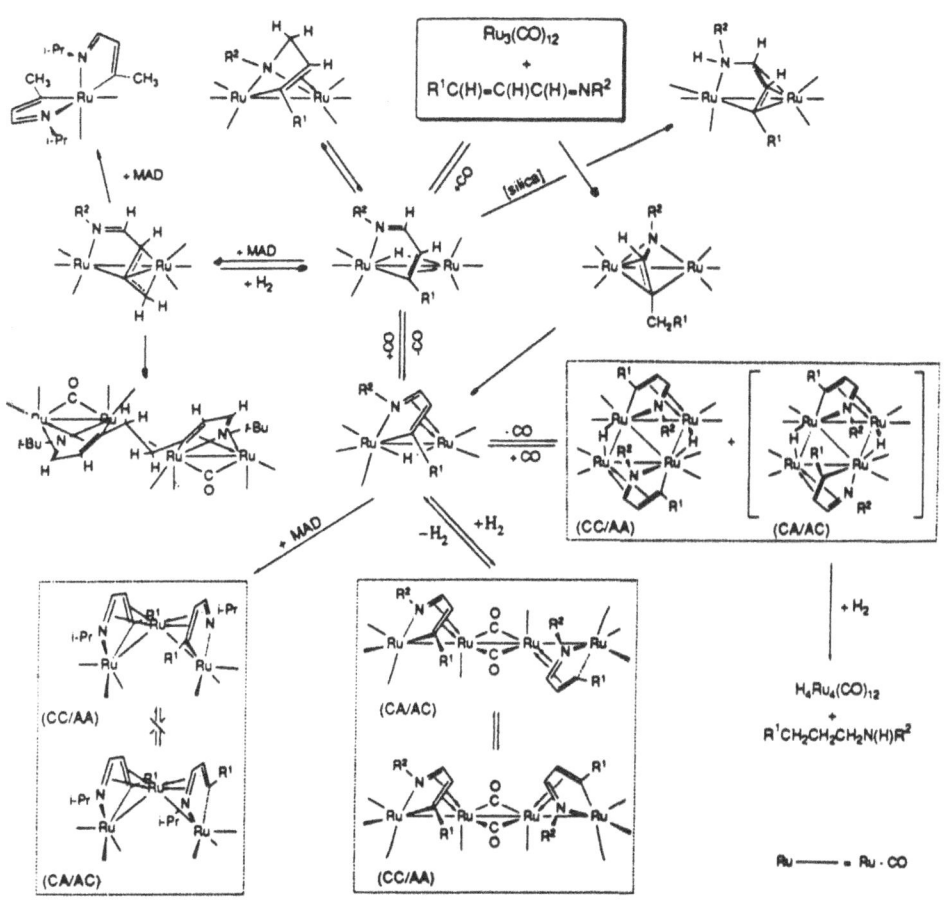

Scheme 1
Selected reactions in the system $Ru_3(CO)_{12}/R,^1$ R^2-MAD: $R,^1$ R^2=CH_3 i-Pr (a);
CH_3, c-Hex (b); t-Bu (c); C_6H_5, i-Pr (d); C_6H_5, t-Bu (e).

enylamido ligand in which the abstracted H-atom has been transferred to the imine C-atom. This hydrogen atom is very mobile and the complex is in equilibrium with $HRu_2(CO)_6[R^2N=C(H)(H)C=CR^1]$. This complex subsequently easily loses a CO molecule, driving the equilibrium towards the crucial intermediate $HRu_2(CO)_5$-$[R^2N=C(H)(H)C=CR^1]$, which is isolobal to $CpRuH(CO)_2$, and in which the MAD-yl ligand has reverted from the 5e- to the 7e-donor mode in order to compensate for the loss of one CO ligand [13,14]. It could be isolated in the case where R^2=iPr and R^1=Ph(X-ray) [15].

Also very interesting is the fact that $Ru_2(CO)_6(R^2NCH_2C(H)=CR^1)$ may be formed back from the tetranuclear complex via the pentacarbonyl hydride and the hexacarbonyl hydride by treatment with H_2 and CO (*Scheme 1*).

3. Structure and Dynamic Behaviour of $Ru_4(CO)_{10}(MAD-yl)_2$

The linear $Ru_4(CO)_{10}(MAD-yl)_2$ is formed as mixtures of two diastereomers CA/AC and CC/AA owing to the presence of two chiral azaruthenacycles [14]. The absolute configuration of the chiral metal centers follows the Brown-Cook-Sloan rules [14,16] which is a modification of the Cahn-Ingold-Prelog (CIP) rules. Since we are only

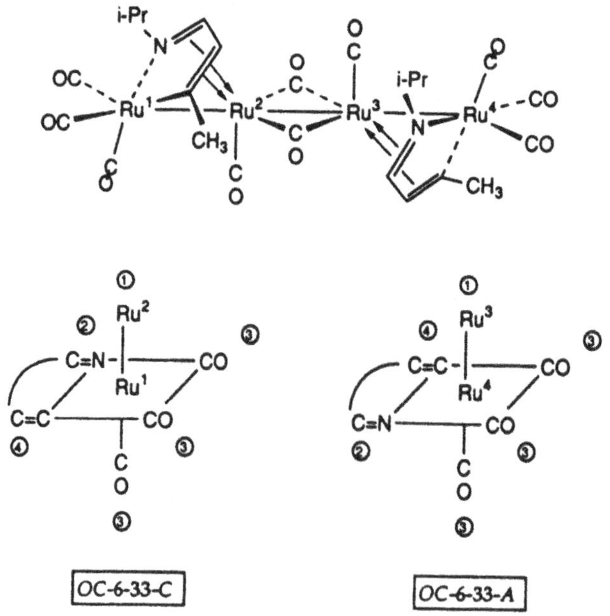

Figure 6
The structure of *trans*-(CA)-Ru$_4$(CO)$_{10}$[CH$_3$C=C(H)C(H)=N-i-Pr]$_2$
and chirality descriptors for Ru(1) and Ru(4).

concerned with enantiomers and diastereomers of the OC-6-33 system and therefore of OC-6-33-C and OC-6-33-A (*Figure 6*) we use as an abbreviation the chirality descriptors C and A throughout.

Furthermore it should be noted that the chirality of Ru(2) depends on those of the next neighbours metal atoms which is also the case for Ru(3). Finally, we prefer to denote the relative geometric stereochemistry by the *cis/trans* notation which is dependent on the mutual position of the terminal carbonyls on Ru(2) and Ru(3). It is therefore sufficient to give the absolute configuration of Ru(1) and Ru(4) i.e. *cis*-(CC) for the stereoisomer with C_2 symmetry and *cis*-(CA) for the one with C_s symmetry. There are four diastereomeric pairs of enantiomers which are therefore denoted as *cis*- and *trans*-(CC/AA) for the chiral $Ru_4(CO)_{10}(MAD-yl)_2$ molecules and *cis*- and *trans*-(CA/AC) for the achiral ones (*Figure 7*).

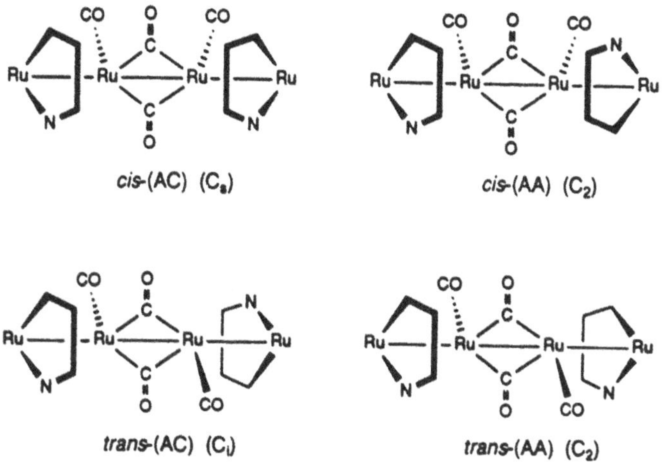

cis-(AC) (C$_s$)   *cis*-(AA) (C$_2$)

trans-(AC) (C$_i$)   *trans*-(AA) (C$_2$)

Figure 7
Configurations of $Ru_4(CO)_{10}(MAD-yl)_2$; symmetry in parentheses.

When we consider the details of the structures of $Ru_4(CO)_{10}(MAD-yl)_2$ depicted in Figure 4 we may mention that both molecules contain three single Ru-Ru bonds with the central Ru(2)-Ru(3) bonds of about 2.77 Å being longer than the Ru(1)-Ru(2) and Ru(3)-Ru(4) bonds (about 2.71 Å) [16]. Since the η^2-C=C and η^2-C=N bonded

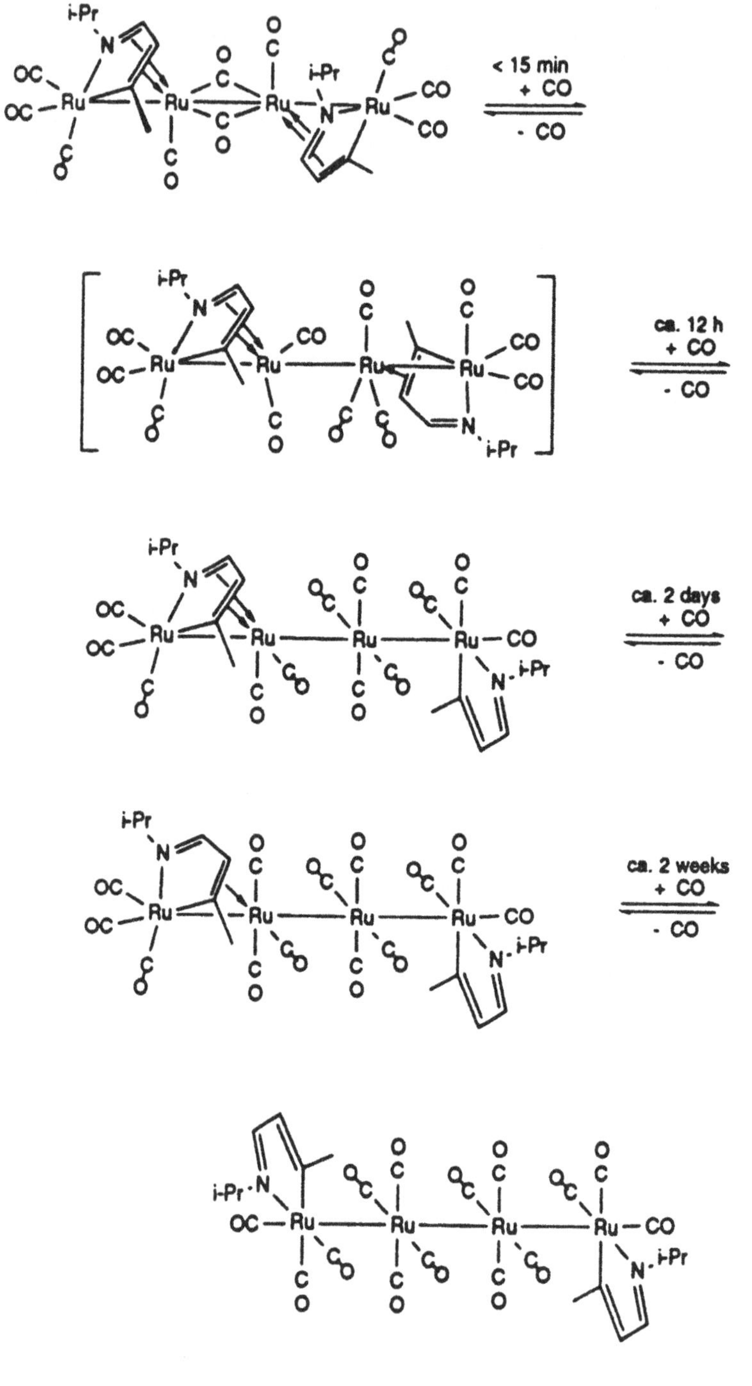

Scheme 2
Stepwise reaction of $Ru_4(CO)_{10}[CH_3C=C(H)(H)C=N-i-Pr]_2$ with CO.

moieties have different π-accepting properties it is not unexpected that the bridging CO groups are asymmetrically bonded. The only real differences are the dihedral angles between the two planes defined by the inner Ru-atoms and the C-atoms of the two bridging CO groups. The angle is about 180° for *trans*-(CA/AC) and 166° for *cis*-(CC/AA), which is analogous to the difference for *trans*- and *cis*-[CpFe(CO)$_2$]$_2$ [17,18].

Conversion between the diastereomers may take place either by raising the temperature above 70°C, or by addition of CO at RT to solutions of each of the diastereomers. The first process is intermolecular and proceeds probably via dinuclear radical intermediates. The CO induced interconversion clearly involves an intramolecular process probably via intermediates containing η^2-bonded MAD-yl ligands (*Scheme 2*), whereby the terminal Ru-η^2-MAD-yl moieties might be involved in a Ray-Dutt twist type of fluxionality [16,19].

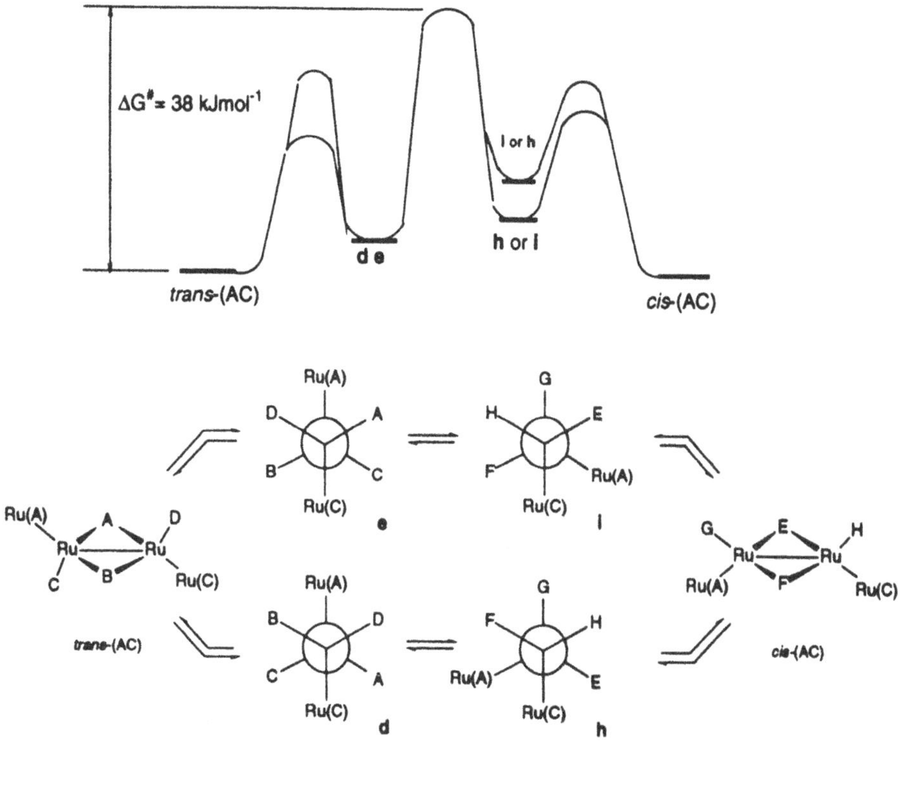

Figure 8

Cis/trans isomerization via non-bridged intermediates of the
trans -(AC) isomer of Ru$_4$(CO)$_{10}$(MAD-yl)$_2$.

104

The most interesting aspect is the investigation of the reaction rates of the bridging/terminal exchange and the *cis/trans* isomerization in each of the diastereomers, which may be compared nicely with the extensive work on [CpFe(CO)$_2$]$_2$ and [CpRu(CO)$_2$]$_2$ [20-25]. In a recent article [16] we have described in detail the studies on the fluxional behaviour of Ru$_4$(CO)$_{10}$ [CH$_3$C=C(H)(H)C=N-i-Pr]$_2$. We therefore restrict ourselves to mentioning the relevant conclusions, which are based on NMR and IR spectroscopy. Owing to the fact that we are dealing with two chiral η^5-Ru-MAD-yl units adjacent to the Ru$_2$(CO)$_4$ core units in each diastereomer, we were able to study separately the *cis/trans* isomerization and the bridge/terminal CO exchage (*Figures 8 and 9*).

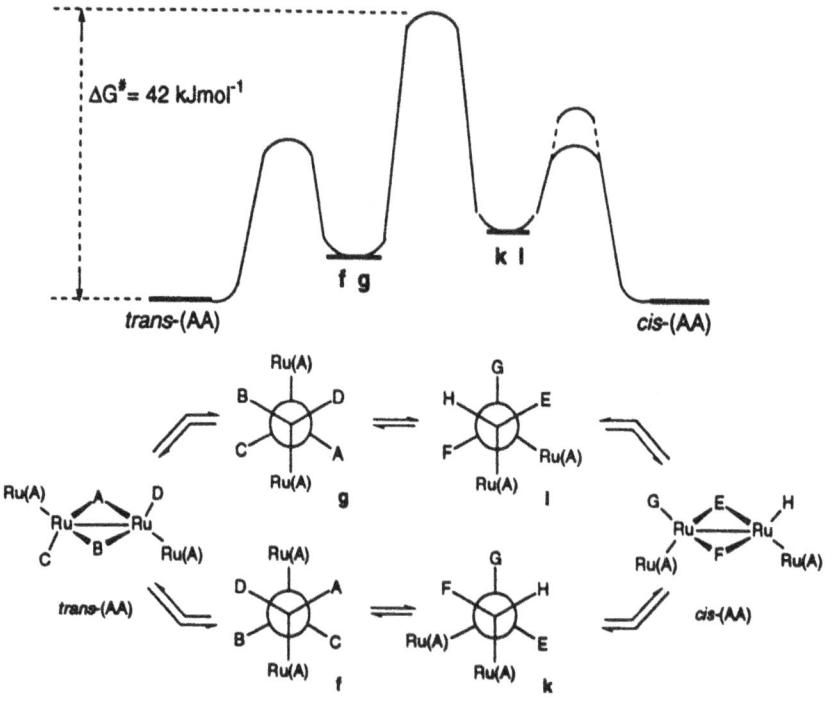

Figure 9

cis/trans isomerization via non-bridged intermediates of
the *trans*-(AA) isomer of Ru$_4$(CO)$_{10}$(MAD-yl)$_2$.

Due to the presence of the two chiral *pseudo*-cyclopentadienyl rings, there will be a kinetic preference for the two *equivalent asymmetric* bridging CO ligands A and

B in the *trans*-AC(or CA) form to migrate to a specific Ru atom with the formation of intermediates d or e. This kinetic preference is caused by the tendency for the bridging asymmetric bonded CO group to move to the nearest Ru atom, which leads to a kinetic preference for intermediate d over e by the pairwise opening of these CO bridges in the *trans*-(CA/AC) form. The synchronous back-formation of the two CO bridges has a similar preference with the CO ligands A and B again acting as the bridging ligands.

Bridge/terminal exchange can take place, but only via the *cis*-(CA/AC) form by rotation of one of the $Ru(CO)_3(\eta^5$-MAD-yl) units in the *trans*-isomer by 120°. Since this rotation will cost extra energy, bridge/terminal exchange does not occur at low temperatures. For *trans*-(AA/CC) there is no preference for the two *nonequivalent symmetric* bridging CO ligands to move to either of the two inner Ru atoms. Thus f and g (*Figure 9*) are formed with equal probability. As a result the four equivalent terminal CO ligands in intermediates f and g may exchange freely, which is responsible for the observed rapid interchange of the bridging and terminal CO ligands in the central $Ru_2(CO)_4$ moiety of the *trans*-(CC/AA) form. At temperatures above 183° K *cis/trans* isomerization occurs for both diastereomers.

In conclusion we may state that the isolobal relationships have been stretched even to the analogous fluxional behaviour of $Ru_4(CO)_{10}(MAD-yl)_2$ and $Cp_2Fe_2(CO)_4$ [16,20-25].

4. Reactivity of $Ru_4(CO)_{10}(MAD-yl)_2$.

4.1. Reactions with CO.

At 20°C the complex $Ru_4(CO)_{10}(MAD-yl)_2$ reacts first rapidly and then (much) slower with four molecules of CO to give first $Ru_4(CO)_{11}(\eta^5$-MAD-yl)(η^3-MAD-yl), in which one π-C=N unit is decoordinated [19]. The second product is $Ru_4(CO)_{12}(\eta^5$-MAD-yl)(η^2-MAD-yl) in which the π-C=C bond is decoordinated instead of the π-C=N moiety of the other MAD-yl ligand. Thereafter $Ru_4(CO)_{13}(\eta^3$-MAD-yl)(η^2-MAD-yl) and $Ru_4(CO)_{14}(\eta^2$-MAD-yl)$_2$ are slowly formed. Interestingly, the latter compound consists of a linear chain of four Ru atoms in which the metals are connected by a Ru-Ru bond only, these are not supported by any bridging ligand whatsoever. The whole reaction sequence which is reversible is shown in Scheme 2.

Under thermolysis and photolysis conditions the reaction of $Ru_4(CO)_{10}(MAD-yl)_2$ leads to cluster breakdown (*Scheme 3*) [19]. Thermolysis affords $Ru_2(CO)_6[CH_2C-C(H)(H)C=N-i-Pr]$ which contains a metal-η^3-allyl bonded unit and the known $Ru_2(CO)_6[CH_3=C(H)(CH_2N-i-Pr]$ (vide infra). It is very likely that in an intermediate stage intramolecular H-transfer takes place from one part of the tetranuclear complex to the other as depicted in Scheme 3. It is postulated that thermal activation causes rupture of the η^2-C=N bonded unit of one of the η^5-MAD-yl groups. The still coordinated η^2-C=C moiety is forced with its CH_3 substituent close to the reactive

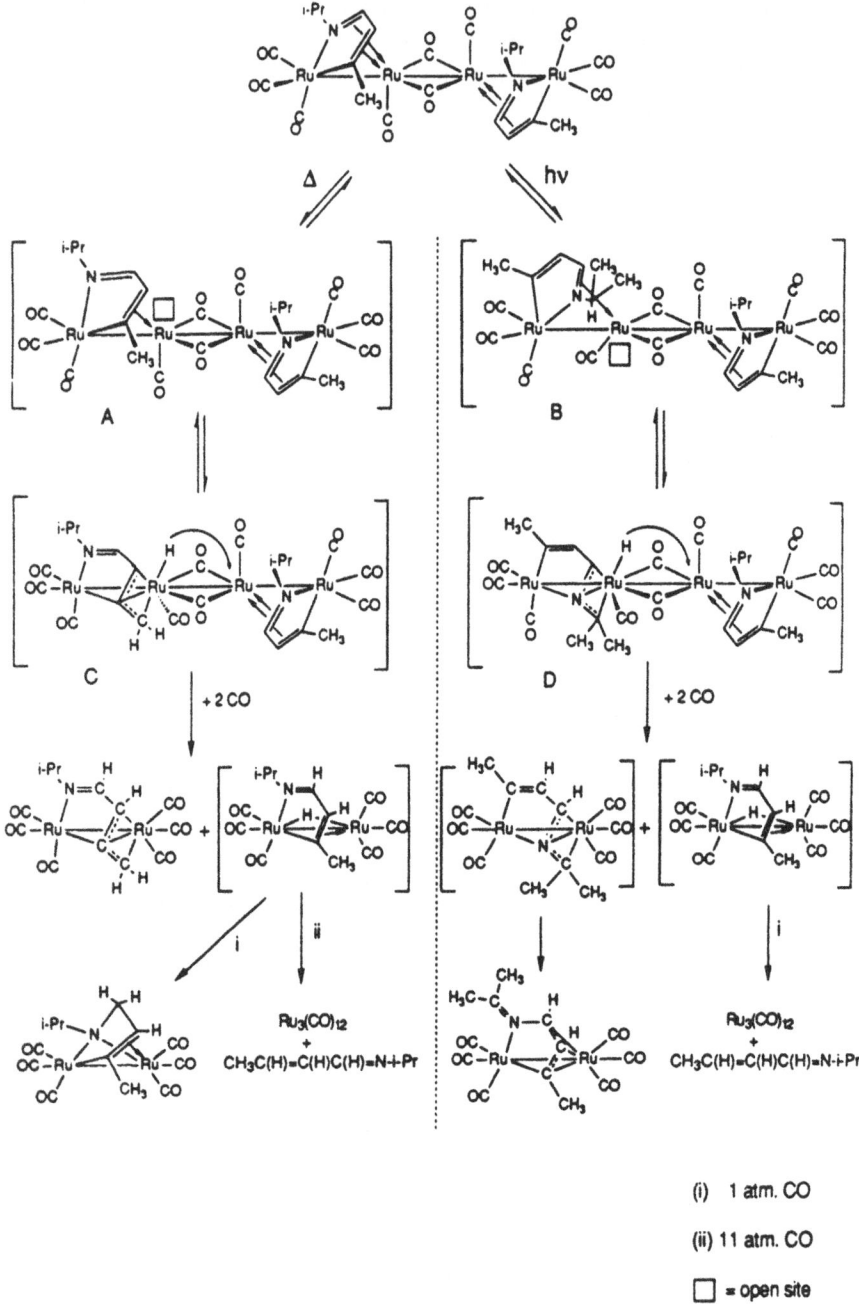

(i)   1 atm. CO

(ii) 11 atm. CO

☐ = open site

Scheme 3
Proposed mechanisms of the thermally and photochemically induced conversions
of $Ru_4(CO)_{10}[CH_3C=C(H)(H)C=N-i-Pr]_2$ in the presence of CO.

16e Ru center after which insertion into the Cγ-H bond may occur. Hydride transfer and CO induced cluster rupture give the final products [19].

In the case of the photochemical reaction we have to assume that now the η^2-C=C bonded moiety, although more strongly bonded to Ru than the η^2-C=N unit, is the one that is dissociating. As a result the isopropyl group comes now close to the 16e Ru center followed by insertion of this Ru atom into the methine C-H bond. In an analogous fashion to the thermolysis reaction, H-transfer and rupture of the central Ru-Ru bond by CO results in the formation of the dinuclear reaction product. In Scheme 3 further details are given, while the reader is referred to the original articles {14,19].

4.2. Reactions of $Ru_4(CO)_{10}(MAD-yl)_2$ with H_2.

At 90°C $Ru_4(CO)_{10}(MAD-yl)_2$ reacts with H_2 to afford the 64e butterfly cluster $H_2Ru_4(CO)_8[R^1C=C(H)(H)C=NR^2]_2$, $H_4Ru_4(CO)_{12}$ and $R^1(CH_2)_3NHR^2$ as the principal products [26]. It should be noted that under the reaction conditions used only the (CC/AA) diastereomer was formed while the other diastereomer (CA/AC) apparently decomposes to the latter two products mentioned above. Both diastereomers, however, can be made from $HRu_2(CO)_5(R^2N=C(H)(H)C=CR^1]$ by loss of CO and subsequent dimerization as shown in Scheme 1.

When applying more moderate temperatures it was found that the two diastereomers of the linear cluster $Ru_4(CO)_{10}(MAD-yl)_2$ react at different temperatures with H_2. At 40°C the (CA/AC) diastereomer reacted with H_2 via the pentacarbonyl-hydride to the products, while the (CC/AA) isomer needed temperatures of 70°C for the reaction to proceed [26] (Scheme 4).

As already mentioned briefly in section 2, the tetranuclear $Ru_4(CO)_{10}(MAD-yl)_2$ affords, when reacted with H_2/CO in a 9 to 1 ratio, two molecules of $Ru_2(CO)_6[R^2NCH_2C(H)=CR^1]$ (Scheme 1). Interestingly this product may also be obtained by reacting the butterfly cluster $H_2Ru_4(CO)_8(MAD-yl)_2$ with CO as depicted in Scheme 1 [26]. Both reactions proceed via the dinuclear intermediates $HRu_2(CO)_5[R^2N=C(H)(H)C=C-R^1]$ and subsequently $HRu_2(CO)_6[R^2N=C(H)(H)-C=C-R^1]$.

5. Rational Synthesis of Heteronuclear MAD-yl Compounds.

From the insights gained during our studies on the di- and tetranuclear ruthenium compounds, we anticipated the feasibility of the rational synthesis of several isostructural M-Ru complexes in general and Fe-Ru compounds in particular [27-30]. As $Ru_4(CO)_{10}[R^2N=C(H)(H)C=CR^1]_2$ can be built up from two molecules of HRu_2-$(CO)_5[R^2N=C(H)(H)C=CR^1]$ (Scheme 1), we staged the synthesis of $HFeRu$-$(CO)_5[R^2N=C(H)(H)C=CR^1]$ with the intention to react it with dinuclear ruthenium complexes and to dimerize it in order to arrive at heterotetranuclear mono-azadienyl

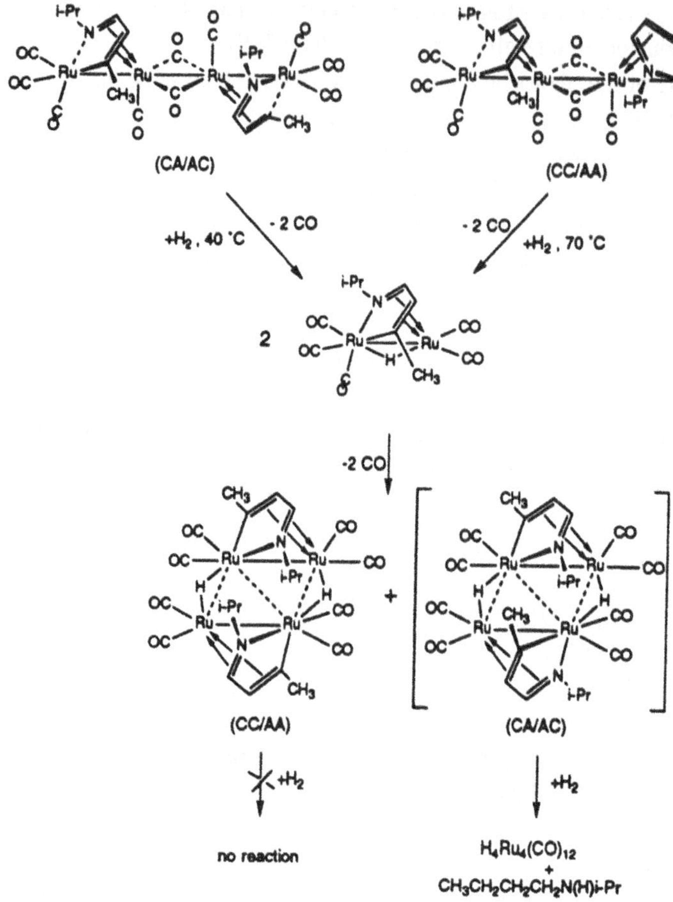

Scheme 4
Hydrogenation of the two diastereomers of $[Ru_4(CO)_{10}[CH_3C=C(H)(H)C=N-i-Pr]_2$.

compounds. The first goal was achieved by photochemical comproportionation between $Ru_2(CO)_6[R^2NCH_2C(H)=CR^1]$ and $Fe_2(CO)_9$ (*Figure 10*) [28]. An X-ray structure of one of these compounds confirmed its structural characterization [27]. $HFeRu(CO)_5[iPrN=C(H)(H)C=CPh]$ was shown to be a precatalyst for the homogeneous hydrogenation of styrene and α-methylstyrene.

Reaction of $HFeRu(CO)_5[R^2=C(H)(H)C=Ph]$ with $Ru_2(CO)_6[R^2N=C(H)(H-C=C=CH_2]$ led to the mixed-ligand tetranuclear $FeRu_3(CO)_{10}[R^2N=C(H)(H)C=CPh]$

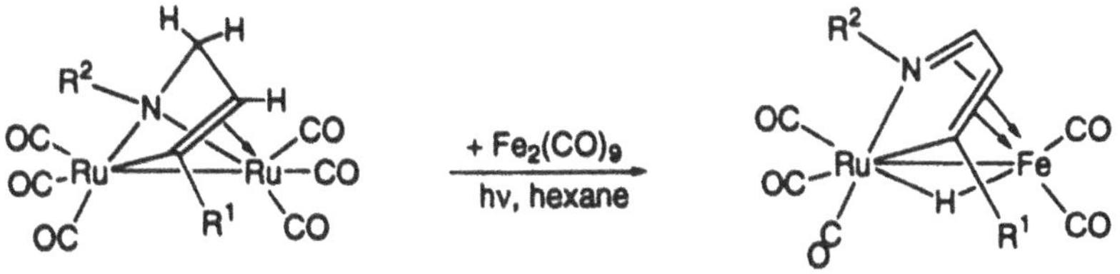

Figure 10
Photochemical synthesis of HFeRu(CO)₅(MAD-yl).

[R²N=C(H)(H)C=CCH₃], whereas the symmetric hetero- tetranuclear compound Fe₂Ru₂(CO)₁₀[R²=C(H)(H)C=CCH₃]₂ was selectively obtained upon dehydrogenation of HFeRu(CO)₅[R²N=C(H)(H)C=CCH₃] in the presence of styrene (*Figure 11*) [29]. Upon treatment with molecular hydrogen, the parent hydrido-iron-ruthenium compound was obtained in good yield again.

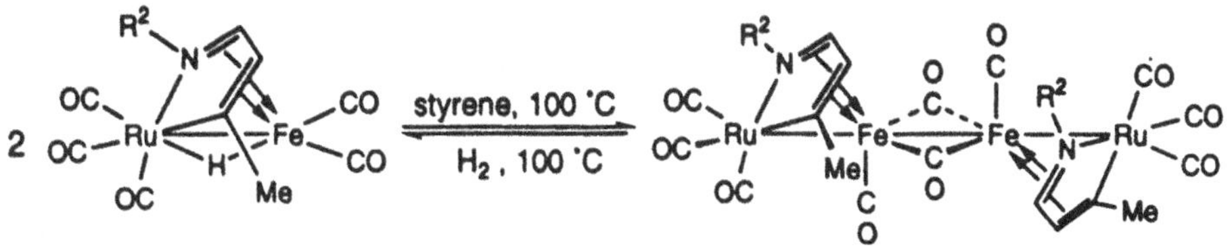

Figure 11
Reversible dehydrogenation of HFeRu(CO)₅(MAD-yl) to give
Fe₂Ru₂(CO)₁₀(MAD-yl)₂.

The NMR of Fe₂Ru₂(CO)₁₀[R²N=C(H)(H)C=CCH₃]₂ at 293 K resembles the low-temperature NMR of Ru₄(CO)₁₀(MAD-yl)₂ and the former occurs in solution as a ca. 1:1 mixture of the *trans*-(CA/AC) and *trans*-(AA/CC) diastereomers. Hence no *trans/cis* isomerization occurs in the case of the Fe₂Ru₂(CO)₁₀(MAD-yl)₂ at 298 K, which was attributed to increased steric interactions in the *cis*-diasteromers of this compound as compared to the tetraruthenium analogues [29], in concert with similar observations for isolobally related Fe-Fe bonded compounds such as (η⁵-C₅H₅)₂-Fe₂(CO)₄ [24,25].

6. Conclusions.

Summarizing our results we may conclude that the present system although more complicated than expected affords a beautiful example of the applicability of isolobal principles not only for structures, but also with regards to dynamic behaviour and reactivity patterns.

7. References.

1.   Van Koten, G., and Vrieze, K. (1982) *Advances in Organometallic Chemistry*, Ed. Stone, F.G.A. and West, R., Academic Press, New York London, **21**, 151-239.
2.   Vrieze, K., and Van Koten, G. (1985) *Inorg. Chim. Acta,* **100**, 79-96.
3.   Vrieze, K. (1986) *J. Organometal. Chem.*, **300**, 307-326.
4.   Elsevier, C.J., Muller, F., Vrieze, K. and Zoet, R. (1988) *New. J. Chem.*, **12**, 571-579.
5.   Muller, F., Van Koten, G., Polm, L.H., Vrieze, K., Zoutberg, M.C., Heijdenrijk, D., Kragten. E., and Stam, C.H. (1989) *Organometallics*, **8**, 1340-1349.
6.   Muller, F., Van Koten G., Kraakman, M.J.A., Vrieze, K., Heijdenrijk, D., and Zoutberg, M.C. (1989) *Organometallics*, **8**, 1331-1339.
7.   Kraakman, M.J.A., De Koning, T.C., De Lange, P.P.M., Vrieze, K., Kooijman, H. and Spek, A.L. (1993) , *Inorg. Chim. Acta,* **203**, 145-155.
8.   Kraakman, M.J.A., Elsevier, C.J., Vrieze, K. and Spek, A.L. (1992) *Organometallics,* **11**, 4250-4260.
9.   Mul, W.P., Elsevier, C.J., Frühauf, H.-W., Vrieze, K., Pein, L., Zoutberg, M.C. and Stam, C.H. (1990) *Inorg. Chem,* **29**, 2336-2345.
10.   Van Baar, J.F., Vrieze, K. and Stufkens, D.J. (1975) *J. Organometal. Chem.*, **97**, 461-472.
11.   Polm, L.H., Mul, W.P., Elsevier, C.J., Vrieze, K., Christophersen, M.J.N. and Stam, C.H. (1988) *Organometallics,* **7**, 423-429.
12.   Humphries, A.P. and Knoz, S.A.R. (1975) *J. Chem. Soc. Dalton,* 1710-1714.
13.   Mul, W.P., Elsevier, C.J., Polm, L.H., Vrieze, K., Zoutberg, M.C., Heijdenrijk, D. and Stam, C.H. (1991) *Organometallics,* **10**, 2247-2259.
14.   Elsevier, C.J., Mul, W.P. and Vrieze, K. (1992)*Inorg. Chim. Acta.* **198-200**, 689-703.
15.   Spek, A.L., Duisenberg, A.J.M., Mul, W.P., Beers, O.C.P. and Elsevier, C.J. (1991) *Acta Cryst. C.*, **C47**, 297-300
16.   Mul, W.P., Ernsting, J.-M., De Lange, W.G.J., Van Straalen, M.D.M., Vrieze, K., Elsevier, C.J., De Wit, M. and Stam, C.H. (1993) , *J. Amer. Chem. Soc.*

115, 980-989.

17. Bryan, R.F., Green, P.T., Newlands, M.J. and Field, D.S. (1970) *J. Chem. Soc. A*, 3068-3074.

18. Bryan, R.F., and Green, P.T. (1970) *J. Chem. Soc. A*, 3064-3068.

19. Mul, W.P., Elsevier, C.J., Vrieze, K., Smeets, W.J.J. and Spek, A.L. (1992) *Organometallics*, **11**, 1891-1901.

20. Manning, A.R. (1968) *J. Chem. Soc. A*, 1319-1324.

21. Bullitt, J.G., Cotton, F.A. and Marks, T.J. (1970) *J. Amer. Chem. Soc.* **92**, 2155-2156.

22. Bullitt, J.G., Cotton, F.A. and Marks, T.J. (1972) , *Inorg. Chem.* **11**, 671-676.

23. Farrugia, L.J. and Mustoo, L. (1992), *Organometallics*, **11**, 2941-2944.

24. Gansow, O.A., Burke, A.R. and Vernon, W.D. (1972) *J. Amer. Chem. Soc.* **94**, 2550-2552.

25. Gansow, O.A., Burke, A.R. and Vernon, W.D. (1976), *J. Amer. Chem. Soc.* **98**, 5817-5826.

26. Mul, W.P., Elsevier, C.J., Van Leijen, M., Vrieze, K., Smeets, W.J.J. and Spek, A.L. (1992) *Organometallics*, **11**, 1877-1890.

27. Beers, O.C.P., Elsevier, C.J., Mul, W.P., Vrieze, K., Häming, L.P. and Stam, C.H. (1990) *Inorg. Chim. Acta.* **171**, 129-131.

28. Beers, O.C.P., Elsevier, C.J., Kooijman, H., Smeets, W.J.J. and Spek, A.L. (1993) *Organometallics,* **12**, 3187-3198.

29. Beers, O.C.P., Elsevier, C.J., Smeets, W.J.J. and Spek, A.L. (1993) , *Organometallics*, **12**, 3199-3210.

30. Beers, O.C.P., Bouman, M.M., Komen, A.E., Vrieze, K., Elsevier, C.J., Horn, E.A. and Spek, A.L. (1993) *Organometallics*, **12**, 315-324.

UNUSUAL LIGAND TRANSFORMATIONS AND REARRANGEMENTS IN HETEROMETALLIC CLUSTERS

Y. Chi, S-J. Chiang and C-J. Su
Department of Chemistry
National Tsing Hua University
Hsinchu 30043, Taiwan, R.O.C.

ABSTRACT The structural behavior of the tetrametal clusters $LWM_3(CO)_{12}H$ (L = Cp and Cp*; M = Os and Ru) has been studied by variation of the ancillary ligands on tungsten atom and the group 8 transition metal atoms. Experimental results clearly indicate that the WOs_3 clusters adopt tetrahedral core arrangements, while the most dominant isomers of the WRu_3 derivatives in solution possess the butterfly arrangement with a unique $\mu_4\text{-}\eta^2\text{-CO}$ ligand. In a parallel study, the treatment of acetylide cluster $CpWRu_2(CO)_8(CCPh)$ with $Ru_3(CO)_{12}$ induces the C-C bond cleavage of the acetylide ligand, affording two high nuclearity carbido-alkylidyne clusters, i.e., $CpWRu_4(\mu_5\text{-C})(CO)_{12}(\mu\text{-CR})$ and $CpWRu_5(\mu_6\text{-C})(CO)_{14}(\mu\text{-CR})$, in high yield.

1. INTRODUCTION

The aim of this presentation is to bring into focus the area of polynuclear heterometallic clusters, as well as showing some novel bonding modes and reactivity pathways of carbonyl and isoelectronic acetylide ligands. Research in this area has been stimulated by two major assumptions i.e. (i) the ligands of transition metal clusters can be used as models to understand the reaction mechanisms of small organic fragments on transition-metal surfaces and metal catalysts, and (ii) a combination of metals, having diverse chemical and electronic properties, within a single compound may induce unique and highly selective chemical transformations that cannot be observed in homonuclear systems.[1]

In the course of studying the generation of the multisite bound $\mu_4\text{-}\eta^2\text{-CO}$ ligand, we have carried out the systematic synthesis of the complexes $LWM_3(CO)_{12}H$

113

L. J. Farrugia (ed.), The Synergy Between Dynamics and Reactivity at Clusters and Surfaces, 113–124.
© 1995 *Kluwer Academic Publishers.*

(L = Cp and Cp*; M = Os and Ru) and $Cp*MoRu_3(CO)_{12}H$. X-ray diffraction studies and their ^1H and ^{13}C NMR data reveal that, for the WOs_3 clusters, only tetrahedral isomers are observed. This is probably due to the fact that the metal-metal interactions between tungsten and osmium (both third-row elements) are relatively stronger than those between tungsten and ruthenium (a second-row metal atom). Moreover, the formation of butterfly isomers with the μ_4-η^2-CO ligand is observed by replacing the osmium atoms with second-row ruthenium atoms.

In the case of the acetylide clusters $CpWRu_2(CO)_8(CCPh)$, the ligated acetylide is analogous to the multisite bound carbonyl ligands. We observed that, upon treatment with $Ru_3(CO)_{12}$, this acetylide cluster produced a pentanuclear square-pyramidal cluster $CpWRu_4(\mu_5$-C)(CO)_{12}(\mu$-CR) and a hexanuclear octahedral cluster $CpWRu_5(\mu_6$-C)(CO)_{14}(\mu$-CR) in sequence. Such a C-C bond cleavage process verifies the intimate relationship between acetylide ligands and alkylidyne-carbide fragments,[2] and provides an illustrative example which parallels the disproportionation of CO ligand [3] or the scission of coordinated isocyanide ligands.[4]

2. GENERATION OF CLUSTERS CONTAINING μ_4-η^2-CO LIGAND

2.1 WOs_3 CARBONYL CLUSTERS

The tetrahedral heterometallic cluster $CpWOs_3(CO)_{12}H$ (1) was first prepared by Shapley and coworkers through a direct condensation of $Os_3(CO)_{10}(CH_3CN)_2$ and $CpW(CO)_3H$.[5] This complex shows two hydride resonances at δ –18.63 and –21.82 (J_{W-H} = 43 Hz) with relative intensities 3 : 1, indicating the presence of two interconverting isomers in solution (Scheme 1).[6] The hydride resonance at δ –18.63, which lacks the J_{W-H} coupling, is assigned to an isomer adopting the solid-state structure, which exhibited the presence of an Os-H-Os hydride ligand. The second isomer possessing a W-H-Os hydride was identified by the assistance of the structural and spectroscopic data of a derivative $Cp*WOs_3(CO)_{12}H$ (2).

In this study, complex 2 was prepared from the reactions with $Cp*W(CO)_3H$ accordingly. Similarly to the solution behavior of 1, the ^1H NMR spectrum exhibited two hydride signals at δ –18.22 and –20.93 (J_{W-H} = 41 Hz) in ratio 1 : 2.1, indicating that the isomer with a W-H-Os interaction becomes the dominant species. X-ray structural determinations on both cluster compounds implied that isomer a possesses a tetrahedral core with only one bridging CO ligand associated with an Os-W edge; whereas isomer b has the less bulky hydride ligand on the W-Os edge. The molecular structure of these two isomers are outlined below.

Scheme 1

Meanwhile, the isomeric distribution of the complex $Cp^{\ddagger}WOs_3(CO)_{12}H$, $Cp^{\ddagger}$ = $C_5H_4CF_3$ was determined, in which the $Cp^{\ddagger}$ ligand is less electron-donating but is similar in size to the Cp^* ligand.[7] Based on this observation, we proposed that the unfavorable steric interaction between the CO and the pentamethylcyclopentadienyl ligands is apparently responsible for the larger reduction of the **2a/2b** ratio compared with that of the **1a/1b** ratio.

2.1 WRu$_3$ AND MoRu$_3$ CARBONYL CLUSTERS

The WRu$_3$ cluster derivative (**3**) with the formula $CpWRu_3(CO)_{12}H$ was similarly obtained from the reaction of $Ru_3(CO)_{12}$ and excess of $CpW(CO)_3H$. [8] An alternative approach would involve using the reaction of $Ru_3(CO)_{12}$ and $[CpW(CO)_3][PPN]$ to produce first the anionic cluster $[CpWRu_3(CO)_{12}][PPN]$, followed by protonation with trifluoroacetic acid. [9] For this WRu$_3$ derivative, the ^1H NMR (400 MHz, CD_2Cl_2, 190 K) spectrum exhibits three hydride resonances at δ -17.03 (**3a**), -18.67 ($J_{W\text{-}H}$ = 53 Hz, **3b**) and -20.60 (**3c**) with a relative intensity ratio of 1 : 1.6 : 3.2, indicating the presence of three interconvertable isomers in solution (Scheme 2).

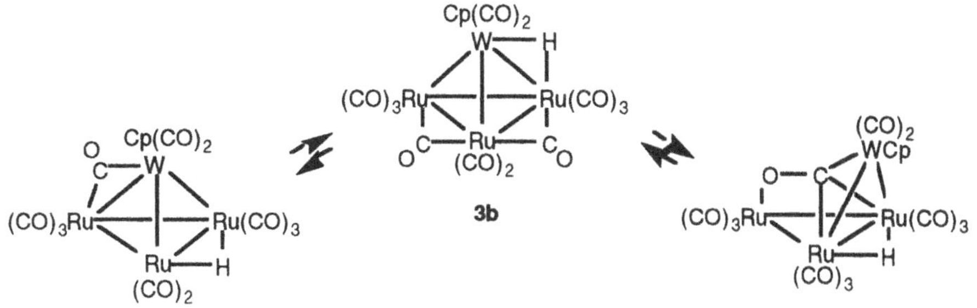

Scheme 2

The variable temperature ^{1}N NMR study indicated that the exchange behavior of the first two isomers **3a** and **3b** is similar to that of the WOs$_3$ complexes **1** and **2** mentioned earlier. Their structures are assigned and can be easily differentiated in the light of the J_{W-H} couplings. The identity of the third isomer (**3c**) was revealed by a single crystal X-ray diffraction, which confirmed the butterfly core arrangement and the occurrence of an unusual μ_4-η^2-CO ligand. The parameters associated with this μ_4-η^2-CO ligand are consistent with those of the Fe_4 butterfly complexes, $[Fe_4(CO)_{13}(\mu$-A$)]^-$, A = H, $AuPEt_3$, $CuPPh_3$, $HgCH_3$, $HgMo(CO)_3Cp$ and $HgFe(CO)_2Cp$,[10] the polynuclear clusters $Cp^*_2W_2Ru_4(CO)_{12}(\mu$-PPh$)_2$,[11] $Cp_2Mo_2Ru_4(CO)_{13}(\mu_4$-CO$)(\mu_4$-S$)$ and $Cp_2Mo_2Ru_5(CO)_{14}(\mu_4$-CO$)_2(\mu_4$-S$)$.[12]

The effects of the ancillary ligand and the transition-metal atom were further investigated by preparing two additional Cp^* complexes $Cp^*WRu_3(CO)_{12}H$ (**4**) and $Cp^*WRu_3(CO)_{12}H$ (**5**). Structural and spectroscopic studies were also carried out and, as expected, both complexes were found to exist as three isomers in solution (Scheme 3). ^1H NMR spectra analysis revealed three hydride resonances at δ –15.26 with J_{W-H} = 51 Hz (**4b**), –20.16 (**4c**) and –20.20 (**4d**) having an intensity ratio of 1 : 1.4 : 4 and at δ –13.75 (**5b**), –20.32 (**5c**) and –20.55 (**5d**) with intensities 1 : 5.9 : 4.2, respectively. The highest frequency hydride signal of each complex is attributed to the tetrahedral isomer **b**, which is related to that of the structurally characterized WOs_3 isomer **b**, due to the observation of J_{W-H} coupling for **4b**. Interestingly, the tetrahedral isomers **a**, which possess a carbonyl ligand bridging across one W-Os edge and a hydride on the opposite Os-Os edge, were not observed. Again, this is apparently due to the extremely unfavorable repulsion effect between a bridging CO ligand and a Cp^* ligand.

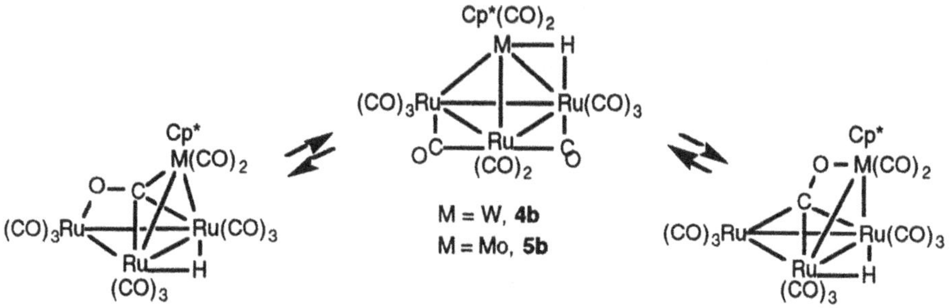

M = W, **4c**
M = Mo, **5c**

M = W, **4b**
M = Mo, **5b**

M = W, **4d**
M = Mo, **5d**

Scheme 3

The structures of the butterfly isomers **c** and **d** were confirmed by X-ray diffraction as well as ^{13}C NMR studies. First, the X-ray study on **4** verifies the key features of derivatives **c**, revealing a butterfly metal core, a μ_4-η^2-CO ligand with its oxygen atom coordinated to the tungsten atom, and Cp^* ligand located in a position syn to the μ_4-η^2-CO ligand. In addition, the X-ray analysis of **5** performed at 110K implies that the cluster contains a similar MoRu$_3$ butterfly framework with a disordered μ_4-η^2-CO ligand, showing the characteristics of both derivatives **c** and **d**.[1] Two molecular drawings are presented to emphasize the interaction between the disordered μ_4-η^2-CO ligand, C(1)O(1) and C(1')O(1'), and the metal core skeleton (Figure 1). The refined occupancy was 65% for C(1)O(1) and 35% for C(1')O(1'). This value differs slightly from that observed in the solution, but agrees with the ratio obtained by a solid-state ^{13}C CPMAS NMR study. In both molecular drawings, the Mo atom occupies the wing-tip position and is linked by two CO ligands and a Cp* ligand. Each of the Ru atoms are coordinated to three terminal CO ligands. The hydride ligand, located by difference Fourier synthesis, is bridging the hinge Ru-Ru bond. The carbon atom C(1) is coordinated to the metal atoms Mo, Ru(1) and Ru(3). The oxygen atom O(1) is tilted to the Ru(2) atom with $\angle$Mo-C(1)-O(1) = 134°. The less abundant C(1') atom is associated with all three Ru atoms and the O(1') is bound to the Mo atom with $\angle$Ru(2)-C(1')-O(1') = 137(2)°. Thus, this solid-state structural data indicates that the isomers **c** and **d** in solution possess essentially identical butterfly core geometries, except for that the oxygen atom of the μ_4-η^2-CO ligand is linked to two opposite wing-tip metal atoms. A summary of the relationship for all three isomers of complexes **4** and **5** is provided in Scheme 3.

Consistent with the presence of pairs of butterfly isomers, the ^{13}C NMR data of ^{13}CO enriched **4** (CD$_2$Cl$_2$, 175K) showed two unique downfield signals at δ 267.5 ($^1J_{W-C}$ = 158 Hz) and 254.7 ($^2J_{C-C}$ = 21 Hz) with ratio 3 : 1 due to the unique μ_4-CO ligand of **4c** and **4d**. Under the same conditions complex **5** showed the μ_4-CO signals at δ 272.6 and 255.0 ($^2J_{C-C}$ = 21 Hz) with ratio 1 : 2. Their chemical shifts correlate with the expected trends of metal carbonyl complexes, i.e., the signal of isomers **c** is more downfield than that of isomers **d** because, in the latter cases, the μ_4-CO is coordinated with the Ru$_3$ triangle, not to the adjacent MRu$_2$ triangle, M = W or Mo. These assignments have been further confirmed by detection of the $^2J_{C-C}$ coupling between the μ_4-CO ligand and its trans CO ligand for isomers **d**, and by the observation of large $^1J_{W-C}$ coupling for isomer **c**. Thus, for the WRu$_3$ complex **4**, the signal at δ 267.5 with $^1J_{W-C}$ = 158 Hz is due to the μ_4-CO ligand of **4c**, and the small $^2J_{C-C}$ coupling of the signal at δ 255.0 for **5** is due to the coupling between the atom C(1') and the trans C(8)O(8) ligand (δ 203.4) on Ru(2).

Furthermore, the Mo-CO ligands of **5d** exhibited signals at δ 239.2 and 228.5, and those of **5c** appeared at δ 226.0 and 222.1. The Mo-CO ligands of **5c** showed signals at highfield positions and underwent fast pairwise CO exchange with respect to those of **5d**. This occurrence could be accounted for by the fact that the Mo atom is not bonded to the oxygen atom of the μ_4-CO, and the coordination

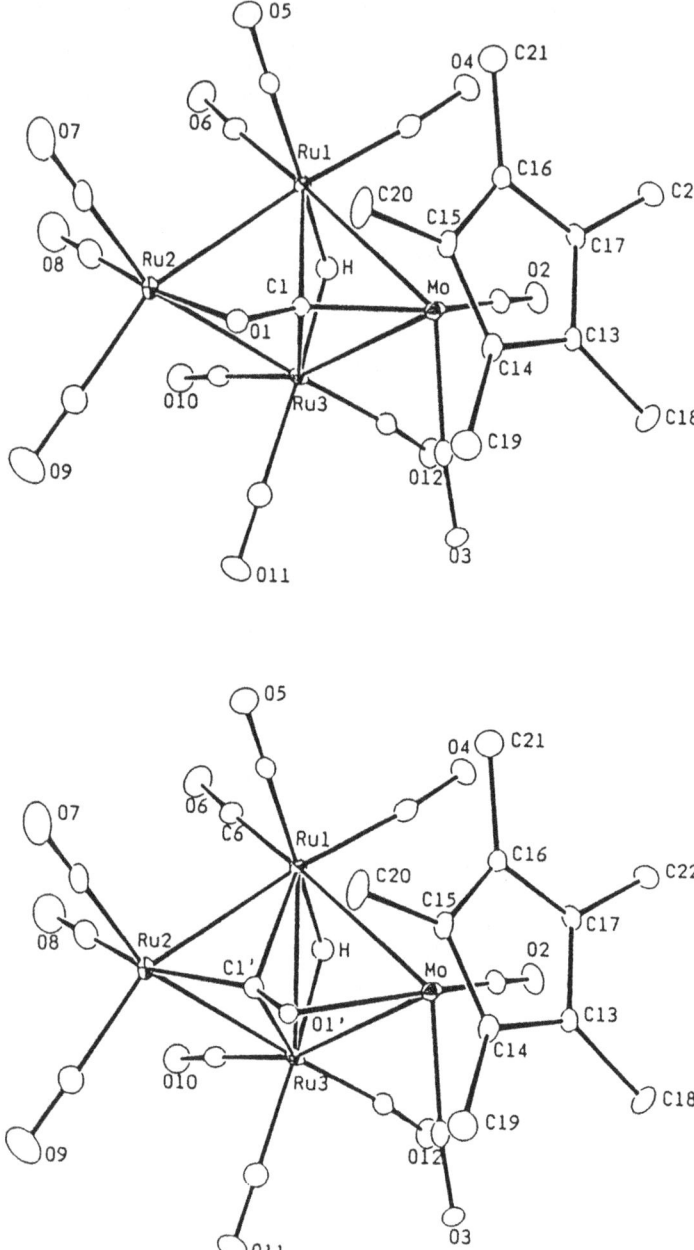

Figure 1. Molecular drawings of the isomers that correspond to isomers **5c** (top) and **5d** (bottom), showing 30% probability thermal ellipsoids.

environment is less sterically congested. The Mo-CO resonances of **5b** were not observed as it may undergo rapid exchange with other CO ligands within the same molecule. This postulation has a precedence in that the ^{13}C NMR spectrum of tetrahedral cluster **2** showed only very broad CO signals at even −190K, indicating the occurrence of fast CO scrambling. [6]

2.3. TETRAHEDRAL TO BUTTERFLY ISOMERIZATION

At least four structural isomers have been observed in the tetranuclear clusters **1 – 5**. The preferred geometry can be moved from tetrahedron to butterfly geometry by simply replacing the osmium atoms with the slightly smaller [14] second-row ruthenium atoms and by using the bulky Cp* ligand. The WRu$_3$ complexes favor the butterfly core arrangement since the W-Ru bond is relatively weaker and since this geometry creates more space for the surrounding CO ligands. Consequently, a CO ligand is transformed to the μ_4-η^2-mode, and provides four cluster valence electrons to compensate the unsaturation generated by opening the W-Ru bond. Thus, the observed preference for isomerization and formation of the μ_4-η^2-CO ligand can be understood in terms of a synergy between steric and electronic effects.

Furthermore, for the Cp* derivatives **4** and **5**, two different butterfly isomers **c** and **d** and a tetrahedral isomer **b** were observed in this study. In isomer **c**, the oxygen atom of the μ_4-η^2-CO ligand is linked to the Ru atom. Whereas, in isomer **d**, the respective oxygen atom is coordinated with the W or the Mo atom instead. The exchange among these three isomers of **5** was also examined. Our experimental results indicated that the tetrahedral isomer **b** is in slow exchange with both butterfly isomers **c** and **d**. This conclusion is strongly supported by the observation of ^1H NMR magnetization transfer between **5b/5c** and **5b/5d** pairs at 294 K by using spin saturation transfer experiment. When the hydride signal due to **5b** was saturated, a decrease in the intensities of the hydride resonances of 39% and 14% was observed for isomers **5d** and **5c**, respectively. In contrast, the direct exchange between **5c** and **5d** is of higher energy. This is in accordance with ^{13}C magnetization transfer NMR experiments, in which one μ_4-CO signal was inverted using a DANTE pulse train at 294 and 304 K. No reduction of intensity of the second μ_4-CO signal was observed at either temperature, indicating that there is no direct exchange occurs between these isomers.

Based on the above discussion, we can conclude that both butterfly isomers **c** and **d** are in equilibrium with the tetrahedral isomer **b** in solution. However, the direct conversion between the butterfly isomers via flipping of the μ_4-CO ligand along the wing-tip metal atoms is not feasible. This is despite the fact that the transition-state structure for this isomerization, which would exhibit a butterfly metal core with

120

a μ_4-η^1-CO ligand,[15] has been observed by X-ray diffraction and studied by EHMO calculations.

2.4. FORMATION OF WRu3 CARBIDO CLUSTERS

After we established that the butterfly WRu3 isomers can interchange with the tetrahedral isomers, an investigation was undertaken to understand whether or not the μ_4-η^2-CO ligand in butterfly isomers can undergo C-O cleavage reaction to yield a carbido cluster as observed in the Fe4 system. [16]

L = Cp, Cp*, **6**

Scheme 4

Unfortunately, hydrogenation of complexes **3** and **4** failed to induce cleavage of the C-O bond but produced instead the elimination of one CO ligand, affording the tetrahedral trihydride clusters LWRu3(CO)11(μ-H)3 (**6**) as two interchangeable isomers in solution.[17] As indicated above (scheme 4), one isomer has three hydrides on the three edges of a WRu2 triangular face, whereas the hydrides of the second isomer are aligned in a zig-zag arrangement. Failure to produce any carbido clusters from **3** and **4** illustrates that the formation of the μ_4-η^2-CO ligand is not essential for the C-O cleavage.

Although C-O bond cleavage could not be induced in these studies, the intended carbido clusters LWRu3(CO)11(μ_4-C)(μ-H), L = Cp, (**7a**) and Cp*, (**7b**), were isolated in low yields by hydrogenation of the related WRu3 methoxymethylidyne clusters LWRu3(CO)11(μ_3-COMe), L = Cp and Cp*, (**8**) in refluxing toluene (Scheme 5). [18] Complexes **8** contain tetrahedral WRu3 metal core with the methoxymethylidyne ligand coordinated to one WRu2 face. The methoxymethylidyne ligand is composed of a CO fragment and methyl group, the latter is isoelectronic with a hydrogen atom. Thus, complex **8** represents a fifth isomer for the tetrametal complexes of formula LWRu3(CO)12H, where the hydrogen (or methyl fragment) is bonded to the oxygen atom of the CO group, rather than a metal-metal edge of the WRu3 core, as observed in the other four isomers **a** – **d**. Hence, the formation of the carbido clusters requires that the precursor must possess a fairly strong bonding with the oxygen atom of CO ligand, e.g., the Me-O interaction of the methoxymethylidyne ligand.

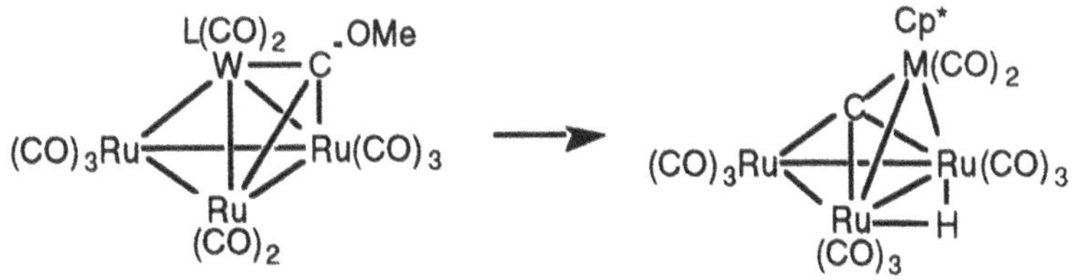

Scheme 5

3. REACTIVITY OF WRu$_2$ ACETYLIDE CLUSTERS

C-C bond cleavage of a coordinated acetylide is relevant to the CO bond cleavage of carbonyl ligands. This study provides, for the first time, an illustrative example of generation of carbido clusters from acetylide clusters upon increase of the nuclearity of the cluster molecules. For these unusual C-C bond cleavage reactions, the starting acetylide cluster is a heterometallic complex (**9**) with formula CpWRu$_2$(CO)$_8$-(C≡CPh), which was prepared by treatment of the mononuclear metal acetylide complex CpW(CO)$_3$(C≡CPh) with approximately 0.67 mole equivalents of Ru$_3$(CO)$_{12}$ in refluxing toluene solution.[19]

For this WRu$_2$ derivative **9**, there are two rotational isomers in solution, which undergo rapid interconversion as determined by variable temperature ^{1}H and ^{13}C NMR studies. In one isomer, the acetylide ligand adopts the asymmetric form with its C-C bond orthogonal to one of the W-Ru bonds. In the second isomer the acetylide possesses the symmetric arrangement, in which the C-C bond bisects with the unique Ru-Ru bond of the WRu$_2$ triangle (Scheme 6). The heterometallic acetylide clusters of formulae CpNiFe$_2$(CO)$_6$(C≡C^{t}Bu) [20] and CoFe$_2$(CO)$_9$-(C≡CSiMe$_3$) [21] have been reported, and these show similar μ_3-η^2-bonding modes for the acetylide ligand. However, the related rotational dynamic process was not observed.

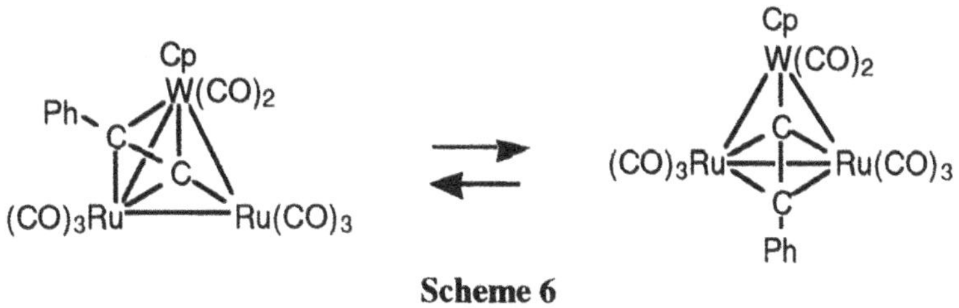

Scheme 6

When **9** was further treated with $Ru_3(CO)_{12}$ in refluxing heptane solution, it was converted into two new high nuclearity cluster compounds.[22] One of these, obtained in 68% yield, was identified as the pentanuclear carbido cluster $CpWRu_4(\mu_5\text{-}C)(CO)_{12}(\mu\text{-CPh})$ (**10**) and the second (formed in 19% yield) as the hexanuclear derivative $CpWRu_5(\mu_6\text{-C})(CO)_{14}(\mu\text{-CPh})$ (**11**). X-ray diffraction studies of **10** and **11** show that they have novel structures as illustrated in scheme 7.

(**10**)   (**11**)

Scheme 7

The metal framework of **10** is best described as a WRu_4 square pyramidal arrangement with one Ru atom occupying the apical position and a carbide atom located at a position slightly below the WRu_3 plane. One CO ligand bridges across a basal W-Ru edge, and an asymmetric alkylidyne is associated with the adjacent W-Ru edge. The overall arrangement resembles that found in many Ru_5 square pyramidal carbido cluster compounds reported in literature. [23] Consistent with the solid-state structure, a carbido and an alkylidyne signal were observed at δ 424.0 ($J_{C\text{-}W}$ = 88 Hz) and 302.3 ($J_{C\text{-}W}$ = 156 Hz) in the ^{13}C NMR spectrum at 294K, in addition to a broad and a sharp CO signal at δ 202.2 and 194.7 respectively, with intensities corresponding to five and three CO ligands. The other CO signals were buried under the baseline, indicating the presence of both exchange between terminal and bridging CO sites and threefold rotation of $Ru(CO)_3$ groups. On the other hand, the hexanuclear cluster **11** possesses a slightly distorted octahedral WRu_5 core with three edge-bridging CO and one alkylidyne on a W-Ru edge. Again, a few ruthenium octahedral carbido clusters with 86 valence electrons have been reported in literature [24] but none possess an unsymmetrical alkylidyne on the ligand sphere.

Complex **10** appears to be an intermediate en route to **11**, as the latter was obtained in 60% yield by heating **10** with $Ru_3(CO)_{12}$ in refluxing toluene. Analysis of the products of the crossover experiment by using 1 : 1 mixture of **9** and $Cp^*WRu_2(CO)_8(CCTol)$ indicates that cluster formation involves no scrambling of the W metal fragment and the alkylidyne unit. Based on these observations and the chemistry developed for the butterfly WOs_3 acetylide clusters, [25] we proposed that the transformation **9** $\rightarrow$ **10** $\rightarrow$ **11** involves the prior formation of a hypothetical acetylide complex $CpWRu_3(CO)_{11}(CCPh)$, followed by cleavage of C-C bond to afford **10** and **11** via an addition of $Ru(CO)_n$ fragments. The presence of W atom appears indispensable, since heating of triruthenium cluster $Ru_3(CO)_9(\mu\text{-}H)(C\equiv C^tBu)$ with $Ru_3(CO)_{12}$ fails to produce the analogous Ru_5 or Ru_6 carbido cluster under similar conditions. Consequently, the resulting alkylidyne ligands in both **10** and **11** always end up coordinated to the W atom, and the related W-Os and W-Ru clusters also facilitate the C-C bond scission of alkyne moieties. [26]

4. ACKNOWLEDGEMENTS

We thank National Tsing Hua University and the National Science Council of the Republic of China for support (Grant No. NSC 83-0208-M007-43).

5. REFERENCES

1.(a) Bullock, R. M.; Casey, C. P (1987). *Acc. Chem. Res.* **20**, 167. (b) Stephan, D. W. (1989) *Coord. Chem. Rev.* **95,** 41.

2. (a) Nucciarone, D.; Taylor, N. J.; Carty, A. J. (1986) *Organometallics* **5**, 1179. (b) Carty, A. J.; Taylor, N. J.; Sappa, E.; Tiripicchio, A.; Tiripicchio Camellini M.(1991) *Organometallics* **10**, 1907.

3. (a) Anson, C. E.; Bailey, P. J.; Conole, G.; McPartlin, M.; Powell, H. R. (1989) *J. Chem. Soc., Chem. Commun.* 442. (b) Bailey, P. J.; Duer, M. J.; Johnson, B. F. G.; Lewis, J.; Conole, C.; McPartlin, M.; Powell, H. R.; Anson, C. E. (1990) *J. Organomet. Chem.* **383**, 441.

4. Adams, R. D.; Mathur, P.; Segmuller (1983) *Organometallics* **2**, 1258.

5. Churchill, M. R.; Hollander, F. J.; Shapley, J. R.; Foose, D. S. (1978) *J. Chem. Soc., Chem. Commun.* 534.

6. Peng, S.-M.; Lee, G.-H.; Chi, Y.; Peng, C.-L.; Hwang, L.-S. (1989) *J. Organomet. Chem.* **371**, 197.

7. Gassman, P. G.; Mickelson, J. W.; Sowa, J. R. Jr., (1992) *J. Am. Chem. Soc.* **114,** 6942.

8. Chi, Y.; Wu, F.-J.; Liu, B.-J.; Wang, C.-C.; Wang, S.-L. (1989) *J. Chem. Soc. Chem. Commun.* 873.

9. Cazanoue, M.; Lugan, N.; Bonnet, J.-J.; Mathieu, R. (1988) *Organometallics* **7**, 2480.

10. (a) Wang, J.; Sabat, M.; Horwitz, C. P.; Shriver, D. F. (1988) *Inorg. Chem.* **27**, 552. (b) Horwitz, C. P.; Holt, E. M.; Brock, C. P.; Shriver, D. F. (1985) *J. Am. Chem. Soc.* **107**, 8316.

11. Wang, J.-C.; Lin, R.-C.; Chi, Y.; Peng, S.-H.; Lee, G.-H. (1993) *Organometallics* **12**, 4061.

12. Adams, R. D.; Babin, J. E.; Tasi, M. (1987) *Angew. Chem., Int. Ed. Engl.* **26**, 685.

13. Su, C. J.; Chi, Y.; Farrugia, L. J.; Peng, S.-M.; Lee, G.-H. (1994) submitted to *Organometallics* for publication.

14. The average Ru-Ru distance in $Ru_3(CO)_{12}$ is about 0.0259 A shorter than the respective Os-Os distance in $Os_3(CO)_{12}$; see (a) Churchill, M. R.; Hollander, F. J.; Hutchinson, J. P. (1977) *Inorg. Chem.* **16**, 2655. (b) Churchill, M. R.; DeBoer, B. G. (1977) *Inorg. Chem.* **16**, 878.

15. Li, P.; Curtis, M. D. (1989) *J. Am. Chem. Soc.* **111**, 8279.

16. (a) Holt, E. M.; Whitmire, K. H.; Shriver, D. F. (1981) *J. Organomet. Chem.* **213**, 125. (b) Whitmire, K. H.; Shriver, D. F. (1981) *J. Am. Chem. Soc.* **103**, 6754.

17. Chi, Y.; Cheng, C.-Y.; Wang, S.-L. (1989) *J. Organomet. Chem.* **378**, 45.

18. Chi, Y.; Chuang, S.-H.; Chen, B.-F.; Peng, S.-M.; Lee, G.-H. (1990) *J. Chem. Soc., Dalton Trans.* 3033.

19. (a) Chi. Y.; Lee, G.-H.; Peng, S.-M.; Liu, B.-J. (1989) *Polyhedron*, **8**, 2003. (b) Hwang, D.-K.; Chi, Y.; Peng, S.-M.; Lee, G.-H. (1990) *Organometallics*, **9**, 2709.

20. Martinetti, A.; Sappa, E.; Tiripicchio, A.; Tiripicchio Camellini, M. (1980) *J. Organomet. Chem.* **197**, 335.

21. Seyferth, D.; Hoke, J. B.; Rheingold, A. L.; Cowie, M.; Hunter, A. D. (1988) *Organometallics*, **7**, 2163.

22. Chiang, S.-L.; Chi, Y.; Peng, S.-M.; Lee, G.-H. submitted to *J. Am. Chem. Soc.*, for publication.

23. (a) Henly, T. J.; Wilson, S. R.; Shapley, J. R. (1987), *Organometallics*, **6**, 2618. (b) Johnson, B. F. G.; Lewis, J.; Nicholls, J. N.; Puga, J.; Raithby, P. R.; Rosales, M. J.; McPartlin, M.; Clegg, W. (1983) *J. Chem. Soc., Dalton Trans.*, 277. (c) Adams, C. J.; Bruce, M. I.; Skelton, B. W.; White, A. H. (1992) *J. Organomet. Chem.*, **423**, 105.

24. (a) Braga, D.; Grepioni, F.; Dyson, P. J.; Johnson, B. F. G.; Frediani, P.; Bianchi, M.; Piacenti, F. (1992) *J. Chem. Soc., Dalton Trans.*, 2565. (b) Braga, D.; Grepioni, F.; Parisini, E., Dyson, P. J.; Johnson, B. F. G.; Reed, D.; Shepherd, D. S.; Bailey, P. J.; Lewis, J. (1993) *J. Organomet. Chem.* **462**, 301. (c) Adams, R. D.; Wu, W. (1992) *Polyhedron*, **16**, 2123.

25. (a) Chi. Y.; Lee, G.-H.; Peng, S.-M.; Wu, C.-H. (1989) *Organometallics*, **8**, 1574. (b) Chi. Y. (1992) *J. Chin. Chem. Soc.* **39**, 591.

26. (a) Park, J. T.; Shapley, J. R.; Churchill. M. R.; Bueno, C. (1983) *J. Am. Chem. Soc.*, **105**, 6182. (b) Stone, F. G. A.; Williams, M. L. (1988) *J. Chem. Soc., Dalton Trans.*, 2467. (c) Chi, Y.; Shapley, J. R. (1985) *Organometallics*, **4**, 1900.

HYDRIDE MOBILITY AND ITS RELATION TO STRUCTURE AND REACTIVITY IN POLYMETALLIC CLUSTERS

EDWARD ROSENBERG
Department of Chemistry
The University of Montana
Missoula, MT 59812

Introduction

Two fundamentally important processes in the chemistry of organic substrates on catalytic surfaces are hydrogenation-dehydrogenation and carbon-hydrogen oxidative addition-reductive elimination. In the course of our studies on the thermal reaction chemistry of organic molecules bound to trimetallic clusters, we have noted that dihydrido metal clusters having face capping organic ligands can exhibit a wide range in the activation energies for hydride site exchange (45-70 kJ/mole). Complexes having relatively mobile hydride ligands tend to undergo hydrogen elimination from the cluster (equation 1) while clusters with relatively rigid hydrides tend to undergo reductive elimination of a carbon hydrogen bond (equation 2)[1]. Since first reporting these observations for ruthenium clusters we have seen similar trends in dihydrido

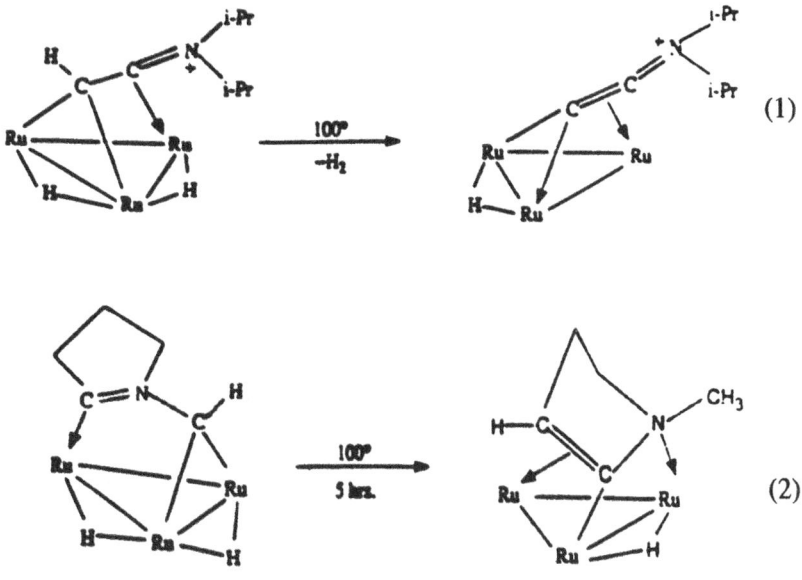

125

L. J. Farrugia (ed.), The Synergy Between Dynamics and Reactivity at Clusters and Surfaces, 125–140.
© 1995 Kluwer Academic Publishers.

trimetallic clusters of osmium. We are now in a position to relate the structure of a particular organometallic cluster to the mobility of its hydrides.

We have also discovered some structurally unusual trihydrido organometallic clusters of osmium, namely

$H(\mu\text{-}H_2)Os_3(CO)_8(\mu_3\text{-}\eta^2\text{-}C=NCH_2CH_2CH_2)$ **(1)** and $(H)(\mu\text{-}H_2)Os_3(CO)_8$-

$(\mu\text{-}\eta^2\text{-}C=NCH_2CH_2CH_2)P(C_6H_5)_3$ **(2)**. In both of these complexes the hydrides are "clustered" around one osmium atom, with the two hydrides bridging edges of the triangle common to the osmium atom bearing the terminal hydride. The reactivity and dynamical properties of these complexes are quite different; **1** being much more rigid than **2**. One and two dimensional NMR techniques have been applied in order to elucidate the mechanism of the dynamical properties of the hydrides and the other ligands in these complexes and related monohydride clusters and a picture emerges in which the exchange processes accessible to the carbonyl ligands have a major influence on hydride mobility[2].

1   **2**   L = PPh₃

We have also been studying the dynamics of the series of complexes $(\mu\text{-}H)_2(\mu\text{-}\eta^2\text{-}$

$C=NCH_2CH_2CH_2(CO)_9(X)(X=Cl, Br, CF_3COO(5))$ where intramolecular hydrogen bonding between X(Cl, Br) and a bridging hydride is indicated and where very subtle structural factors govern the dynamical processes[3].

1.   **Geometry and Dynamics in Dihydrido Trimetallic Clusters with μ_3-Capping Ligands.**

The dynamical properties of dihydrido trimetallic clusters have been well investigated[4]. The most common mechanism for the exchange of two magnetically

Scheme 1

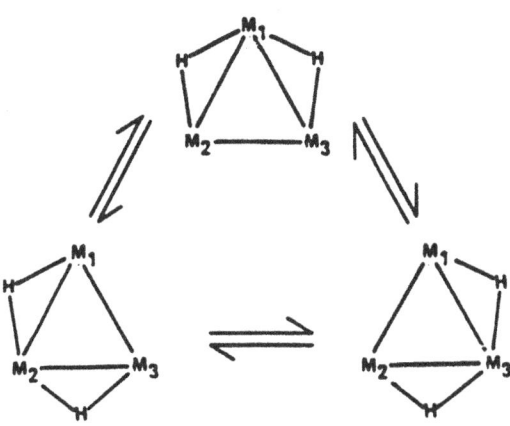

Table 1

Correlation of Hydride Ligand Exchange and Free Energies of Activation with the Dihedral Angle Difference in Dihydride Clusters

	$\Delta G^{\ddagger}$ kJ mol^{-1}	θ_H (deg)
Fluxional		
$H_2Os_3[Me_3CNCHCHCHC](CO)_9$	46.1(2)	43
$H_2Ru_3[(i-Pr)_2NCCH](CO)_9$	46.2(2)	43
$H_2Os_3[MeCCMe](CO)_9$	46.1(2)	44
$H_2Os_3[MeCCNMe_2](CO)_9$	57.5(2)	50
$H_2Ru_3[MeOCCMe](CO)_9$	60.1(2)	54
Static		
$H_2Os_3[CCH_2CH_2CH_2N](CF_3COO)(CO)_9$	~65.0	58
$H_2Os_3[C_4H_7N](CO)_9$ red	>72	64
$H_2Ru_3[MeCHNCMe](CO)_9$	66,8(2)	65
$H_2Os_3CHCNCH_2CH_2CH_2](CO)_9$	65.3(2)	71

$$\theta = \alpha_{H1} - \alpha_{H2}$$

unequivalent hydrides is a two stage process involving simple edge hopping of one hydride followed by a similar motion of the other hydride via a symmetrical intermediate. Collapse of this intermediate to the ground state structure effectively exchanges the two hydrides (Scheme 1). The second stage of this process is normally associated with other ligand motions such as localized carbonyl scrambling and/or rotation of the face capping ligand. [4] The barriers for this can vary over a fairly large range ($\Delta G^{\ddagger}$ = 45 to > 70 kJ/mole) and can be greater in certain second row dihydrido trimetallic clusters than for third row complexes of related geometries (Table 1). We have found that the barriers for hydride site exchange can be divided into two groups: one in which the tuck angle (see diagram associated with Table 1) is <55° where the hydrides are fluxional on the NMR time scale at room temperature ($\Delta G^{\ddagger}$ = 45-60 kJ/mole) and one in which the tuck angle is >55 where the hydrides are static on the NMR time scale at room temperature. Although this generalization applies only to a limited range of trimetallic clusters with face capping ligands, it relates to a more general conclusion presented at this workshop based on calculations which states that two hydrides on a metal cluster need to migrate to the same face of a cluster polyhedron in order to undergo H_2 desorption[5]. Indeed we have found that the dihydride cluster complexes with smaller tuck angles and more fluxional hydrides are more prone to undergo H_2 elimination.

2.   **Dynamics of $H(\mu\text{-}H)_2(\mu_3\text{-}\eta^2\text{-}C\overline{=NCH_2CH_2CH_2})Os_3(CO)_9$, and $H(\mu\text{-}H)_2$**

$\text{-}(\mu\text{-}\eta^2\text{-}C\overline{=NCH_2CH_2CH_2})Os_3(CO)_9P(C_6H_5)_3$

The reaction of $(\mu\text{-}H)(\mu_3\text{-}\eta^2\text{-}C\overline{=NCH_2CH_2CH_2})Os_3(CO)_9$ with one atmosphere of dihydrogen in refluxing octane gives one major product whose infrared ^{1}H-NMR and elemental analysis indicated the molecular formula $H(\mu\text{-}H)_2Os_3(CO)_8$ $(\mu_3\text{-}\eta^2\text{-}$

$C\overline{=NCH_2CH_2CH_2})$(**1**, eq 3) in 46% yield. Compound **1** is also obtained in similar

(3)

1

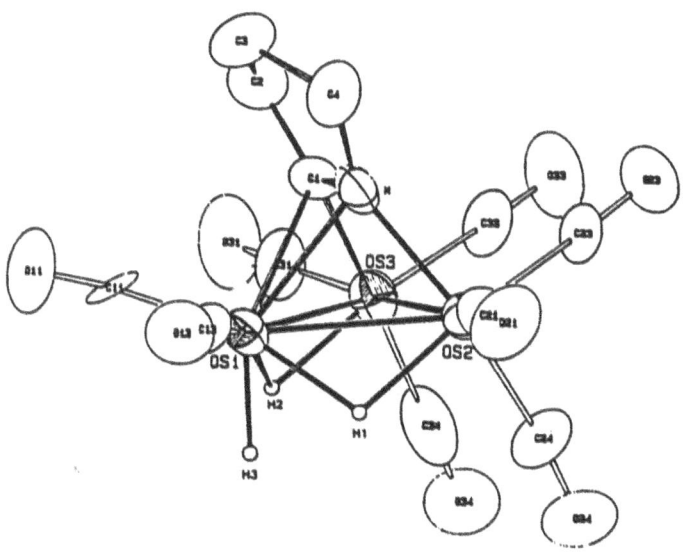

Figure 1   ORTEP drawing of H(μ-H)$_2$(μ_3-η^2-C=NCH$_2$CH$_2$CH$_2$)Os$_3$(CO)$_8$ (1) showing the calculated positions of hydrides.

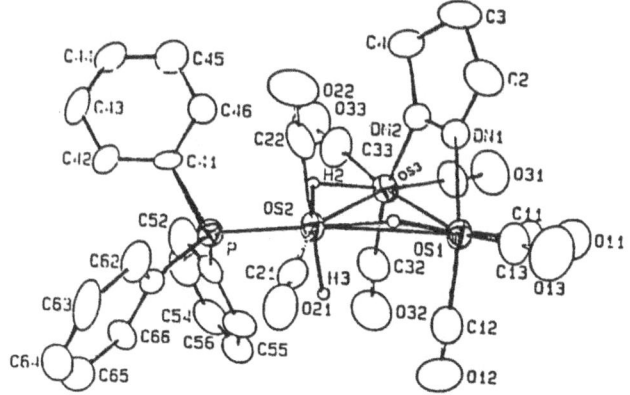

Figure 2   ORTEP drawing of H(μ-H)$_2$(μ-η^2-C=NCH$_2$CH$_2$CH$_2$)Os$_3$(CO)$_8$PPh$_3$ (2) showing the calculated positions of the hydrides.

yield (47%) by photolysis of the same compound in the presence of one atmosphere of dihydrogen in hexane for one hour. Irradiation of **1** under these conditions for longer periods of time resulted only in decomposition of **1**.

It was apparent from the ^{1}H-NMR data that **1** contained one terminal (-9.65 ppm) and two bridging (-15.31 and -15.89 ppm) hydrides but because of the low symmetry of this cluster we could not definitively assign the structure shown (eq 3). We therefore undertook a solid state structural investigation of **1**. The solid state structure of **1** is shown in Figure 1. Compound **1** exists as a triangular core of osmium atoms with three nearly equivalent metal-metal bonds (Os(1)-Os(2) = 2.836 (1) Å, Os(1)-Os(3) = 2.857 (2) Å and Os(2)-Os(3) = 2.817 (1) Å). The organic ligand serves as a five electron donor by donating one electron through a sigma bond from C(1) to Os(3) (C(1)-Os(3) = 2.10 (3) Å), two electrons by a donor sigma bond between N and Os(2) (N-Os(2) = 2.11 (2) Å) and two additional electrons *via* a π-bond from the double bond between C(1) and N (C(1)-N = 1.39 (3) Å) to Os(1) (N-Os(1) = 2.29 (2) Å and C(1)-Os(1) = 2.35 (2) Å). All but the last two metal ligand bond lengths are very similar to other μ_3-imidoyl clusters[6]. The positions of the hydrides were calculated using the program HYDEX[2] where the program was allowed to search for potential energy minima along all three metal edges for bridging hydrides and all three osmium atoms for terminal hydrides. Two bridging hydrides and one terminal hydride positions were found. The bridging hydrides were found along the edges of the cluster not bridged by the organic ligand and to be tucked down below the plane of the metals (1.09 Å for H(1) and 0.95 Å for H(2)). Their positions as *trans* to CO(11) and CO(23) for H(1) and *trans* to CO(33) and CO(13) for H(2) verify these calculated positions[5]. The terminal hydride was found to be bonded to the rear metal atom, Os(1), where a CO ligand is located in the parent compound[6].

Although the hydrides in **1** are clearly rigid on the NMR time scale with respect to bridge-terminal exchange, the chiral nature of the cluster makes the carbonyl groups on Os(1) diastereotopic. It is therefore possible that the $(CO)_2H$ grouping on Os(1) is undergoing tripodal motion. The proton decoupled ^{13}C-NMR spectrum of **2** at room temperature shows the expected eight resonances. The observation of two complex multiplets for two carbonyls allows partial assignment of these resonances to the radial carbonyl groups on Os(1) and confirms that these ligands are rigid on the NMR time scale. The rigidity of the ligands on Os(1) in **1** is in sharp contrast to the three carbonyl groups in the starting material which are undergoing tripodal motion on the NMR time scale even at -60°C, where the "windshield wiper" of the hydride and the imidoyl ligand is slow on the NMR time scale[6]. This difference is attributable either to the presence of the bridging hydrides, or to the different bonding properties of the terminal hydride in **1**.

Since μ_3-imidoyl complexes undergo facile reactions with two electron donors with apparent displacement of the C=N π-bond[6], we thought it would be interesting to see if the stereochemically rigid **1** would exhibit the same type of

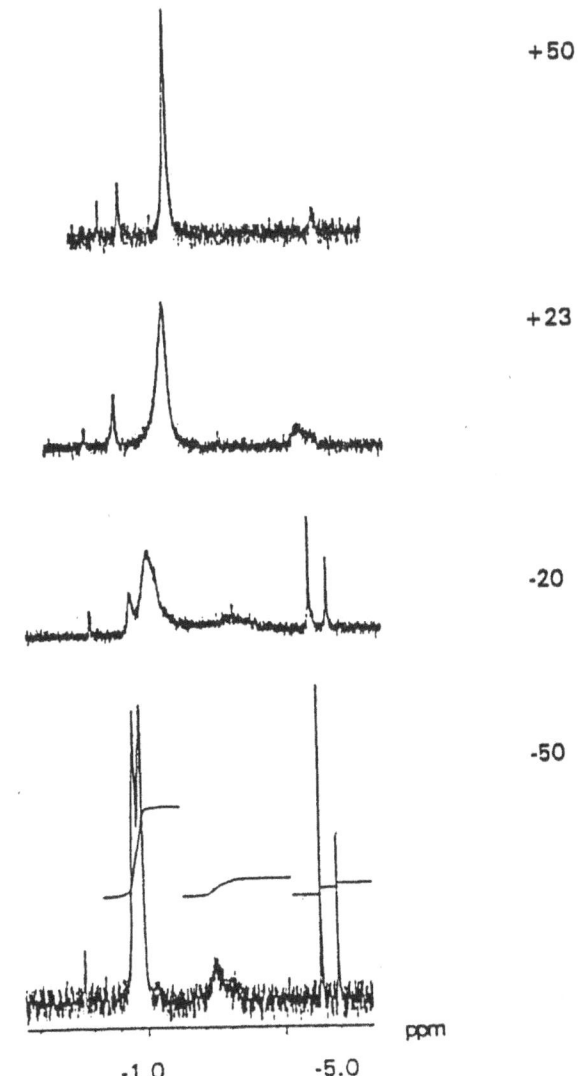

+50

+23

-20

-50

ppm

-1.0   -5.0

Figure 3   VT ^{31}P{H} NMR spectra of 2 at 161 MHz in CDCl$_3$.

reactivity. Indeed, treatment of **1** with triphenylphosphine at ambient temperatures for twenty-four hours leads to conversion to a single product in 86% isolated yield whose ^{1}H and ^{31}P NMR, infrared and elemental analysis are consistent with the

formula $H(\mu\text{-}H)_2(\mu\text{-}\eta^2\text{-}C\text{=}NCH_2CH_2CH_2)Os_3(CO)_8PPh_3$ (**2**) where the phosphine has coordinated to the osmium atom formerly π-bound to the C=N bond (eq 4).

(4)

Here again, as for **1**, although it was clear what the approximate structure of **2** was (eq 4), the exact disposition of hydride ligands was unclear and so a solid state structure of **2** was undertaken. The solid state structure of **2** is shown in Figure 2. Although the connectivity of the hydrides in **2** is similar to **1**, phosphine addition results in significant changes in several aspects of the structure. Both metal-metal bonds bearing the bridging hydrides are elongated by ~0.2Å relative to **1** and this is accompanied by a dramatic change in the location of the bridging hydrides which HYDEX finds on the same face of the Os_3 triangle as the imidoyl ligand. This location is again supported by the disposition of the phosphine and CO(21) which are distinctly tilted toward the opposite face. The overall geometry at Os(2), the metal atom bonded to all three hydrides, is octahedral (excluding the metal-metal vectors) and the $Os(2)P(CO)_2H_3$ moiety tilts away from the μ-imidoyl on the Os(1)-Os(3) edge. Thus, the terminal hydrides makes an angle of 77.20(2)Å with the metal triangle while the corresponding angle in **1** is 93.43(5)Å.

We have been able to measure an approximate rate for the formation of **2** by following the disappearance of **1** by ^{1}H-NMR in the hydride region. It was not possible to monitor the rate of appearance of **2** since its hydride resonances are broadened by exchange and because it exists as several isomers in solution (*vide infra*). Using the rate of decay of the resonances of **2** we can estimate a second order rate constant of $2.0 \pm 1.2 \ M^{-1}m^{-1}$; over a hundred times faster than the rate constant for triphenylphosphine addition to the starting compound $(0.012 \pm 0.04 \ M^{-1}m^{-1})$[7]. This

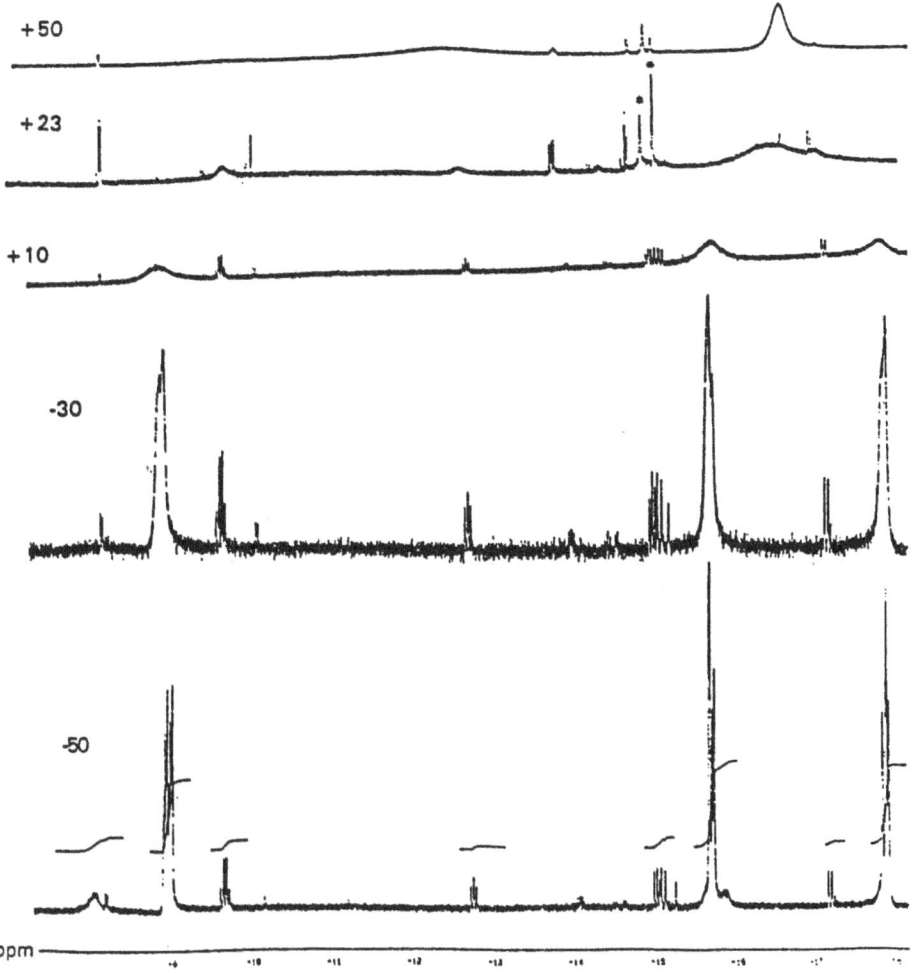

Figure 4   VT ¹H-NMR spectra of 2 at 360 MHz

result clearly indicates that the relatively high reactivity of μ_3-imidoyls is due to the facile displacement of the C=N π-bond from the metal it is coordinated to, rather than partial opening of coordination site brought about by ligand motion over the face of the cluster. We suggest that the much faster rate of reaction of **1** with triphenylphosphine is a consequence of a more electron rich Os(1) in **1** which leads to a weak C=N-Os(1) interaction. This weaker interaction is manifest in the C(1)-Os(1) and N-Os(1) bond lengths which are both 2.25(1)Å in the starting compound and 2.35(2) and 2.29(2)Å respectively in **1**.

The structural changes observed on going from **1** to **2** have significant consequences for the dynamical properties of the latter. Thus the $^{31}P\{^1H\}$ NMR spectrum of **2** at -50°C shows one major set of two resonances at -1.65 and -1.35 ppm relative to ext. H_3PO_4; (rel. int. $\approx$ 2:1), a set of broadened resonances at -2.35 and -2.95 ppm (rel. int. $\approx$ 2:1), a set of sharp resonances at -6.31 and -6.90 ppm (rel. int. $\approx$ 2:1 (Figure 3). The overall relative intensity of these sets of resonances is 7.2:1.4:1.0. In addition, there are two very minor resonances at +0.25 and 1.15 ppm. The resonance at +0.25 ppm shows a significant chemical shift dependence and appears to have shifted under the major set of resonances at -50°C. Neither of these resonances shows significant line broadening with decreasing temperature to -50°C. At -80°C in methylene chloride, we can observe the resonance at 0.25 ppm as a shoulder on the major resonance set, but resolution of the latter is lost. The resonances at -2.95 and -2.35 ppm sharpen slightly at -80°C.

The ^1H-NMR of **2** at -50°C consists of three sharp complex multiplets centered at -8.94, -15.74 and -17.97 ppm (rel. int. = 1:1.2:1.1). In addition, we observe a series of sharp multiplets ranging in relative intensity from 0.02:1 to 0.13:1 with respect to the major isomers at -8.18, -9.66, -12.76, -15.08 and -17.28 ppm. Furthermore, there are broad resonances at -8.01 and -15.90 ppm each in a relative intensity of 0.14:1 with respect to the major resonances (Figure 4). As the temperature is increased to room temperature we observe a broadening of the major resonances while the sets of sharp minor resonances remain sharp until +10°C and the two initially broad resonances have merged with the baseline. At room temperature, we observe the averaging of the three sets of bridging hydride resonances and the onset of broadening of the resonances at -9.66 and -12.16 ppm while the terminal hydride resonance due to the major isomer has merged into the baseline. At +50°C, a narrowing resonance appears at -16.75 ppm, the approximate average of the two bridging hydrides of the major isomers. In the room temperature spectrum a broad shoulder can be seen on the merging resonances at -17.20 ppm, which is the approximate average of the broad minor bridging hydride at -15.90 ppm (partially overlapping with the major bridging hydride resonance at -15.74 ppm) observed at -50°C and a second bridging hydride which is apparently hidden under the major bridging hydride resonance at -17.97 ppm. This proposal is consistent with the fact that this bridging hydride resonance integrates 11% higher than the terminal hydride resonance of the major isomer which is well separated from its minor isomer counterparts. As for the $^{31}P\{^1H\}$ NMR data there appears a set of isomers which is more fluxional than the major isomer set (the broadened resonances in the -50°C spectrum at -8.01, -15.90 and -17.97 ppm) and a set of minor isomers which is slightly more rigid than the major isomer set (the resonances at -9.66, -15.08, and a companion bridging hydride under the major resonance at -15.74 ppm; the latter integrates about

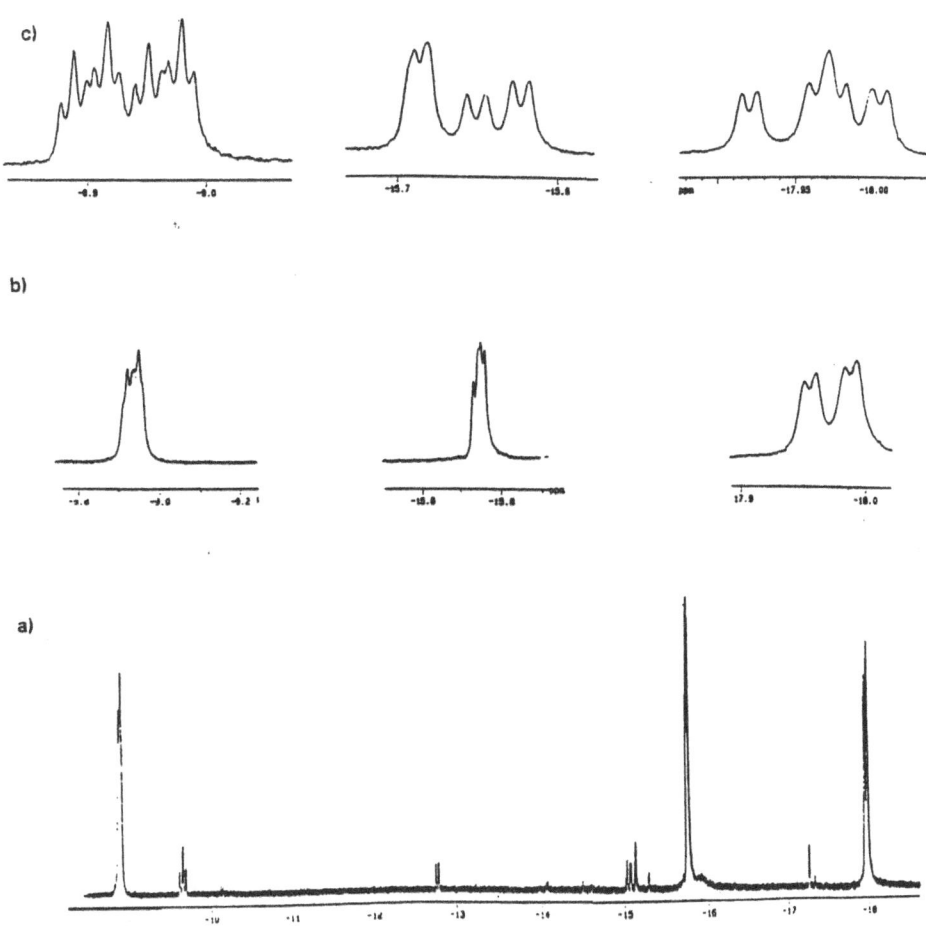

Figure 5   Phosphorus decoupled ^1H-NMR spectrum of **2** (a) entire hydride region, (b) expansion of major isomer hydride resonances and (c) expansion of major isomer resonances with phosphorus coupling.

20% higher than the terminal hydride). The lower intensity resonances remain rigid throughout the temperature range examined except for the one at -12.76 ppm which begin to broaden at ambient temperatures.

From the above ^{1}H-NMR data, we cannot conclude whether the bridging and terminal hydrides are actually exchanging with each other. However, a 2D-EXSY experiment conducted at -50°C with a mixing time of 0.5 sec reveals that this is indeed the case. Furthermore, the complexity of the hydride multiplets observed for the major isomer does not clearly show that each resonance represents an isomer set of two. This is, however, clearer from the phosphorus decoupled ^{1}H-NMR spectrum at -50°C (Figure 5) where each of the bridging hydride resonances appear as a partially overlapping set of doublets. Given that coupling between bridging hydrides is usually <1 Hz the spectrum is clearly assignable to two isomers. As expected, the terminal hydride which is coupled to both μ-hydrides appears as a more complex multiplet.

There are in all twelve possible isomers for **2**. If one excludes the isomers with PPh$_3$ in the axial positions this reduces to six (Scheme 2). The variable temperature ^{1}H and ^{31}P-NMR spectra show that there are three sets of isomers (which are present in comparable relative intensities for both nuclei) averaging at different rates. We suggest that these three sets differ by the location of the terminal hydride. We suggest that these three sets differ by the location of the terminal hydride. We further propose that its position controls the rate of tripodal rotation at the osmium atom bearing the terminal hydride and that this process energetically overlaps with bridge-terminal hydride exchange. There are additional resonances which are extremely low in intensity which show up in the long term accumulation at room temperature. These may be persistent trace impurities or other isomer sets. They will not be considered further. In our previous work on the dynamics of a range of imidoyl complexes of the type $(\mu$-H$)(\mu$-η^2-RC=NR')Os$_3$(CO)$_9$L (L=CO, PR$_3$) where L is located on the unbridged osmium atom, analogous to **2**, we found that tripodal motion first involves three ligands only, excluding the axial ligand on the same face of the triangle as the imidoyl ligand[6]. On this basis we can assign the slowest set of exchanging hydrides to the minor isomer set 2 (Scheme 2). The major isomer set can be assigned to set 1 based on the solid state structure where the terminal hydride is axial but on the opposite face of the metal triangle to the imidoyl ligand. The most rapidly exchanging isomer is probably set 3 where the hydride and the phosphine are both in radial positions. Of course, the assignment of the minor isomers must be taken as tentative, but the more important aspect of these results is the rationale behind the much greater mobility of the hydrides in **2** relative to **1**. We have previously observed that phosphine substitution leads to a lowering of the barrier to tripodal motionat the unbridged osmium atom in the complexes

$(\mu$-H$)(\mu$-η^2-$\overline{\text{C=NCH}_2\text{CH}_2\text{CH}_2})Os_3(CO)_9$L (L=CO,PR$_3$) [6]. In **2**, however, this process apparently overlaps with bridge-terminal hydride exchange. This is probably a result of the elongation of the metal-metal bonds relative to **1** which would result in

Scheme 2

L = PPh₃

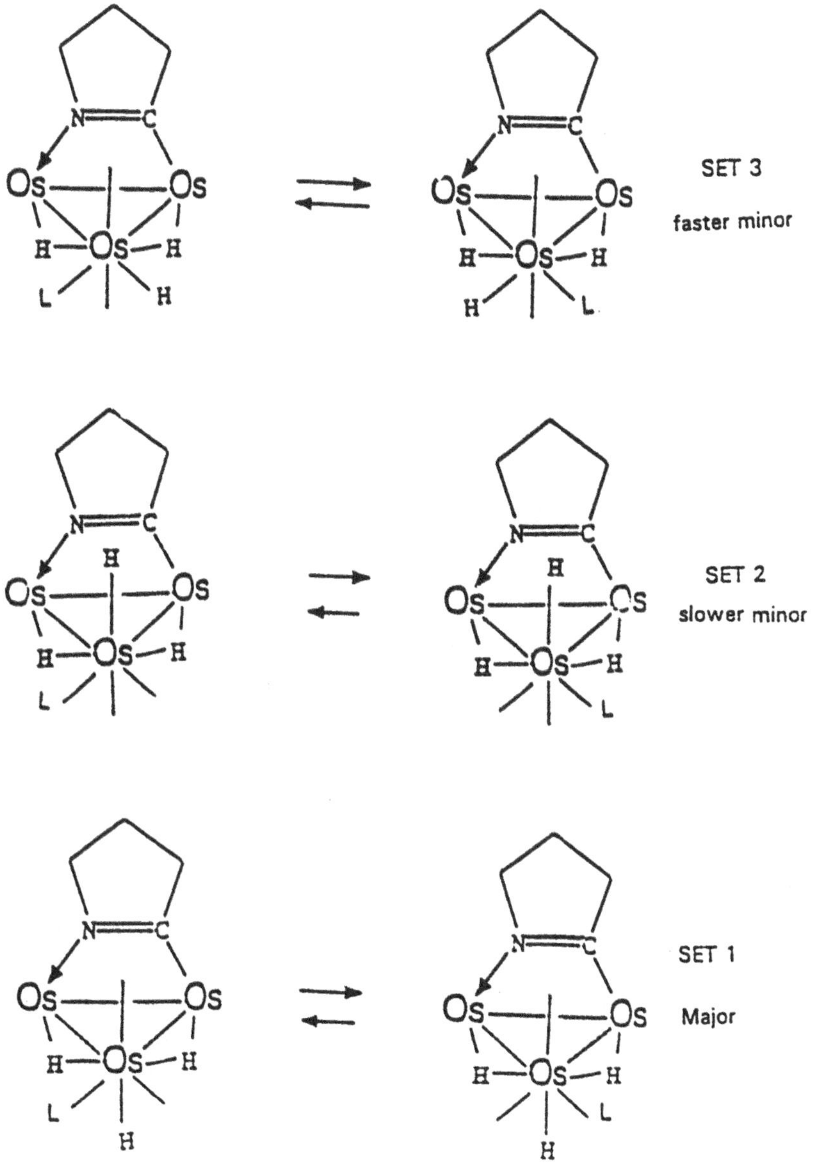

138

longer, weaker hydrogen-metal bridges. The greater mobility may be the result of significant asymmetry in the hydride bridges induced by phosphine substitution but this is not evident since our version of HYDEX places all bridging hydrides symmetrically. It is clear, however, that the proximity of the bridging and terminal hydrides is not related to their rate of exchange since they are much closer to each other in **1** than in **2**.

3.   **Hydride Dynamics of HBr Complexes of μ_3-Imidoyls**

We recently reported the solid state structures of the complexes $(\mu\text{-H})_2(\mu\text{-}\eta^2\text{-}$

$\overline{\text{C}=\text{NCH}_2\text{CH}_2\text{CH}_2)\text{Os}_3(\text{CO})_9\text{X}}$ (X=Cl,Br(**3**), CF$_3$COO)[3,8]. More recently we studied the thermolysis of the bromide derivative **3** and found that it undergoes an unusual isomerization in addition to loss of HBr and dissociation of the organic ligand (eq 5).

The solid state structure of **3** and its isomer **4** show evidence of intramolecular hydrogen bonding (i.e., short H-Br distances 2.96 Å in **3** and 2.78 Å in **4**, sum of van der Waals is 3.20 Å, see reference 8 and figure 6). Despite these overall similarities the dynamical properties of the hydrides in **3** and **4** are remarkably different with direct hydride exchange having a $\Delta G^{\ddagger} = 76 \pm 2$ kJ/mole in **3** and 54 ± 2 kJ/mole in **4** as estimated from the coalescence of the two hydride resonances in the major isomers of **3** and **4** (figure 7). This large difference in the activation energy for exchange cannot be accounted for by differences in hydride geometry between **3** and **4**. The hydrides in both complexes were located by taking successive slices in the Fourier maps (15) and show that the differences in the hydride location (error in hydride location is ± 0.04 Å) are relatively small. Indeed the short H-Br distance in **4** is maintained by a distinct tilting of the bromide compared with **3** where the bromide is perpendicular to the Os$_3$ plane (figure 6). It is possible the relocation of the hydride to the opposite face of the cluster as that occupied by the imidoyl ligand in **4** allows for the formation of a doubly bridged or diterminal intermediate which is inaccessible for steric reasons in **3**.

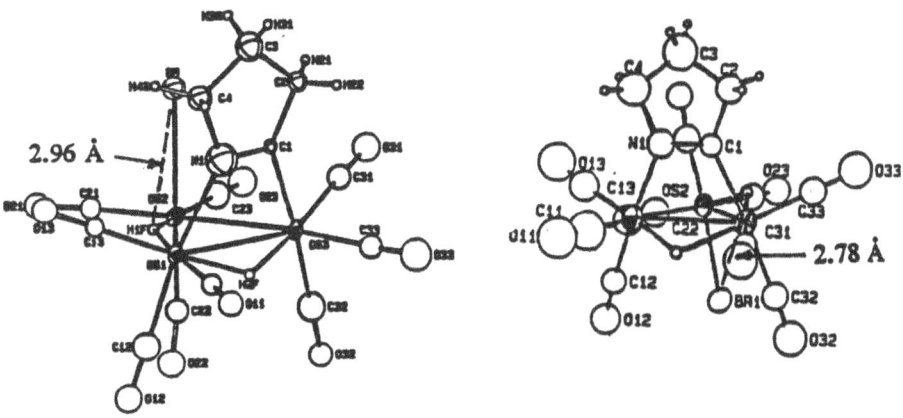

Figure 6 Solid state structure of **3** and **4** showing the located positions of the hydrides and the bromine-hydrogen distance

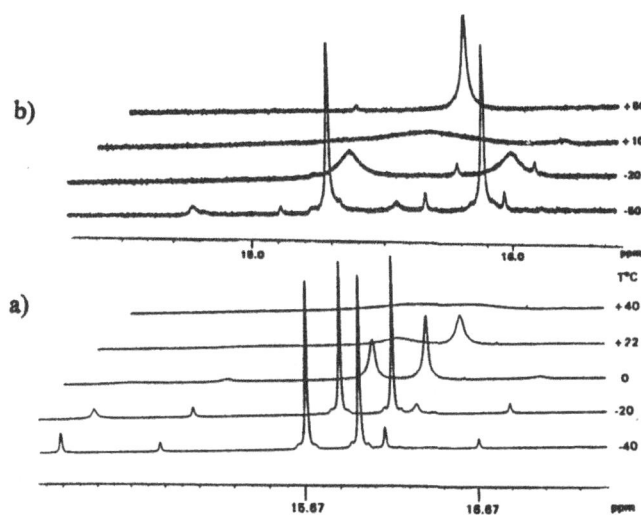

Figure 7 VT ¹H NMR spectra of (a) **3** and (b) **4** at 360 MHz in the hydride region

Conclusions

Although the rates of hydride exchange in dihydrido-trimetallic clusters containing μ_3-capping ligands appear to fall into a consistent pattern where the rate of exchange depends on the "tuck angle" between the hydrides, the picture is much less clear for polyhydrido complexes outside of this category. From our studies, it would appear that asymmetry in the hydride bridges, and the overall steric environment, including metal metal bond lengths and the relative disposition of bulky ancillary ligands play a role. The internuclear distance between hydrides as determined from HYDEX[8] and the actual location of the hydrides appears to have little bearing on the rate of hydride exchange as seen by the comparisons between 1 and 2 and 3 and 4 reported here.

Acknowledgements

We gratefully acknowledge the support of the National Science Foundation for support of this research (CHE9319062). The solid state structures reported here were performed at the California State University, Northridge, by Professor Kenneth Hardcastle and coworkers and the synthetic work was carried out by Professor Shariff Kabir

References

1. Rosenberg, E., Day, M., Hardcastle, K.I., Milone, L., Wolf, E., Kabir, S.E., McPhillips, T., Gobetto, R., Osella, D. and Irving, M. (1991) *Organometallics*, **10**, 2743-2751.

2. Rosenberg, E., Kabir, S.E., Day, M., and Hardcastle, K.I. (1994) *Organometallics*, in press.

3. Rosenberg, E., Kabir, S.E., Yin, M., Nishimura, N., Hardcastle, K., Milone, L., Gobetto, R. and Osella, D. (1994) *Organometallics*, submitted.

4. Rosenberg, E., Hajela, S., Lend, M., Zhang, M., Bracker-Novak, J., Gobetto, R., Osella, D. and Milone, L. (1990) *Organometallics*, **9**, 1379-1382 and references therein.

5. Jellinek, J. and Güvenc, Z.B. (1994) See article in this volume.

6. Rosenberg, E., Day, M., Espitia, D., Hardcastle, K., Kabir, S.E., McPhillips, T., Gobetto, R., Milone, L. and Osella, D. (1993) , *Organometallics*, **12**, 2309-2324.

7. Rosenberg, E., Freeman, W., Carlos, Z., Hardcastle, K., Yoo, Y.J., Milone, L. and Gobetto, R. (1992) *J. Cluster Sci.*, **3**, 439-457.

8. Hardcastle, K.I. and Irving, M. (1993) *J. Cluster Sci.*, **4**, 77-88.

STRUCTURAL VARIATIONS IN TETRANUCLEAR PLATINUM-RUTHENIUM CLUSTERS

L. J. Farrugia, D. Ellis and A. M. Senior
Department of Chemistry,
University of Glasgow,
Glasgow,
G12 8QQ, U.K.

Introduction

While the well known electron counting rules [1,2] are of considerable utility in rationalising the metal core structures of many transition metal cluster compounds, there are problems with their use for those elements at the end of the transition series. This is of course well known for the Group 11 elements, especially gold [3], but this also holds true for Group 10 metals such as platinum [4,5]. The underlying causes for this behaviour are reasonably well understood from a molecular orbital viewpoint, and in the case of Pt-containing clusters, [5] it is possible to consider (at least as a crude approximation) the Pt atoms as being either 16 or 18 electron centres. One result of the electronic "ambivalence" of Pt is that there are often several structures of closely similar energy for cluster species with the same molecular formulae. In this paper we discuss some of our recent unpublished results relating to the structural variability of tetranuclear Pt-Ru clusters.

We have previously noted [5, 6] that there are two common metal skeletal geometries for 60 cluster valence electron (CVE) triosmium-platinum clusters, *i.e.* the essentially regular tetrahedral skeleton, found for example in $Os_3Pt(\mu\text{-}H)_2(CO)_{10}(COD)$ (1) [7] or $Os_3Pt(\mu\text{-}H)_2(\mu\text{-}SO_2)(CO)_{10}(PCy_3)$ (2) [8] and the non-planar butterfly skeleton, *e.g.* in $Os_3Pt(\mu\text{-}H)_2(CO)_{10}(PR_3)_2$ (3) [9], where one Pt-Os bond is elongated (generally ~3.5 Å) compared with the other Pt-Os separations, and may be considered as non-bonding. Braunstein and coworkers [10] have also shown that the 58 CVE clusters $Pt_2M_2(CO)_6(PR_3)_2Cp_2$ (M=Mo, W) have two isomeric forms

141

L. J. Farrugia (ed.), The Synergy Between Dynamics and Reactivity at Clusters and Surfaces, 141–157.
© 1995 *Kluwer Academic Publishers.*

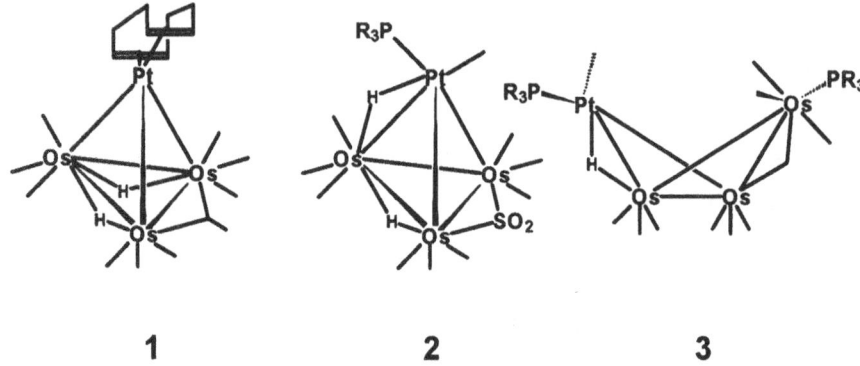

1   **2**   **3**

in the solid state, one with a planar butterfly skeleton, the other with a tetrahedral metal core. Depending on the steric bulk of the phosphine, these isomers may co-exist in solution and in this case the isomers undergo a facile interconversion. These studies led us to consider whether similar structural variablility would be apparent in related 60 CVE tri-ruthenium platinum clusters.

We have already shown [11] that the 62 CVE hydrido-alkynyl *spiked-triangular* cluster $Ru_3Pt(\mu\text{-}H)(\mu_4\text{-}\eta^2\text{-}C\equiv CBu^t)(CO)_9(dppe)$ (**4**) is in thermodynamic equilibrium with *butterfly-* vinylidene tautomer $Ru_3Pt\{\mu_4\text{-}\eta^2\text{-}C=C(H)Bu^t\}\text{-}(CO)_9(dppe)$ (**5**). This unusually facile skeletal interconversion in solution has been followed *directly* [12] through observation of the $Pt_{L_{III}}$ edge EXAFS spectra.

1. Synthesis of 60 CVE triruthenium-platinum clusters

The methodology of cluster synthesis is now on a firm footing [13], with a variety of routes being available. In terms of the synthesis of heteroplatinum clusters, one of the most successful routes has been the utilisation of low (usually zero-valent) Pt complexes such as $Pt(COD)_2$, or phosphine-olefin complexes such as $Pt(C_2H_4)_2(PR_3)$ [5]. These compounds react by ready loss of (usually) one COD ligand, or of all the ethene ligands. One conceptually simple method of preparing 60 CVE Ru_3Pt clusters is to utilise the phosphine-olefin complexes as synthons which can add the 12 electron fragment "$Pt(PR_3)$" directly to preformed 48 electron Ru_3 clusters. The preparation of these reagents involves several steps, including the preparation of $Pt(COD)_2$ and the unstable $Pt(C_2H_4)_3$, while the olefin complex $Pt(nb)_3$ (nb=bicyclo[2.2.1]hept-2-ene), can be prepared easily in good yield. [14] We have recently found that $Pt(nb)_2(PR_3)$, prepared by *in situ* reaction of this tris-olefin complex with one mole equivalent of phosphine, also acts as a good source of the "$Pt(PR_3)$" fragment, by loss of the bulky nb ligands.

1.1 Synthesis structure and dynamics of $Ru_3Pt(\mu\text{-}H)(\mu_3\text{-}COMe)(CO)_{10}(PR_3)$.

Some time ago, Stone and coworkers [15] reported the synthesis of $Fe_3Pt(\mu_3\text{-}H)(\mu_3\text{-}COMe)(CO)_{10}(PPh_3)$ (6) from the reaction of $Pt(PPh_3)(C_2H_4)_2$ with $Fe_3(\mu\text{-}H)(\mu\text{-}COMe)(CO)_{10}$. The analogous reaction of $Pt(PCy_3)(C_2H_4)_2$ with $Os_3(\mu\text{-}H)(\mu\text{-}COMe)(CO)_{10}$ reported somewhat later [16], did not give the corresponding Os_3Pt cluster, but instead afforded the pentanuclear carbido species $Os_3Pt_2(\mu\text{-}H)(\mu_5\text{-}C)(\mu\text{-}OMe)(CO)_{10}(PCy_3)_2$ as the sole isolable product. In these studies, it was also stated [15] that treatment of $Ru_3(\mu\text{-}H)(\mu\text{-}COMe)(CO)_{10}$ (7) with $Pt(PPh_3)(C_2H_4)_2$ did not afford any analogous Ru-Pt clusters. We have reinvestigated this synthetic procedure, [17] and have found that the reaction of $Pt(PR_3)(nb)_2$ (R=Cy or ^iPr) with $Ru_3(\mu\text{-}H)(\mu\text{-}$

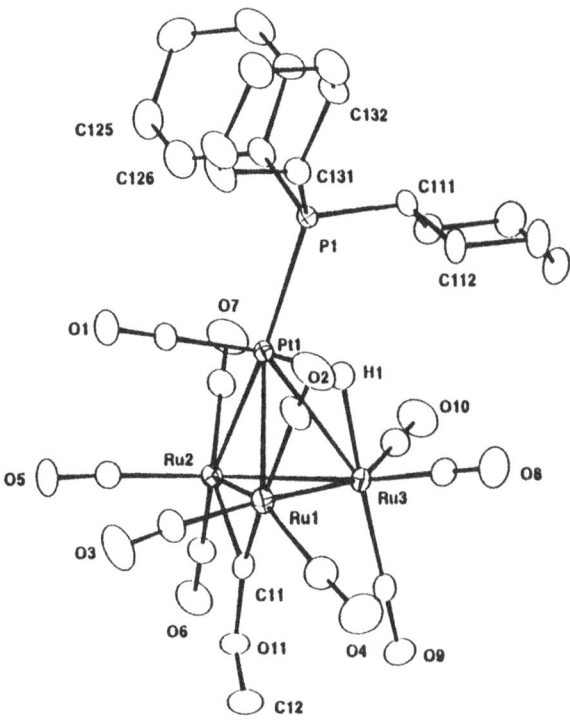

Figure 1. Molecular structure of $Ru_3Pt(\mu\text{-}H)(\mu_3\text{-}COMe)(CO)_{10}(P^iPr_3)$ (7)
Important bond lengths (Å): Pt-Ru(1) 2.821(1), Pt-Ru(2) 2.760(1), Pt-Ru(3) 2.903), Ru(1)-Ru(2) 2.972(1), Ru(1)-Ru(3) 2.767(1), Ru(2)-Ru(3) 2.754(1), Pt-P 2.359(2), Ru-(1)-C(11) 1.973(6), Ru(2)-C(11) 1.990(6), Ru(3)-C(11) 2.597(6).

COMe)(CO)$_{10}$ does indeed result in the high yield formation of the desired product Ru$_3$Pt(μ-H)(μ_3-COMe)(CO)$_{10}$(PR$_3$) (7). The reaction appears quite sensitive to the nature of the phosphine ligand and, in line with the previous work [15], we were unable to isolate any Ru-Pt clusters using Pt(nb)$_2$(PPh$_3$).

The X-ray structure of cluster 7, shown in Figure 1, revealed a tetrahedral metal skeleton, with the COMe ligand asymmetrically bridging the tri-ruthenium face. Pt L$_{III}$-edge and Ru K-edge EXAFS spectra on both solid and THF solution samples of 7 are consistent with the X-ray determination, and show that the same skeletal geometry is retained in both phases. The overall ligand disposition in 7 is similar to that previously observed for 6 [15], with one important exception. This concerns the orientation of the "T-shaped" Pt(μ-H)(CO)(PR$_3$) unit relative to the Ru$_3$ triangle. The atoms Pt1, P1, C1, O1 and H1 are virtually coplanar, and lie in the molecular pseudo-mirror plane, such that the hydride ligand bridges the Pt-Ru(3) vector. The orientation of the Pt(μ-H)(CO)(PR$_3$) units in cluster 6 (Figure 2) and 7 (Figure 1) are related by a rotation of *ca* 60° about the pseudo-C_3 axis of the M$_3$ triangles. This results in the hydride ligand bridging a PtFe$_2$ face in 6, rather than a Pt-Fe edge as in 7. This relationship between the structures of 6 and 7 provides an insight into the fluxional behavior of 7.

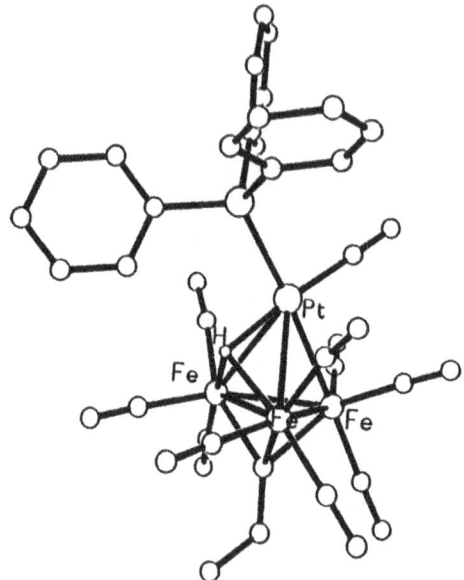

Figure 2. The molecular structure of Fe$_3$Pt(μ_3-H)(μ_3-COMe)(CO)$_{10}$(PPh$_3$) (6)
Taken from coordinates obtained from Cambridge Crystallographic Data Base

The variable temperature ^{13}C NMR spectra of a ^{13}CO enriched sample of **7** show that at the lowest temperature measured (208 K), all the Ru-bound carbonyl ligands are equivalent, giving rise to a sharp singlet at δ 199.1. The Pt-bound carbonyl shows up as a doublet resonance at δ 182.3. The equivalence of all the Ru-bound carbonyls *could* arise from intermetallic CO scrambling, but we suggest that rotation of the $Pt(\mu-H)(CO)(PR_3)$ unit about the *pseudo-C_3* axis of the Ru_3 triangle, coupled with rapid tripodal rotation of the individual $Ru(CO)_3$ groups is responsible. We have previously unambiguously characterised [18] a similar rotation of a $Pt(\mu-H)(CO)(PR_3)$ unit about the *pseudo-C_3* axis of the Os_3 triangle in $Os_3Pt(\mu-H)_2(CO)_9(CNCy)(PCy_3)$. In addition, earlier EHMO calculations by Hoffmann and Schilling [19] have suggested a very low activation barrier to the rotation of the PtL_2 unit in related $PtFe_3$ clusters.

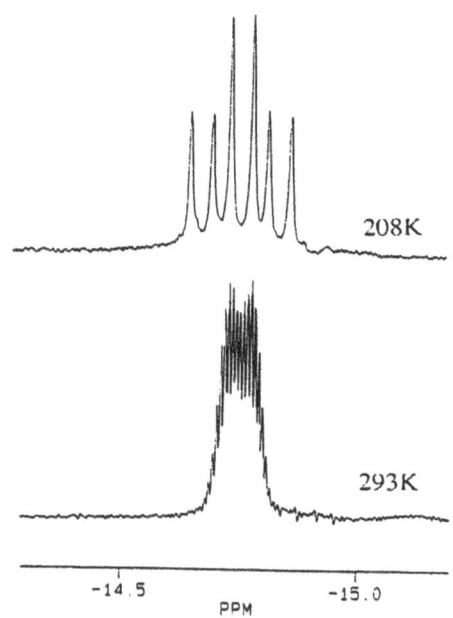

Figure 3. The variable temperature ^1H NMR spectrum of cluster 7 in the hydride region.

The two ^{13}C signals for the Ru-CO ligands and the Pt-CO ligand coalesce on warming indicating complete intermetallic carbonyl scrambling. This process is also observable in the ^1H NMR spectrum in the hydride region, shown in Figure 3. At 208 K this signal is a doublet resonance due to coupling to the ^{31}P nucleus, and the intensity of the ^{13}C satellites indicates that the statistical level of ^{13}C enrichment at each site is *ca* 50%. At this temperature there is only one observable ^1H-^{13}C coupling, due to the carbonyl on the Pt atom $C(1)$, which is *trans* to the hydride ligand. On warming to 293 K there is no shift in the resonance position, but the coupling pattern becomes more complex. At the level of 50% ^{13}CO enrichment, the most abundant isotopomer of **7** is that one containing five ^{13}CO ligands and five ^{12}CO ligands. The two isotopomers with four ^{13}CO ligands and six ^{13}CO ligands respectively will have similar abundances. The abundances of other isotopomers is much less. If rapid exchange occurs between the Pt-bound CO ligand and the Ru-bound CO ligands, then the hydride ligand in these isotopomers will experience an *averaged* coupling to either four, five or six ^{13}CO ligands, in addition to the ^{31}P coupling. The observed signal will therefore be a superposition of a doublet of quintets, a doublet of sextets, and a doublet of septets; hence the complexity of the coupling pattern.

1.2 Synthesis, structure and dynamic behaviour of $Ru_3Pt(\mu\text{-}H)(\mu_3\text{-}\sigma\text{-}\pi\text{-}CMe.CH.CMe)(\mu\text{-}CO)(CO)_8(P^iPr_3)$

The reaction of the σ-π-allyl cluster $Ru_3(\mu\text{-}H)(\mu_3\text{-}\sigma\text{-}\pi\text{-}CMe.CH.CMe)(CO)_9$ with $Pt(nb)_2(PR_3)$ in a 1:1 molar ratio affords the dark green cluster $Ru_3Pt(\mu\text{-}H)(\mu_3\text{-}\sigma\text{-}\pi\text{-}CMe.CH.CMe)(\mu\text{-}CO)(CO)_8(PR_3)$ (**8**) in *ca* 50-60% yield, together with a second purple, as yet unidentified product. The X-ray structure of **8** (Figure 4) shows that the metal skeleton has a distorted tetrahedral geometry. Effectively, the $Pt(PR_3)$ fragment has added to the Ru_3 face of $Ru_3(\mu\text{-}H)(\mu_3\text{-}\sigma\text{-}\pi\text{-}CMe.CH.CMe)(CO)_9$, normal to the Pt-P vector, so that we may say that the $Pt(PR_3)$ fragment has "landed" on the Ru_3 surface. The most significant pertubation to the ligand disposition of $Ru_3(\mu\text{-}H)(\mu_3\text{-}\sigma\text{-}\pi\text{-}CMe.CH.CMe)(CO)_9$, which occurs on addition of the $Pt(PR_3)$ fragment , concerns the *axial* terminal CO ligand on the unique Ru atom which is trans with respect to the π-bonded allyl ligand. In cluster **8** this CO ligand bridges one of the Pt-Ru vectors, in an essentially symmetric mode [Pt-C = 1.955(7), Ru-C = 2.090(7)Å]. The Pt atom lies on the opposite face of the Ru_3 triangle to the σ-π-allyl ligand, so that there is no interaction between the Pt atom and this ligand. The hydrocarbyl ligand is bonded to the Ru_3 triangle in a very similar fashion to that found [20] in the precursor Ru_3 cluster. There is considerable asymmetry in the solid state structure. As can clearly be seen from Figure 4, the bridging CO ligand is substantially displaced from the *pseudo*-mirror plane defined by Pt-Ru1 and H1. In addition, the bond distances Pt-Ru2 = 2.746(1)Å and Pt-Ru3 = 2.959(1)Å are significantly different, though there is no obvious reason for this. Nevertheless, in solution the molecule possesses an effective mirror plane in the ^{13}C NMR spectrum, even at the lowest temperature measured (183 K), resulting in pairwise equivalence of corresponding carbonyl groups on Ru2 and Ru3.

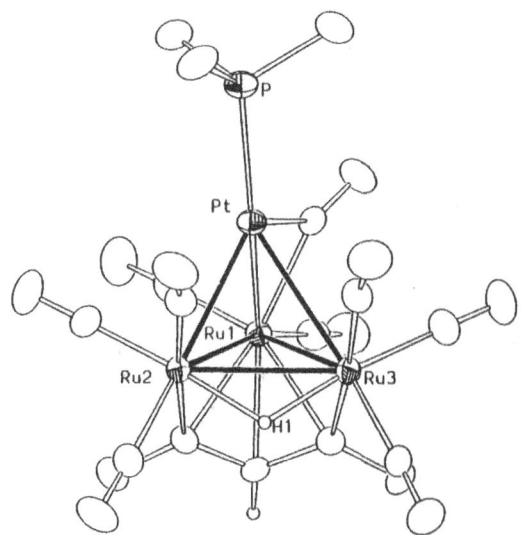

Figure 4. Structure of Ru$_3$Pt(μ-H)(μ_3-σ-π-CMe.CH.CMe)(μ-CO)(CO)$_8$(PR$_3$) (8)
Important bond lengths (Å): Pt-Ru(1) 2.752(1), Pt-Ru(2) 2.746(1), Pt-Ru(3) 2.959(1), Ru(1)-Ru(2) 2.899(1), Ru(1)-Ru(3) 2.882(1), Ru(2)-Ru(3) 2.906(1), Pt-P 2.274(2), Ru(1)-C(allyl) 2.232(7), 2.227(7), 2.261(7), Ru(2)-C(allyl) 2.071(7), Ru(3)-C(allyl) 2.070(7)

Interestingly, Pt L$_{\mathrm{III}}$ EXAFS spectra of **8** in solution at room temperature are consistent with an instantaneous structure similar to that found in the solid phase, with two Pt-Ru bonds of 2.75Å and one longer bond of 2.93Å. This implies a very low energy fluxional process which is fast on the NMR timescale, but slow on the much shorter EXAFS timescale, and which results in a time averaged molecular mirror plane. We have observed two carbonyl exchange processes; the lowest energy process results in the mutual exchange of all three CO ligands bonded to Ru1, while a second, higher energy process involves tripodal rotation of the apparently equivalent Ru(CO)$_3$ on Ru2 and Ru3. From a ^{13}C EXSY spectrum at 260 K we estimate an activation barrier $\Delta G^{\ddagger} = 64\pm1$ kJ mol^{-1} for this second process. We were unable to "freeze out" the lower energy process (it is still rapid at 183 K), and it probably has an activation barrier considerably less than ~40 kJ mol^{-1}.

The corresponding fluxional exchanges in $Ru_3(\mu\text{-}H)(\mu_3\text{-}\sigma\text{-}\pi\text{-}CMe.CH.CMe)(CO)_9$ have been examined in detail [21], and the exchange barriers are ~40 and 76 kJ mol^{-1} for the tripodal rotations of the $Ru(CO)_3$ groups on the unique and equivalent pair of Ru atoms respectively. Thus the general pattern of carbonyl fluxionality is very similar to that found in **8**, though in this latter cluster we have found that the barriers are somewhat lower. There is no evidence for any mobility of the hydrocarbyl fragment. For the low energy exchange process in **8** involving the $Ru(CO)_3$ group bonded to Ru1, a sequential formation and breaking of CO bridges to the Pt atom must occur. In contradiction to simplistic expectations, the extra steric bulk of the $Pt(PR_3)$ group does not appear to play a major role in deciding the activation barriers in these two related clusters.

1.3 Synthesis and structure of $Ru_3Pt(\mu_3\text{-}PhCCPh)(\mu\text{-}CO)_2(CO)_8(P^iPr_3)$

Reaction of $Pt(nb)_2(P^iPr_3)$ with $Ru_3(\mu\text{-}CO)(CO)_9(\mu_3\text{-}PhCCPh)$ [22] in a 1:1 molar ratio results in reasonable yields of the cluster $Ru_3Pt(\mu_3\text{-}PhCCPh)(\mu\text{-}$

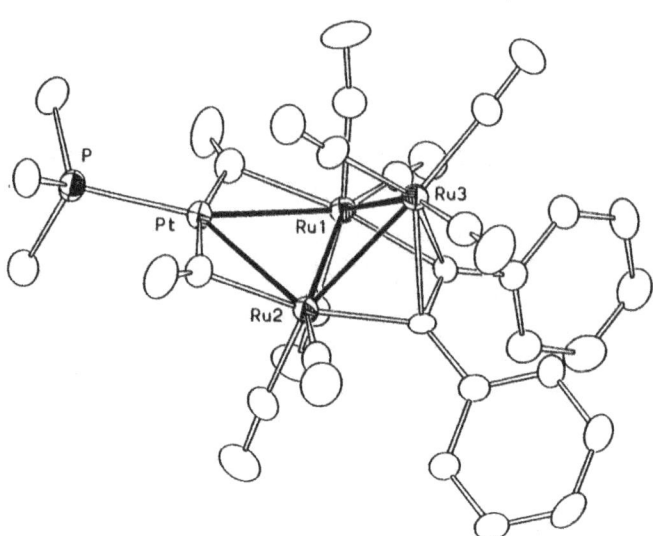

Figure 5. The molecular structure of $Ru_3Pt(\mu_3\text{-}PhCCPh)(\mu\text{-}CO)_2(CO)_8(P^iPr_3)$ (9) **Important bond lengths (Å):** Pt-Ru(1) 2.723(1), Pt-Ru(2) 2.696(1), Pt.....Ru(3) 4.037(1), Ru(1)-Ru(2) 2.822(1), Ru(1)-Ru(3) 2.829(1), Ru(2)-Ru(3) 2.694(1), Pt-CO(bridge, av) 2.04(1), Ru-CO(bridge, av) 2.17(1), Pt-P 2.297(2), Ru(1)-C(alkyne) 2.198(7), Ru(2)-C(alkyne) 2.168(7), Ru(3)-C(alkyne,av) 2.18(1)

$CO)_2(CO)_8(P^iPr_3)$ (9), as well as significant quantities of $Ru_3(CO)_9(\mu_3\text{-PhCCPh})$- (P^iPr_3) (10). Both complexes have been characterised by spectroscopic methods, and also by X-ray diffraction studies. The structure of 9, shown in Figure 5, contains a non-planar butterfly metal skeleton. It is closely related to that of the starting cluster $Ru_3(\mu\text{-CO})(CO)_9(\mu_3\text{-PhCCPh})$, which is presumed [22] to have a similar structure to the crystallographically characterised dimethyl derivative. The "Pt(PR$_3$)" fragment has effectively been inserted into the symmetric Ru(μ-CO)Ru bridge found in $Ru_3(\mu$- $CO)(CO)_9(\mu_3\text{-PhCCPh})$, and a second carbonyl bridge also forms to the Pt atom. This results in somewhat asymmetric structure, since the atom Ru2 bears two terminal CO ligands, while the atom Ru1 bears three terminal CO ligands. This is also reflected in the distances Ru(1)-Ru(3) = 2.829(1) and Ru(2)-Ru(3) = 2.694(1)Å. There is no interaction between the alkyne and the Pt atom , since these are on opposite faces of the Ru$_3$ triangle, and the alkyne is bonded to the Ru$_3$ triangle in the familiar μ_3-η^2-|| mode. We have previously reported [11] a closely related cationic alkyne cluster $[Ru_3Pt(\mu\text{-H})(HCC^tBu)(CO)_9(dppe)]^+(BF_4^-)$ (11), which is unstable and has only been characterised by spectroscopic methods. Neither of the alkyne carbon atoms in 11 display any coupling to [195]Pt, hence we suggested [11] that there were no bonding interactions between Pt and the alkyne ligand. In view of the structure we have recently obtained for cluster 9, it seems plausible that the metal-ligand disposition in both these complexes is similar. If the Pt atom and alkyne ligand in 11 lie over *opposite* faces of the Ru$_3$ triangle, this is consistent with the spectroscopic evidence.

The formation of the phosphine-substituted Ru$_3$ cluster 10 indicates that the reaction pathway is not simple. Indeed in many of the reactions of Pt(nb)$_2$(PR$_3$) with triruthenium clusters, e.g. with $Ru_3(\mu\text{-H})(\mu\text{-SR})(CO)_{10}$ or $Ru_3(\mu\text{-H})(\mu\text{-}C,N\text{-}$ $C_5H_4N)(CO)_{10}$, we have observed that the phosphine substituted Ru$_3$ cluster is the *only* isolable product. The formation of these Ru$_3$ complexes may arise from phosphine loss from the Pt(nb)$_2$(PR$_3$) reagent, or through decomposition of unstable Ru-Pt cluster intermediates. One point of interest about the structure of 10 concerns the carbonyl arrangement, *vis-a-vis* terminal or bridging modes. Deeming and Senior [23] have compared a number of structures of the type $M_3(CO)_{10}$(alkyne) (M=Ru,Os) and their phosphine substituted derivatives. They suggest the observed structures cover a spectrum between two principal structural types, one of which contains one symmetric bridging CO, and the other two weakly semi-bridging CO ligands. Cluster 10 appears to be an example of the latter type, similar in structure to $Os_3(CO)_9(\mu_3\text{-}$ EtCCEt)(PPh$_3$). [24]

1.4 Structure and reactivity of Ru₃Pt(μ-CO)₂(CO)₉(PⁱPr₃)₂

The successful addition of the "Pt(PR₃)" to several 48 CVE triruthenium clusters containing hydrocarbyl ligands, led us to investigate whether this reaction would also work for the simplest 48 CVE triruthenium cluster, i.e. Ru₃(CO)₁₂. Addition of Pt(nb)₂(PⁱPr₃) to Ru₃(CO)₁₂ in a 1:1 molar ratio gave moderate yields of the intense purple 60 CVE cluster Ru₃Pt(μ-CO)₂(CO)₉(PⁱPr₃)₂ (12) as the sole isolable product. In accordance with the product stoichiometry, the yield is substantially improved if the bis phosphine complex Pt(nb)(PⁱPr₃)₂ is used instead. The X-ray structure of 12 (Figure 6) reveals a planar butterfly metal skeleton, and the overall ligand disposition is very similar to that observed for the related osmium complex Os₃Pt(μ-CO)₂(CO)₉(PPh₃)₂ reported by Farrar et al. [25]. The Pt L$_{III}$ edge EXAFS spectra of 12 indicate that a similar structure is maintained in solution, at least in terms of the bonding metal-metal distances.

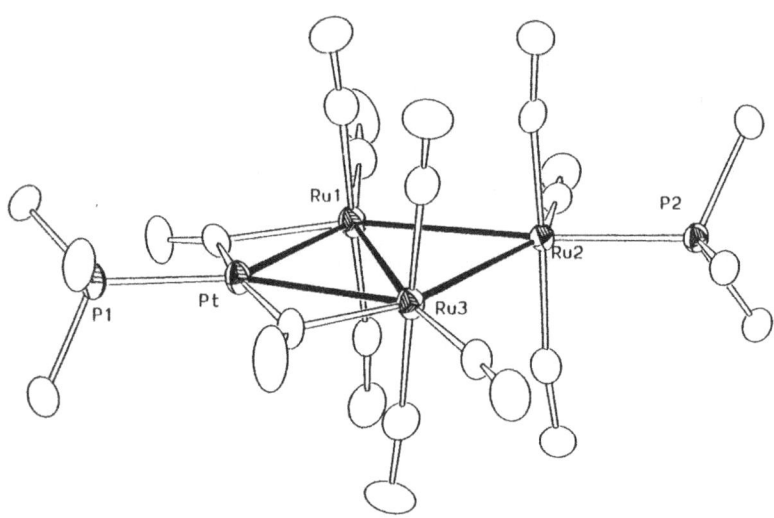

Figure 6. The molecular structure of Ru₃Pt(μ-CO)₂(CO)₉(PⁱPr₃)₂
Important bond lengths (Å): Pt-Ru(1) 2.689(1), Pt-Ru(3) 2.670(1), Pt...Ru(2) 4.782(1), Ru(1)-Ru(2) 2.905(1), Ru(1)-Ru(3) 2.906(1), Ru(2)-Ru(3) 2.903(1), Pt-P 2.286(3), Ru(2)-P 2.379(3), Pt-CO(bridging, av) 2.05(1), Ru-CO(bridging, av) 2.11(1)

Since cluster **12** could be prepared in high yield we have investigated some reactions with small molecules. Reaction of **12** with HCC^{t}Bu or Ph$_2$PCCPPh$_2$ leads to a complex mixture of products which are still under investigation. Treatment with 2 atm pressure of carbon monoxide results in a slow degradation to give Ru$_3$(CO)$_{11}$(P^{i}Pr$_3$) and Pt$_3$(μ-CO)$_3$(P^{i}Pr$_3$) as the only tractable products. Reaction with dihydrogen afforded three complexes, the major product being Ru$_3$Pt(μ-H)$_2$(μ-CO)$_2$(CO)$_8$(P^{i}Pr$_3$)$_2$ (**13**) , with small amounts of Ru$_2$Pt$_2$(μ-H)$_2$(CO)$_8$(P^{i}Pr$_3$)$_2$ (**14**) and Ru$_4$(μ-H)$_2$(μ-CO)(CO)$_{10}$(P^{i}Pr$_3$) (**15**). Cluster **13** is a simple substitution product of **12**, in which one CO ligand has been replaced with a pair of hydride ligands, so that the electron count remains the same. Clusters **14** and **15** are degradation products involving exchange of metal atoms according to the equation:

$$2Ru_3Pt(\mu\text{-}CO)_2(CO)_9(P^iPr_3)_2 \quad + 2H_2 \quad \Rightarrow \quad Ru_2Pt_2(\mu\text{-}H)_2(CO)_8(P^iPr_3)_2$$
$$+ \quad Ru_4(\mu\text{-}H)_2(CO)_{11}(P^iPr_3)_2$$
$$+ \quad 3CO$$

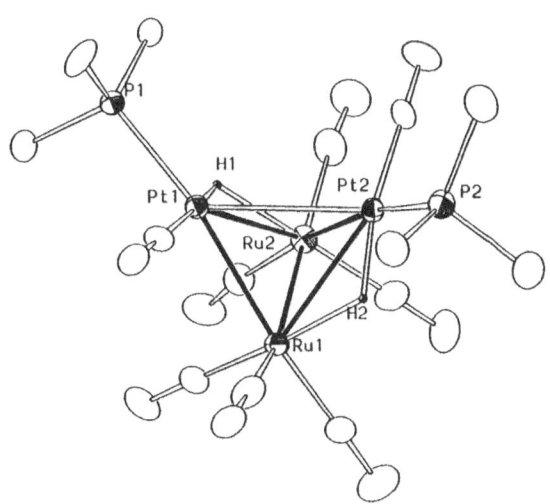

Figure 7. The molecular structure of Ru$_2$Pt$_2$(μ-H)$_2$(CO)$_8$(P^{i}Pr$_3$)$_2$ (14)
Important bond lengths (Å): Pt(1)-Pt(2) 3.096(1), Ru(1)-Pt(1) 2.700(2), Ru(1)-Pt(2) 2.823(1), Ru(1)-Ru(2) 2.741(2), Ru(2)-Pt(1) 2.818(2), Ru(2)-Pt(2) 2.704(1), Pt(1)-P(1) 2.310(4), Pt(2)-P(2) 2.305(7)

Cluster **15** is a phosphine derivative of the well known $Ru_4(\mu\text{-H})_2(CO)_{13}$. The unsubstituted species contains two weakly semi-bridging carbonyls [26], while **15** has one (crystallographically constrained) symmetric bridging carbonyl.

Cluster **14** (Figure 7) is the ruthenium analogue of the previously known species $M_2Pt_2(\mu\text{-H})_2(CO)_8(PR_3)_2$ M=Fe [27] and M=Os[28], and thus completes the series. All three clusters have the same non-planar butterfly metal skeleton, with the Pt atoms occupying the wingtip positions. The ligand dispositions are essentially identical, and the molecules possess idealised C_2 symmetry, which is crystallo-graphically defined for $Os_2Pt_2(\mu\text{-H})_2(CO)_8(PPh_3)_2$ [28a]. The most interesting structural feature of these clusters, in the context of this article, lies in the Pt..Pt separation which is rather long and may be considered non-bonding, or only weakly bonding. For M=Fe this distance is 2.988(2)Å, while for M=Os there have been two reported X-ray structures, with PPh₃ [Pt...Pt = 3.206(1)Å] and with PCy₃ [Pt...Pt = 3.230(1)Å]. The corresponding Pt...Pt separation in **14** is 3.096(1)Å, which is intermediate in value between the Fe and Os compounds. The covalent radii of Ru and Os differ by only 0.02Å, and this is reflected in the close similarity of all the other metal-metal distances . Hence, one might expect the Ru and Os analogues to have

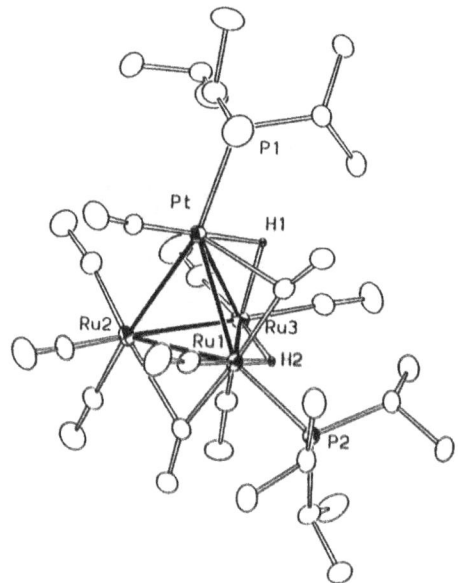

Figure 8. The molecular structure of $Ru_3Pt(\mu\text{-H})_2(\mu\text{-CO})_2(CO)_8(P^iPr_3)_2$ (13)
Important bond lengths (Å): Pt-Ru(1) 2.879(1), Pt-Ru(2) 2.735(1), Pt-Ru(3) 2.880(1), Ru(1)-Ru(2) 2.878(1), Ru(1)-Ru(3) 2.975(1), Ru(2)-Ru(3) 2.878(1), Pt-P(1) 2.331(2), Ru(1)-P(2) 2.368(2).

essentially identical metrical parameters. The significant difference in the Pt...Pt separations is taken as an indicator of the flexibility of the M_2Pt_2 butterfly skeleton.

The cluster $Ru_3Pt(\mu\text{-}H)_2(\mu\text{-}CO)_2(CO)_8(P^iPr_3)_2$ (**13**) also has an analogue in osmium chemistry in the species $Os_3Pt(\mu\text{-}H)_2(CO)_{10}(PR_3)_2$ [9]. In this case however, there are substantial structural differences. The skeletal core in the Os_3Pt clusters is a non-planar butterfly [9] with a long non-bonding wingtip Os...Pt separation of ~3.5Å. All the carbonyl ligands are terminal. In contrast, the metal skeleton in **13** (Figure 8) is tetrahedral, and there are two semi-bridging carbonyl ligands lying across the Ru1-Pt and Ru1-Ru2 bonds. The structure of **13** may be conceptually related to its Os analogue by a breaking of the Pt-Ru1 bond and the carbonyl bridges to Pt and Ru2, and a "tilting" of the $Ru(CO)_3(PR_3)$ group on Ru1 about its *pseudo*-twofold axis, so that these ligands adopt more regular axial and equatorial positions.

This relationship between the crystal structure of **13** and its Os analogues provides a fascinating insight into the fluxional behaviour of $Os_3Pt(\mu\text{-}H)_2(CO)_{10}$-$(PR_3)_2$ (**3**). Variable temperature ^{13}C and EXSY studies on $Os_3Pt(\mu\text{-}H)_2(CO)_{10}$-$(PCy_3)_2$ [29] have shown an unusual carbonyl exchange, which can be interpreted as occuring *via* two different mechanisms. One mechanism involves a planar butterfly intermediate or transition state (mechanism A), while the other involves a double rotation of the $Pt(H)(CO)(PR_3)$ group about its *pseudo*-twofold axis coupled with a rotation of the wingtip $Os(H)(CO)_3(PR_3)$ group about the *pseudo*-twofold axis (mechanism B). Both mechanisms involve inversion of chirality of the cluster, which can in principal be detected using a derivative containing a phosphine such as PMe_2Ph with diastereotopic groups. We have prepared such a derivative of **3**, and shown by ^1H NMR studies [6] that the two Me groups of the PMe_2Ph ligand are inequivalent at low temperature, but coalesce on warming, confirming that enantiomerisation of the cluster occurs. Unfortunately, quantitative studies relating the rate of enantiomerisation to the carbonyl exchange rates were not possible, since this particular derivative exists as two exchanging isomers in solution.

Variable temperature ^{31}P NMR studies on **13** have shown that, in solution, this complex also exists as two exchanging isomers. At room temperature these two isomers are present in roughly equal proportions. Presumably one isomer is similar to that observed in the solid state, while the other isomer may resemble cluster **3**. Our recent results therefore indicate that mechanism B (or some similar mechanism involving an analogue of **13**) is a distinct possibility in the fluxional behaviour of **3**, and further work in this area is required to elucidate the mechanisms of these interesting dynamic processes.

154

2. Structures of 62 CVE triruthenium-platinum clusters

The 60 CVE clusters mentioned above have skeletal geometries ranging from regular tetrahedral, through distorted tetrahedral (with one elongated Pt-Ru bond) to angular butterfly and planar butterfly. The local geometry around the Pt centres changes from non-planar in the tetrahedral molecules, to planar in the butterfly species, and these latter structures formally contain one less metal-metal bond. A few 62 CVE Ru$_3$Pt clusters are also known, [5,6] such as the above mentioned alkynyl cluster Ru$_3$Pt(μ-H)(μ_4-η^2-C$\equiv$CBu^t)(CO)$_9$(dppe) (4) [11] which has a spiked

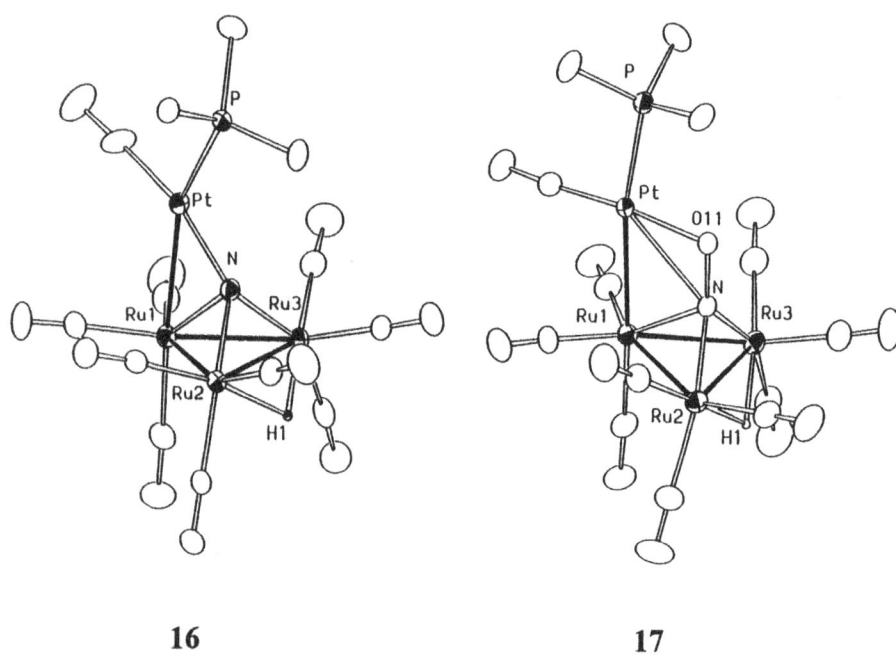

16   **17**

Figure 9. The molecular structures of **Ru$_3$Pt(μ-H)(μ_4-N)(CO)$_{10}$(P^{i}Pr$_3$) (16)** and
Ru$_3$Pt(μ-H)(μ_4-η^2-NO)(CO)$_{10}$(P^{i}Pr$_3$) (17)

Important bond lengths (Å). Cluster 16: Pt-Ru(1) 2.718(1), Ru(1)-Ru(2) 2.729(1), Ru(1)-Ru(3) 2.729(1), Ru(2)-Ru(3) 2.821(1), Pt-N 1.934(7), Ru(1)-N 2.073(6), Ru(2)-N 2.011(7), Ru(3)-N 2.016(7), Pt-P 2.296(2).

Cluster 17: Pt-Ru(1) 2.710(1), Ru(1)-Ru(2) 2.724(1), Ru(1)-Ru(3) 2.727(1), Ru(2)-Ru(3) 2.840(1), Pt-N 2.597(4), Ru(1)-N 2.121(4), Ru(2)-N 1.988(4), Ru(3)-N 1.999(4), Pt-O(11) 1.993(4), N-O(11) 1.365(6), Pt-P 2.319(2).

triangular metal skeleton. This species contains the five electron donor μ_4-η^2-alkynyl ligand, and has one less metal-metal bond than the butterfly clusters, and a planar coordination geometry at the Pt atom. In a simplistic sense, non-planar geometries at the Pt atom may be associated with 18 electron counts, while planar geometries are associated with 16 electron counts.

We have recently prepared two further examples 62 CVE Ru_3Pt clusters, involving the relatively unusual μ_4-N and μ_4-η^2-NO ligands, which also act as five electron donors. The reaction of $Ru_3(\mu$-H$)(\mu$-NO$)(CO)_{10}$ with one mole equivalent of $Pt(nb)_2(P^iPr_3)$ affords, in addition to significant amounts of $Ru_3(\mu$-H$)(\mu$-NO$)(CO)_9(P^iPr_3)$, low yields of two Ru-Pt clusters $Ru_3Pt(\mu$-H$)(\mu_4$-N$)(CO)_{10}(P^iPr_3)$ (16) and $Ru_3Pt(\mu$-H$)(\mu_4$-η^2-NO$)(CO)_{10}(P^iPr_3)$ (17). The structures of complexes 16 and 17 are quite similar and are shown in Figure 9. Both contain a spiked-triangular Ru_3Pt skeleton, with the Pt atoms having an essentially square planar coordination geometry. A number of clusters containing the four-coordinate nitrido-ligand are known [30], but these examples primarily involve metal clusters with a butterfly geometry. On the other hand, clusters containing the μ_4-η^2-NO ligand are very rare, the only other example [31] of which we are aware is $Mo_2Co_2(\mu_4$-η^2-NO$)\{C_6H_4(P^tBu_2)_2\}(CO)_6Cp_2$ The N-O bond length in this latter species is 1.27(7)Å [31], but in cluster 17 this bond is considerably longer, cf N-O = 1.365(6)Å. This suggests a high degree of "activation" and bond weakening, and it is tempting to view cluster 17 as an intermediate for 16. However we have no evidence for this, and solutions of 17 appear quite stable, and do not show any NMR signals for 16, even on standing for several days.

Conclusions

It is clear that prediction of the metal skeletal geometry for 60 CVE Ru_3Pt clusters is not a straight-forward matter, even allowing for the small sample of clusters presented in this article. The exact nature and number of the peripheral ligands plays a crucial role in determining this geometry, but as yet there are no clear patterns emerging. Although semi-empirical MO calculations may provide some clues in this direction, it is important to realise that the energy differences between differing skeletal isomers may be very small.

Acknowledgements

This work has been carried out in collaboration with Professor A. G. Orpen (Bristol), whose group was responsible for the EXAFS results. We also thank the EPSRC for financial support.

References

1. Mingos, D. M. P. (1984) *Acc. Chem. Res.* **17**, 311-319

2. Owen, S. M. (1988) *Polyhedron*, **7**, 253-283

3. Hall, K. P. and Mingos. D. M. P. (1984) *Prog. Inorg. Chem.*, **32**, 237-325

4. Mingos, D. M. P. and Wardle, R. W. M. (1985) *Transition Met. Chem. (N.Y.)* **10** 441-459

5. Farrugia , L. J. (1990) *Adv. Organomet. Chem.*, **31**, 301-391.

6. Farrugia, L. J. (1992) *J. Cluster Sci.*, **3**, 361-383.

7. Ewing, P. and Farrugia, L. J. (1988) *J. Organomet. Chem.* **347**, C31-C34.

8. Ewing, P. and Farrugia, L. J. (1989) *Organometallics*, **8**, 1665-1673.

9. (a) Farrugia, L. J., Howard, J. A. K., Mitrprachachon, P., Stone, F. G. A. and Woodward, P. (1981) *J. Chem. Soc. Dalton Trans.*, 162-170. (b) Farrugia, L. J. (1988) *Acta. Crystallogr. Sect. C.* , **C44**, 1307-1309. (c) Farrugia, L. J. (1991) *Acta Crystallogr. Sect. C*, **C47**, 1310-1312.

10. Braunstein, P., de Meric de Bellefon, C., Bouaoud, S.-E., Grandjean, D., Halet, J.-F., Saillard, J.-Y. (1991) *J. Am. Chem. Soc.*, **113**, 5282-5292

11. Ewing, P. and Farrugia, L. J. (1989) *Organometallics*, **8**, 1246-1260.

12. Dent, A. J., Farrugia, L. J., Orpen, A. G. and Stratford, S. E. (1992) *J. Chem. Soc., Chem. Commun.*, 1456-1457.

13. (a) Roberts, D. A. and Geoffroy, G. L. in *Comprehensive Organometallic Chemistry* , Eds, Wilkinson, G., Stone, F. G. A., and Abel, E. (1982) Pergamon, Oxford , Vol 6, Chapter 40. (b) Geoffroy, G. L. in *Metal Clusters in Catalysis*: Eds., Gates, B. C., Guczi, L., Knozinger, H. (1986) Elsevier, New York, Chapter 1. (c) Adams, R. D. in *The Chemistry of Metal Cluster Complexes* Eds., Shriver, D. F., Kaesz, H. D., and Adams, R. D. (1990) VCH, New York, Chapter 3.

14. (a) Green, M., Howard, J.A. K., Spencer, J. L., and Stone, F. G. A. (1977) *J. Chem. Soc., Dalton Trans.*, 271- 277. (b) Crascall, L. E., Spencer, J. L. *Inorg.Synth.*, **1990**, *28*, 126-129.

15. (a) Green, M., Mead, K. A., Mills, R. M., Salter, I. D., Stone, F. G. A. and Woodward, P. (1982) *J. Chem. Soc., Chem. Commun.*, 51- 53. (b)Salter, I. D. (1983) *Ph. D. Thesis, University of Bristol.*

16. Farrugia, L. J., Miles, A. D., and Stone, F. G. A. (1985) *J. Chem. Soc., Dalton Trans.*, 2437-2447.

17. Ellis, D., Farrugia, L. J., Wiegeleben, P., Crossley, J. G., Orpen, A. G., and Waller, P. N. (1995) *Organometallics*, **14**, in press.

18. Farrugia, L. J. (1989) *Organometallics*, **8**, 2410-2417.

19. Schilling, B. E. R., and Hoffmann, R., (1979) *J. Am. Chem. Soc.*, **101**, 3456-3467.

20. Rosenberg, E., Hursthouse, M., Randall, E. W., Milone, L., Valle, M., and Evans, M., (1972) *J. Chem. Soc., Chem. Commun.*,, 545-546

21. Hawkes, G. E., Lian, L. Y., Randall, E. W., Sales, K. D., and Aime, S., (1985) *J. Magn. Reson.*, **65**, 173-177.

22. Rivomanana, S., Lavigne, G., Lugan, N. and Bonnet, J.-J. (1991) *Organometallics*, **10**, 2285-2297.

23. Deeming, A. J. and Senior, A. M. (1992) *J. Organomet. Chem.* **439**, 177-188.

24. Rosenberg, E., Bracker-Novak, J., Gellert, R. W., Aime, S., Gobetto, R. and Osella, D. (1989) *J. Organomet. Chem.* **365**, 163-185.

25. Farrar, D. H., Gukathasan, R. R. and Lunniss, J. A. (1991) *Inorg. Chim. Acta* **179**, 271-274.

26. Rheingold, A. L., Haggerty, B. S., Geoffroy, G. L. and Han, S.-H. (1990) *J. Organomet. Chem.* **384**, 209-216.

27. Farrugia, L. J., Howard, J. A. K., Mitrprachachon, P., Stone, F. G. A. and Woodward, P. (1981) *J. Chem. Soc., Dalton Trans.*, 1134-1145.

28. (a) Farrugia, L. J., Howard, J. A. K., Mitrprachachon, P., Stone, F. G. A. (1981) *J. Chem. Soc., Dalton Trans.*, 1274-1277. (b) Farrugia, L. J. (1988) *Acta Crystallogr., Sect C.*, **C44**, 818-820.

29. Ewing, P., and Farrugia, L. J., (1988) *Organometallics*, **7**, 859-870.

30. Gladfelter, W. L., (1985) *Adv. Organomet. Chem.*, **24**, 41-86.

31. Kyba, E. P., Kerby, M. C., Kashyap, R. P., Mountzouris, J. A., and Davis, R. E. (1990) *J. Am. Chem. Soc.*, **112**, 905-907.

INTRAMOLECULAR EXCHANGE IN d⁹ METAL CARBONYL CLUSTERS

R. ROULET
Institut de Chimie Minérale et Analytique de l'Université
BCH, 1015 Lausanne
Switzerland

1. Introduction

In recent years the migration of small molecules over polymetallic aggregates containing metal–metal bonds has gained as much attention as the stereochemical nonrigidity of monometallic coordination compounds had in the sixties. The possible relationship of such migrations to those occurring during the chemisorption of small molecules to a metal surface is an additional incentive for understanding the pathways by which movement of coordinated molecules occurs relative to the framework of a metallic cluster compound.

The bulk of the studies on the fluxional behaviour of cluster compounds have dealt with the migration of carbon monoxide over tri– and tetrametallic carbonyl clusters and have led to two theories of the mechanism of carbonyl scrambling. The first proposal was made by Cotton in 1966 [1], and involves bridge opening and closing of the carbonyl ligands to move then around the cluster in concerted steps.

The underlying assumption of this theory was that the interconversion of ligand positions is about a rigid metal cluster core. Subsequently, Johnson and Benfield proposed in 1978 that the scrambling processes of carbonyl ligands in $M_3(CO)_{12}$ (M = Fe, Ru, Os) and $M_4(CO)_{12}$ (M = Co, Rh, Ir) and related clusters could be explained in terms of 'the initial icosahedral arrangement of ligands rearranging along a well–defined reaction co–ordinate *via* a cubo–octahedral transition state' [2]. This rearrangement conserves all antipodal relationships between the ligands. However, it is known that total CO scrambling is observed in certain systems with a ground state symmetry lower than C_{3v}. This means that these antipodal relationships must be destroyed and Johnson *et al.* [3] have recently revised their Ligand Polyhedral Model accordingly by proposing rearrangements *via* an icosahedron ↔ anticubeoctahedron ↔ icosahedron mechanism.

Most of the normal modes of a cluster involve some coupling of modes which are largely ligand motion and of others which are largely metal motion. Therefore, the motions of ligand and metal atoms are interdependent and should not be separated from one another. Nevertheless, we will adopt in this report the classical description of the

159

L. J. Farrugia (ed.), The Synergy Between Dynamics and Reactivity at Clusters and Surfaces, 159–173.
© 1995 *Kluwer Academic Publishers.*

fluxional behaviour in terms of bond breaking and forming and will address the question of the relative motion of metal nuclei when examining the few experimental results available on reaction volumes and activation volumes.

Numerous studies have demonstrated the importance of including pressure as a kinetic parameter in the elucidation of inorganic reaction mechanisms [4] [5]. The only data on intramolecular processes involving carbonyl complexes are those concerning the isomerisation of $[RuCl_2(CO)_2(PR_3)_2]$ [6]. Derivatives of $M_4(CO)_{12}$ (M = Ir, Rh) and other cluster compounds are suitable for variable–pressure studies if two conditions are met. First, whether the compound exists as a single species in solution or as an equilibrium mixture of two geometrical isomers, the dynamic connectivities of the intramolecular exchange(s) must be established accurately using magnetisation transfer experiments (quantitative evaluation of 2D–EXSY spectra [7] or DANTE sequences [8]). Second, the compound should preferably be an uncharged molecule. In this case, there is no charge creation or annihilation on forming the transition state or the product. The electrostriction can therefore be neglected, and both the activation volume $\Delta V^{\ddagger}$ and the reaction volume ΔV_0 can be deduced from the variable–pressure NMR spectra using eqs. (1) and (2) where k_0 and K_0 are the rate constant of the site exchange process

$$\ln k = \ln k_0 - (\Delta V^{\ddagger}P/RT) \qquad (1)$$
$$\ln K = \ln K_0 - (\Delta V_0 P/RT) \qquad (2)$$

and the isomerisation equilibrium constant at 0.1 MPa, respectively. The resulting activation and reaction volumes reflect molecular volume changes and can be directly compared to characterise the transition state. If only one species is observed, a positive value of $\Delta V^{\ddagger}$ will reflect a volume expansion on going from the ground state to the transition state. In certain cases, the value of ΔV_0 can be compared to the value obtained from variable–pressure IR spectra or from crystallographic data.

2. Fluxional behaviour of $M_4(CO)_{12}$

2.1 HOMO– AND HETEROMETALLIC DODECACARBONYLS

$Co_4(CO)_{12}$ and $Rh_4(CO)_{12}$ are disordered in the solid [9]. IR [10] and ^{59}Co–NMR [11] [12] measurements indicate that the structure of $Co_4(CO)_{12}$ is of C_{3v} symmetry with three edge–bridging carbonyl groups as in the crystal, but no quantitative, dynamic information could be extracted from its variable–temperature ^{13}C–NMR spectra due to intensity distortions [13], nor from those of most other systems containing cobalt [14].

The ground state geometry of $Rh_4(CO)_{12}$ in solution has also C_{3v} symmetry [15]. A recent simulation of its variable–temperature and –pressure ^{13}C–NMR spectra [16] (*Fig. 1a*) has shown that the fluxional process leading to complete CO scrambling is the merry–go–round of six CO's about any one of the triangular faces of an unbridged transition state (the Cotton mechanism).

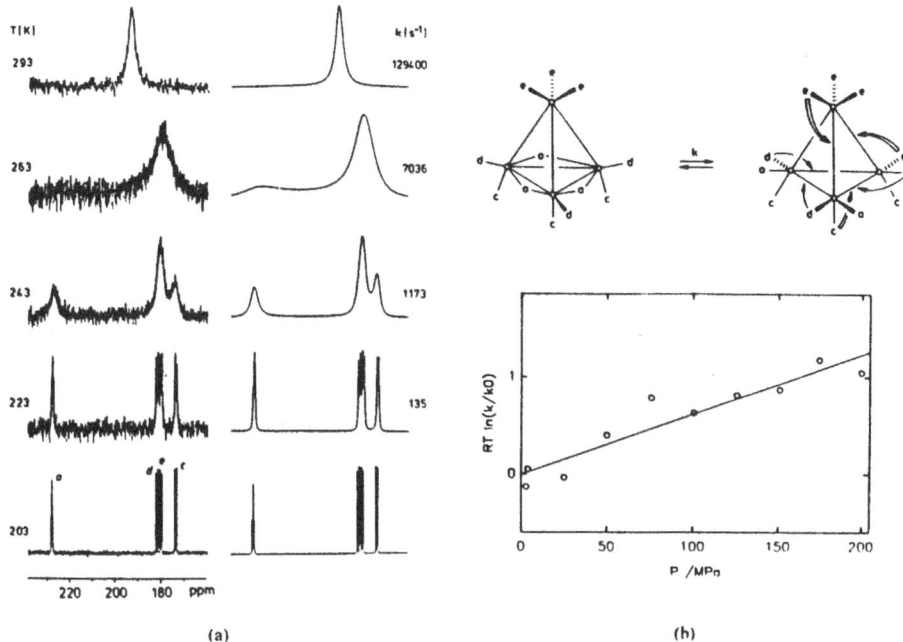

Figure 1. a) Experimental and calculated ^{13}C–NMR spectra of $Rh_4(CO)_{12}$ in CD_2Cl_2 ($\Delta G^{\ddagger} = 43.8 \pm 0.4$ kJ mol^{-1} at 298 K). *b)* Rate dependence on pressure ($\Delta V^{\ddagger} = -6 \pm 1$ cm^3 mol^{-1})

The dependence of the rate constant on pressure gave a negative value for the activation volume (*Fig. 1b*), indicating that the transition state has a smaller molar volume than that of the bridged ground state. Since the opening of bridges should lead to an increase in volume (see part 3), the negative value of $\Delta V^{\ddagger}$ suggests a substantial shortening of the unbridged M–M bonds relative to the bridged ones. There is to date no example of a Rh_4 cluster with terminal ligands only. However, a statistical study of X–ray diffraction data has shown that the differences between the mean bond lengths for unbridged and CO–bridged M–M bonds have values of –0.027 and –0.019 Å for Rh and Ir, respectively [17].

$Ir_4(CO)_{12}$ is also disordered in the solid, but definitely has T_d symmetry with terminal ligands only [18]. This compound is practically insoluble in common organic solvents. However, it is somewhat soluble (0.02 M) in N,N'–dimethyl–N,N'–propylene urea (DMPU) without loss of CO [19]. The IR spectrum of the yellow solution has a band at 1825 cm^{-1} with an absorbance ratio of 1:3 relative to those in the terminal CO region (2069–1967 cm^{-1}), indicating that the ground state structure in DMPU is similar to that of $Rh_4(CO)_{12}$ in CD_2Cl_2 [16]. CO site exchange is still rapid on the NMR time scale at 200 K, the lowest temperature accessible in DMPU/THF. This is a first indication that a process requiring a $\mu_2 \leftrightarrow \eta_1$ switch in the bonding modes of CO is slower in a Rh cluster than in the isostructural Ir cluster.

$CoRh_3(CO)_{12}$ and $CoIr_3(CO)_{12}$ are thermally unstable and the high quadrupole moment of ^{59}Co has probably precluded so far the determination of the dynamic connectivities in $Co_3Rh(CO)_{12}$, $Co_3Ir(CO)_{12}$ and $Co_2Ir_2(CO)_{12}$ [20–22]. In solution $Co_3Rh(CO)_{12}$ has a ground state geometry with three CO's bridging the Co_2Rh basal face. The low temperature ^{13}C–NMR spectrum is consistent with the ground state structure if three apical carbonyls give a single resonance; the $Co(CO)_3$ unit appears to be rotating even at 168 K. This fluxional process localised at one metal centre is also observed in other M_4 clusters (see part 3). Over an intermediate temperature range, there is a 10:2 pattern of CO equivalence with two distinct CO's bonded to Rh [23]. $Ir_3Rh(CO)_{12}$ can now be obtained in the pure state [24], but its low solubility in common organic solvents has prevented any NMR study.

$Ir_2Rh_2(CO)_{12}$ is disordered in the solid. The ground state geometry in solution has C_s symmetry with three CO's edge–bridging one $IrRh_2$ face [25], and is similar to that found for $Co_2Rh_2(CO)_{12}$ in the solid state [21], with Ir–atoms replacing the two Co–atoms. It is fluxional above 230 K with the dynamic connectivities $b \leftrightarrow f \leftrightarrow g$, $d \leftrightarrow c$ and $e \leftrightarrow h$ (*Scheme 1*).

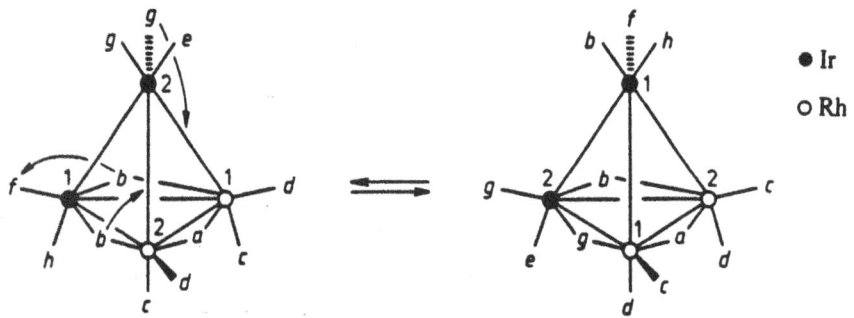

Scheme 1. The 'change of basal face' in $Ir_2Rh_2(CO)_{12}$ ($\Delta G^{\ddagger} = 51.2 \pm 0.4$ kJ mol^{-1} in CD_2Cl_2 at 298 K). Hereafter, labels *a* or *b* designate bridging CO's, *d* or *f* radial CO's, *c* or *h* axial CO's and *e* or *g* apical CO's.

Carbonyl *a* remains unaffected, excluding site exchanges by a merry–go–round of six CO's, as proposed for $Rh_4(CO)_{12}$. Since restricted axial–basal site exchanges are observed, this fluxional process may be called a 'change of basal face' (Rh(1)–Rh(2)–Ir(1) to Rh(1)–Rh(2)–Ir(2)). This new process which has a positive volume of activation [19] is also encountered in other systems and will be further described in part 3.2.

$IrRh_3(CO)_{12}$ has been recently obtained in the pure state by a redox cocondensation of $[Ir(CO)_3]^{3-}$ and $[Rh(COD)_2(THF)_2]^+$ (1:3) in THF followed by saturation with CO [19]. Its ground state geometry in solution has three CO's edge–bridging the Rh_3 basal face. Basal–axial site exchanges occur above 260 K, but still have to be characterised quantitatively.

3. Kinetic pathways of carbonyl scrambling

3.1 THE MERRY–GO–ROUND

When the ground state of a tetrahedral cluster has C_{3v} symmetry, the merry–go–round of terminal and symmetrically bridging CO's about a triangular face is believed to take place in concerted steps [1]. An IR and NMR study of $[Ir_4(CO)_9(\mu_3-1,3,5-trithiane)]$ gave the first quantitative evidence for such a mechanism [26] [27]. Two isomers of this cluster have been isolated in the solid state: the unbridged isomer (A) and the bridged one (B, *Scheme 2*).

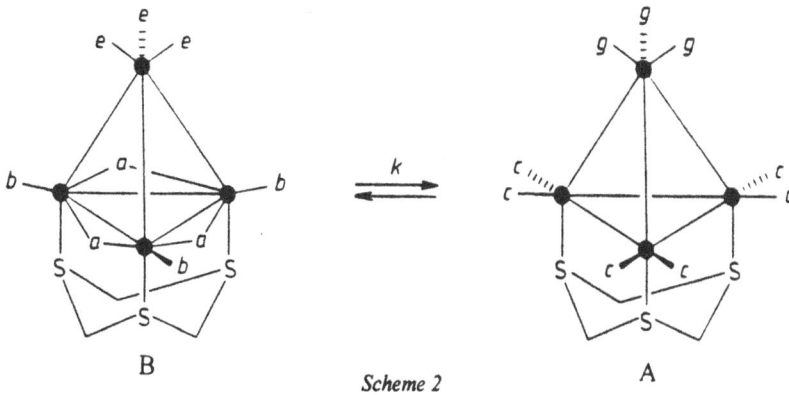

Scheme 2

An endothermic isomerisation equilibrium B = A takes place in solution and involves the mutual exchange of the carbonyl ligands more or less coplanar with the Ir_3 face capped by the tridentate ligand. The dependence of the population ratio of the two isomers and that of the rate constant k (B → A) on pressure gave values of 15.4 ± 0.4 and 8.3 ± 0.5 cm^3mol^{-1} for ΔV_0 and $\Delta V^\ddagger$, respectively. Since the activation volume is about half the reaction volume, the merry–go–round is a concerted process and a transition state with three semi–bridging CO's of C_3 symmetry can be proposed. The value of the reaction volume is in good agreement with the value calculated from the crystallographic data. This value results from the opening of three CO bridges, the synergic contraction of the metallic cage, but also from the elongation of the three Ir–S bonds on going from B to A. In this case, the latter factor is preponderant in determining the positive sign of ΔV_0 which is therefore not characteristic of a merry–go–round of CO's. As observed for $Rh_4(CO)_{12}$ and other cluster compounds exempt of bridging, non–CO ligands (see below), a merry–go–round of carbonyl ligands seems to be associated with negative values of ΔV_0 and $\Delta V^\ddagger$ when the ground state is the CO–bridged species B.

The bulk of kinetic data available to date (*Table 1*) concerns the monosubstituted derivatives $[Ir_4(CO)_{11}L]$ (L = PEt_3 [28], $PMePh_2$ [29], PH_2Ph, $PHPh_2$ [30], PPh_3 [31] [32], PAr_3 [32], $P(OMe)_3$, $P(OPh)_3$, $P(OCH_2)_3CEt$ [33], t–BuNC [34] [35], ArNC [35],

COCH$_2$CH$_2$O (dioxycarbene) [36], Br$^-$ [37], I$^-$, CNS$^-$, NO$_2^-$ [32]. These compounds exist in solution as a single species (A or B) or as an equilibrium mixture of two or three isomers (A + B + C, *Fig. 2*). When L is a σ and π–donor ligand such as Br$^-$, the bridged ground state B is observed, probably because bridging CO's are better π–acceptors than terminal CO's. The unbridged ground state A is usually observed when L is a better π–acceptor than CO. Isomer C is always the minor one, as a radial substitution causes more steric crowding than an axial substitution [29].

TABLE 1. Kinetic data on [Ir$_4$(CO)$_{11}$L] clusters (references given in the text)

| L | Ground state(s) | $\Delta G^\ddagger$ [kJ mol^{-1}] at 298 K | |
		1st merry-go- round	2nd merry-go-round
PEt$_3$	B + C	44.3	44.3
PH$_2$Ph	A + B + C	41.8 (A → B)	43.2 (A → C)
PHPh$_2$	A + B + C	39.7 (A → B)	42.6 (A → C)
PPh$_3$	B	45.6 ± 0.4	46.3 ± 0.5
PAr$_3$	B	45 – 47	47 – 49
P(OPh)$_3$	A + B	37.7 (A → B)	–
P(OMe)$_3$	A + B + C	36.5 (A → B)	58.6 (A → C)
P(OCH$_2$)$_3$CEt	A + B + C	31.4 (A → B)	23.4 (A → C)
COCH$_2$CH$_2$O	A + B	42.3 ± 0.9	48 ± 2
t–BuNC	A	47.5 ± 0.4	48.6 ± 0.4
PhNC	A	49.4 ± 0.4	49.9 ± 0.4
ArNC	A	49 – 51	50 – 53
Br$^-$	B	37.0 ± 0.6	37 ± 3
I$^-$	B	36.8 ± 0.4	38.6 ± 0.4
NCS$^-$	B	33 ± 2	–
NO$_2^-$	B	40.7 ± 0.9	44 ± 2

A clear picture of the various exchange processes results from the magnetisation transfer experiments effected on the clusters with L = PHPh$_2$, PH$_2$Ph, P(OMe)$_3$, P(OCH$_2$)$_3$CEt and P(OPh)$_3$ which all behave similarly. The observed isomers merely correspond to three minima on the kinetic pathways of CO scrambling (*Fig. 2*), whose free enthalpies differ at most by *ca.* 9 kJ mol^{-1} at 298 K.

For L = P(OPh)$_3$, only equilibrium B = A is observed (ΔG^0 = 1.8 ± 0.1 kJ mol^{-1} at 298 K [33]). The dependence of the population ratio of the two isomers and of the rate constant k$_1$ (B → A) on pressure gave values of –8.3 ± 0.2 and –9.4 ± 1.1 cm^{3}mol^{-1} for ΔV_0 and $\Delta V^\ddagger$, respectively [19]. The negative value of ΔV_0 indicates that the unbridged isomer A has a smaller volume than the bridged one (B), and therefore that unbridged Ir–Ir bonds are shorter than bridged Ir–Ir bonds (as observed in the solid state [17]). Since $\Delta V^\ddagger \cong \Delta V_0$, the transition state should have a bridged–like geometry. This finding backs up an earlier proposal that the unbridged species A is a poor representation of the transition state of a merry–go–round in clusters of C$_s$ symmetry with basic ligands such as L = Br$^-$ [37] or PEt$_3$ [28].

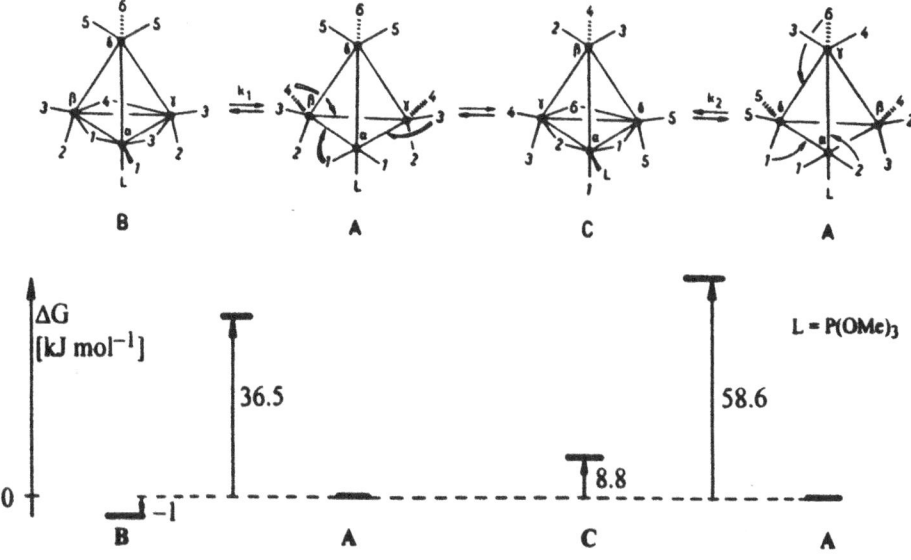

Figure 2. Relative stabilities of the isomers of $[Ir_4(CO)_{11}P(OMe)_3]$ in solution and free enthalpies of activation of the two CO site exchange processes.

One kinetic step of a merry–go–round only involves the CO ligands which are more or less coplanar with a given triangular face of the cluster compound. Better coplanarity should thus lower the activation energy of a merry–go–round. A clear evidence for this is given by the comparison of the $\Delta G^{\ddagger}$'s at 298 K for $[Ir_4(CO)_{10}(\mu_2-dppm)]$ (38.0 ± 0.5 kJ mol^{-1}) and $[Ir_4(CO)_{10}(\mu_2-dppp)]$ (53.9 ± 0.4 kJ mol^{-1} [32]) (*Scheme 3*). The dppm ligand has a smaller bite than dppp and crystallographic data [19] show that carbonyls a, b, d, and f are roughly coplanar with the basal face of the dppm complex, whereas in the dppp complex these CO's deviate from that plane towards the apical Ir–atom.

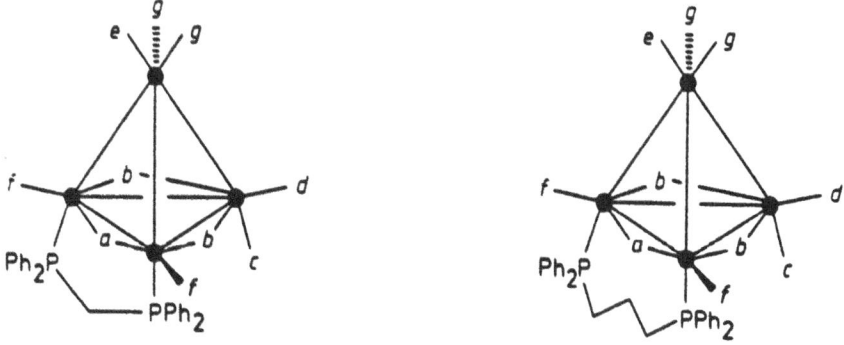

Scheme 3

3.2 THE CHANGE OF BASAL FACE

[Ir$_4$(CO)$_{10}$(η^2–diarsine)] has a ground state structure with three edge–bridging CO's and the bidentate ligand chelating one basal Ir–atom (*Scheme 4*).

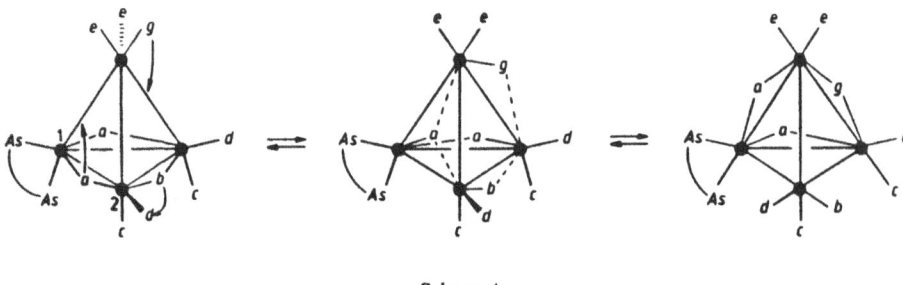

Scheme 4

Two CO–bridges are asymmetrical with shorter bonds towards the electron rich Ir–atom (Ir(1)–C$_a$O 2.012, Ir(2)–C$_a$O 2.257 Å [38]). The variable–temperature ^{13}C–NMR spectra indicated that carbonyl *a* was unaffected while basal–apical exchanges were observed [39]. This was later confirmed by 2D–EXSY measurements giving the exchange matrix elements (*b*, *g*) = k, (*c*, *d*) = (*c*, *e*) = (*d*, *e*) = k/2 and by line shape analysis of the VT ^{13}C–NMR spectra. A consequence of this CO site–exchange is that the two Me$_2$As groups swap positions and this was confirmed by simulation of the VT ^{1}H–NMR spectra [40]. As for Ir$_2$Rh$_2$(CO)$_{12}$, the exchange mechanism is clearly distinct from the merry–go–round and Shapley *et al.* [39] proposed that the carbonyl exchange takes place *via* a face–bridged mechanism (*Scheme 4*).

[Ir$_4$(CO)$_{11}$(μ_2–SO$_2$)] displays the same fluxional behaviour as Ir$_2$Rh$_2$(CO)$_{12}$ and [Ir$_4$(CO)$_{10}$(η^2–diarsine)] [41]. Recently, the dependence of the rate of CO–exchange on pressure has been determined, giving a positive value for the activation volume (+ 15 ± 1 cm^{3}mol^{-1} [19]) of the change of basal face (*Fig. 3*).

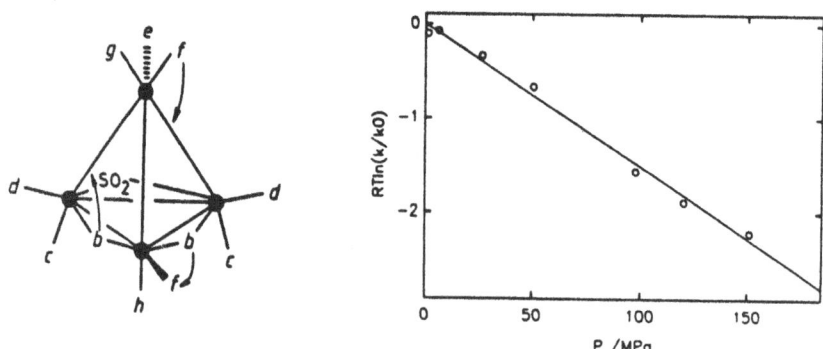

Figure 3. Dependence of the rate of change of basal face in [Ir$_4$(CO)$_{11}$(μ_2–SO$_2$)] on pressure.

A positive activation volume ($+ 6.0 \pm 0.1$ cm^{3}mol^{-1} [19]) has also been obtained for the change of basal face in $[Ir_4(CO)_8(\mu_2-(EtO)_2POP(OEt)_2)_2]$. However, more data are needed before one can rely on the sign of the activation volume for distinguishing a merry–go–round (involving three metal atoms) from a change of basal face (involving four metal atoms).

3.3 ROTATION OF CARBONYL LIGANDS AT ONE METAL CENTRE

The rotation of a $M(CO)_3$ group about a local C_3 axis is a common process in mono-metallic complexes. This rotation is usually the slowest site exchange process observed in Ir_4 cluster compounds, and can thus seldom be characterised by 2D–EXSY measurements. In $[Ir_4(CO)_8(\eta^4$–diene$)_2]$ (*Scheme 5, left*), the only exchange observed is $e \leftrightarrow g$ and, therefore, the rotation of three CO's about the apical Ir–atom is the only site exchange process present in these tetrasubstituted clusters [42]. The main factor affecting the activation energy of the rotation of apical CO's in an Ir_4 cluster is the relative bulkiness of the ligand(s) in radial position. Line shape analysis of the variable–temperature ^{13}C–NMR spectra of $[Ir_4(CO)_8(\eta^4$–COD$)_2]$ (COD = 1,5–cyclooctadiene), $[Ir_4(CO)_8(\eta^4$–COD$)(\eta^4$–NBD$)]$ and $[Ir_4(CO)_8(\eta^4$–NBD$)_2]$ (NBD = norbornadiene) gave values of 62.2 ± 0.3, 53.2 ± 0.4, and 46.7 ± 0.6 kJ mol^{-1} for the free enthalpies of activation at 298 K, respectively [42]. The sequence of decreasing $\Delta G^{\ddagger}$'s is due to a decrease of steric hindrance caused by the C=C bonds in radial positions on going from a 1,5– to a more strained 1,4–diene.

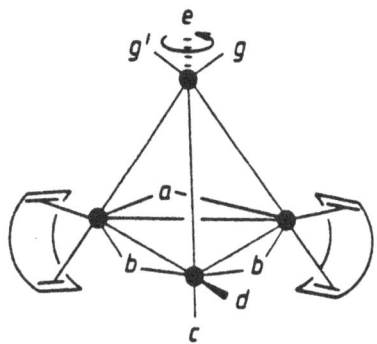 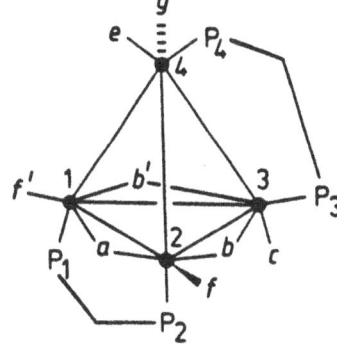

Scheme 5

The rotation of a $L(CO)_2$ moiety about one metal centre is also observed in Ir_4 clusters. A ^{31}P– and ^{13}C–NMR study of $[Ir_4(CO)_8(\mu_2$–dppm$)_2]$ [42] has shown that a first CO site exchange, which is the change of basal face Ir(1)–Ir(2)–Ir(3) to Ir(2)–Ir(3)–Ir(4) (*Scheme 5, right*), averages the P–atoms pairwise (P$_1$–P$_4$ and P$_2$–P$_3$). Then, above 270 K, complete averaging of the P–atoms results from the restricted rotation of P$_4$, e and g ($\Delta G^{\ddagger} = 62.2 \pm 0.6$ kJ mol^{-1} at 298 K).

3.4 SUBSTITUTION OF Ir BY Rh IN THE METAL FRAMEWORK

Since the discovery of new synthetic routes to pure $Ir_3Rh(CO)_{12}$ [24], $Ir_2Rh_2(CO)_{12}$ [25] and $IrRh_3(CO)_{12}$ [19], a large number of substituted derivatives of these mixed clusters have been obtained. All clusters have a ground state geometry with three edge–bridging CO's defining the basal face (Ir_2Rh, $IrRh_2$ and Rh_3, respectively). The apical position is always occupied by an Ir–atom. The same CO scrambling processes as for the Ir_4 clusters are generally observed. There is however a notable exception concerning $[Rh_4(CO)_9(\mu_3–1,3,5–trithiane)]$ which has a butterfly structure in the solid state and in solution at very low temperature [43].

TABLE 2. CO scrambling in homo– and heterometallic M_4 cluster compounds

Compound	$\Delta G^{\ddagger}$ [kJ mol^{-1}] at 298 K		
	Ir_4	Ir_3Rh	Ir_2Rh_2
Merry–go–round			
$[M_4(CO)_{11}Br]^-$	37 ± 1	52.4 ± 0.5	> 68
$[M_4(CO)_{11}I]^-$	38.6 ± 0.3	51.7 ± 0.6	> 68
$[M_4(CO)_9(\mu_3–1,3,5–trithiane)]$	38.0 ± 0.5	48.4 ± 0.5	> 60
Change of basal face			
$[M_4(CO)_{10}(\eta^4–NBD)]$	36.8 ± 0.8	52.8 ± 0.8	57.1 ± 0.5
$[M_4(CO)_{10}(\eta^4–COD)]$	43.8 ± 0.4	55 ± 1	57.1 ± 0.6
$[M_4(CO)_{10}(\eta^2–diarsine)]$	29.8 ± 0.5	52.3 ± 0.6	
Rotation of apical CO's			
$[M_4(CO)_{10}(\eta^4–NBD)]$	54 ± 1	49.4 ± 0.6	46.6 ± 0.6
$[M_4(CO)_{10}(\eta^4–COD)]$	> 70		46.9 ± 0.4
$[M_4(CO)_8(\eta^4–NBD)_2]$	46.7 ± 0.6	43.0 ± 0.3	43.9 ± 0.6
$[M_4(CO)_8(\eta^4–COD)(\eta^4–NBD)]$	53.3 ± 0.2	50.5 ± 0.2	46.7 ± 0.3
$[M_4(CO)_8(\eta^4–COD)_2]$	62.2 ± 0.3	59.9 ± 0.3	56.6 ± 0.3

The results of kinetic studies on several series of isostructural Ir_4, Ir_3Rh and Ir_2Rh_2 clusters are collected in Table 2. These results clearly show that the effect of substituting Ir by Rh in the basal face is twofold: it slows the merry–go–round and the change of basal face, that is the two processes which require debridging of CO's, and it accelerates the rotation of CO's about the apical Ir–atom. One can rationalise this by a ground state effect: substituting Ir by Rh causes a shift of electron density towards the basal metal atoms, thus stabilising the bridged structure since μ_2–CO's are better π–acceptors than terminal CO's. It therefore seems that Rh has a greater tendency to maintain M–M bonds with edge–bridging CO's in a cluster than does Ir. This is the same trend as that found in d^8 metal clusters (e.g. $Fe_3(CO)_{12}$ has bridged M–M bonds, whereas $Os_3(CO)_{12}$ has a structure with all terminal ligands).

4. Non–CO ligand mobility

4.1 HYDRIDE LIGAND MOBILITY

The metal cluster hydride $[Ir_4(CO)_{11}H]^-$ [44] has a ground state structure analogous to that of $[Ir_4(CO)_{11}(SO_2)]$ with the hydride replacing the SO_2 as the μ_2–bridging ligand. However, the hydride ligand may have a terminal bonding mode accessible at moderate energy; indeed in $[Ir_4(CO)_{10}H_2]^{2-}$ the hydride ligands appear to occupy terminal positions even in the ground state [45]. A quantitative analysis of 2D–EXSY ^{13}C–NMR spectra of $[Ir_4(CO)_{11}H]^-$ has shown that the lowest energy fluxional process is an opening of all bridging ligands to give an unbridged intermediate, followed by reforming of the bridges about the face defined by the Ir–atoms α, γ and δ. A merry-go–round process would lead to a structure with three bridging CO's and a terminal hydride, which is not that of the ground state.

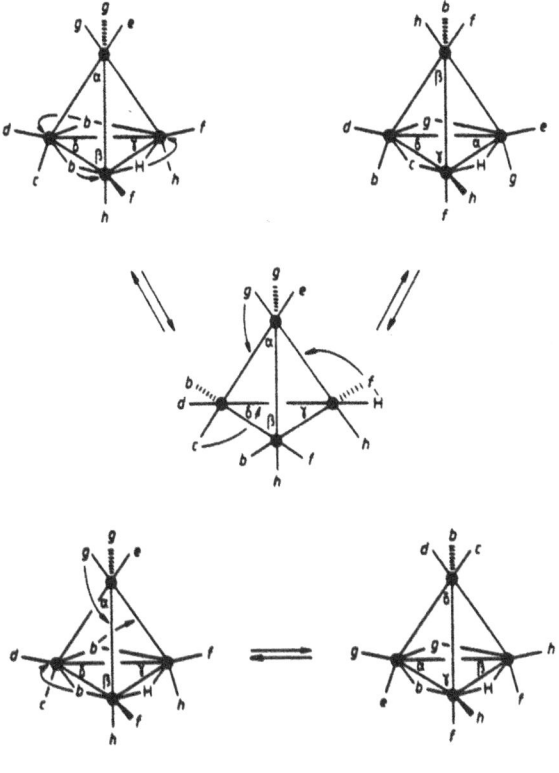

Scheme 6

The process would need to be repeated to reform the ground state structure and exchanges between f and e and also between h and c would be expected in the EXSY spectrum, but are not observed. Rebridging of the unbridged intermediate about the face defined by the Ir–atoms α, β and γ would also lead to a structure with a terminal hydride and is not observed. This mechanism is notable since it was the first time that the substituted ligand had been seen to participate in the fluxional process of an Ir_4 cluster, presumably because the H-atom has, like CO, several possible bonding modes. The next lowest energy process is a change of basal face, which is the same as the lowest energy process in $[Ir_4(CO)_{11}(SO_2)]$ (*Scheme 6*) [46].

4.2 MOBILITY OF LIGANDS WITH A P–DONOR ATOM

The cluster $[Ir_4(CO)_{10}(Ph_2PCH=CHPPh_2)]$ [47] exists in solution as an equilibrium mixture of two isomers (*Scheme 7*) [48]. Substitution of Br^- by *cis*–$Ph_2PCH=CHPPh_2$ in $[Ir_4(CO)_{11}Br]^-$ at low temperature gives the kinetically favoured isomer A which is fluxional (k_{213} = 55 s^{-1} for the merry–go–round *a–f–b–d*). Isomer A slowly converts to B until equilibrium is reached (k_1 = 2.3·10^{-7}, k_{-1} = 1.63·10^{-7} s^{-1} at 213 K). Isomer B is also fluxional (k_{213} = 670 s^{-1} for the change of basal face). The total concentrations of

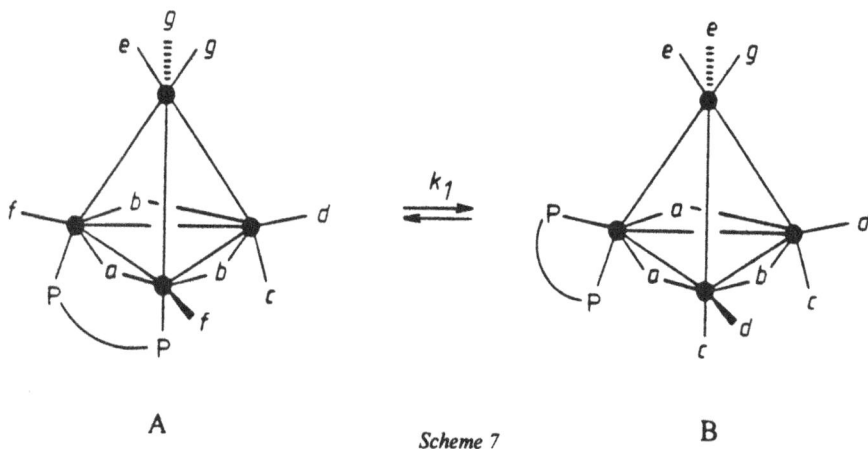

A   *Scheme 7*   B

Ir_4 and of bidentate ligand remain constant and no other species is observed during this isomerisation. One observes therefore a migration of a P–atom of the bidentate ligand to a site left vacant by the faster CO scrambling processes. As expected, the activation parameters of this isomerisation are much higher than those reported in Table 1 ($\Delta G^{\ddagger}_1$ = 79.9 and $\Delta G^{\ddagger}_{-1}$ = 82.4 kJ mol^{-1} at 298 K [48]).

More interesting is the new kinetic process encountered in the mixed cluster $[Ir_2Rh_2(CO)_{11}PPh_3]$ [49] (and in $[Ir_3Rh(CO)_{11}PPh_3]$ which behaves similarly [50]). The reaction of $Ir_2Rh_2(CO)_{12}$ with one mol–equiv. of PPh_3 yields $[Ir_2Rh_2(CO)_{11}PPh_3]$ as a mixture of two isomers with the phosphine ligand axially bound either to one basal Rh–atom in the kinetically preferred isomer A or to one basal Ir–atom in the thermodynamically preferred isomer B (*Scheme 8*). Both isomers are fluxional on the ^{13}C–NMR time scale at low temperature due to CO scrambling. Around room temperature, a new process starts to operate which is responsible for the isomerisation A = B, *i.e.* the migration of the PPh_3 ligand from one metal centre to another. No other species, in particular no free PPh_3, was detected in the equilibrium mixture of A and B.

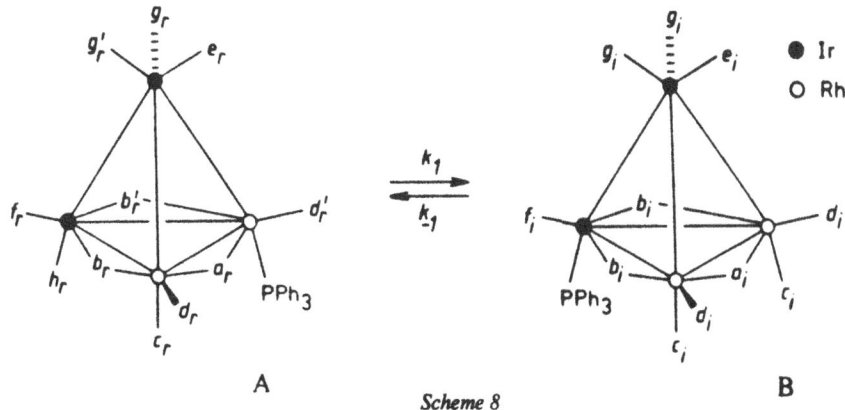

Scheme 8

Adding free PPh_3 to a solution of A caused a rapid substitution of CO giving $[Ir_2Rh_2(CO)_{10}(PPh_3)_2]$ whose ^{13}C–NMR signals are clearly distinguishable from those of A and B. The half-time of isomerisation A = B was found to be independent of the initial concentration of A and did not vary in presence of H_2O_2, which would selectively oxidise traces of free PPh_3 to $OPPh_3$. The ^{31}P–NMR signal of A is a doublet and the Rh–P coupling is maintained during the isomerisation. These observations indicate that the migration of PPh_3 between two metal centres is an intramolecular process. The isomerisation was further studied as function of pressure (1–1840 bar) and gave the following volumes: $\Delta V_0 (A \rightarrow B) = -3 \pm 2$, $\Delta V^{\ddagger}_1 = 10.1 \pm 1.5$, and $\Delta V^{\ddagger}_{-1} = 13.2 \pm 2.1 \text{ cm}^3 \text{ mol}^{-1}$. As expected, the reaction volume is close to zero. The activation volumes are equal within experimental error and positive, in agreement with the positive entropy values obtained from a line shape analysis of the variable–temperature ^{31}P–NMR spectra. The exchange of PPh_3 therefore takes place *via* breaking a M–P bond in the rate–determining step, the phosphine ligand then migrating within the solvent cage to an axial co-ordination site left vacant on an adjacent metal atom by the more rapid CO scrambling process. These results show that a phosphine ligand, which is reputedly inert in a homometallic compound, may be mobile in a heterometallic cluster compound at room temperature. They also suggest that the μ_2–bonding mode of PPh_3 between heterometallic centres is energetically accessible, although there is no structural evidence to date for a μ_2–PPh_3 ligand in a cluster compound.

The majority of the results described in this report have been obtained by Katya Besançon, Giacomo Bondietti, Chris Hall, Gábor Laurenczy, Tito Lumini, Alessandra Orlandi, Andrés Strawczynski and Gianfranco Suardi at the University of Lausanne, with financial support from the Swiss National Science Foundation.

172

5. References

1. Cotton, F.A. (1966), *Inorg. Chem.* **5**, 1083–1085.
2. Johnson, B.F.G. and Benfield, R.E. (1978), *J. Chem. Soc., Dalton Trans.*, 1554–1568.
3. Johnson, B.F.G. and Roberts, Y.V. (1994), *Inorg. Chim. Acta*, in press.
4. Van Eldik, R., Ed. (1986) *Inorganic High Pressure Chemistry: Kinetics and Mechanisms*, Elsevier, Amsterdam: Chapters 2–4 and references cited therein; Van Eldik, R., Asano, T. and le Noble, W.J. (1989), *Chem. Rev.* **89**, 549–688.
5. Merbach, A.E. (1987), *Pure Appl. Chem.* **59**, 161–172; Merbach, A.E. and Akitt, J.W. (1990), *NMR Basic Princ. Prog.* **24**, 189–232.
6. Krassouski, D.W., Nelson, J.H., Brower, K.R., Hamerstein, D. and Jacobson, R.A. (1988), *Inorg. Chem.* **27**, 4294–4307.
7. Bodenhausen, G., Kogler, H. and Ernst, R.R. (1984), *J. Magn. Reson.* **58**, 370–388.
8. Morris, G.A. and Freeman, R. (1978), *J. Magn. Reson.* **29**, 433–462.
9. Wei, C.H., Wilkes, G.R. and Dahl, L.F. (1967), *J. Am. Chem. Soc.*, **89**, 4792–4793; Wei, C.H. (1969), *Inorg. Chem.* **8**, 2384–2397; Carré, F.H. , Cotton, F.A. and Frenz, B.A. (1976), *Inorg. Chem.* **15**, 380–387.
10. Bor, G., Sbrignadello, G. and Noack, K. (1975), *Helv. Chim. Acta* **58**, 815–832.
11. Aime, S., Botta, M., Gobetto, R. and Hauson, B.E. (1989), *Inorg. Chem.* **28**, 1196–1198; Hauson, B.E. and Lisic, E.C. (1986), *Inorg. Chem.* **25**, 715–716.
12. Lucken, E.A.C., Noack, K. and Williams, D.F. (1967), *J. Chem. Soc. A*, 148–154; Haas, H. and Sheline, R.K. (1967), *J. Inorg. Nucl. Chem.* **29**, 693–698.
13. Evans, J., Johnson, B.F.G., Lewis, J. and Matheson, T.W. (1975), *J. Am. Chem. Soc.* **97**, 1245–1246; Cohen, M.A., Kidd, D.R. and Brown, T.L. (1975), *J. Am. Chem. Soc.* **97**, 4408–4409.
14. Evans, J., Johnson, B.F.G., Lewis, J., Matheson, T.W. and Norton, J.R. (1978), *J. Chem. Soc., Dalton Trans.*, 626–634.
15. Cotton, F.A., Kraczynski, L., Shapiro, B.L. and Johnson, L.F. (1972), *J. Am. Chem. Soc.* **94**, 6191–6193; Evans, J., Johnson, B.F.G., Lewis, J. and Norton, J.R. (1973), *J. Chem. Soc., Chem. Comm.*, 807–808.
16. Besançon, K. Lumini, T. and Roulet, R. (1993), *J. Organomet. Chem. Conf.'93*, Munich, november 4–5, Abstract, p. 75; ibid, unpublished results.
17. Braga, D. and Grepioni, F. (1987), *J. Organomet. Chem.* **336**, C9–C12.
18. Churchill, M.R. and Hutchinson, J.P. (1978), *Inorg. Chem.* **17**, 3528–3535.
19. Besançon, K. (1994), PhD thesis, UNIL, Lausanne.
20. Martinengo, S., Chini, P., Albano, V.G., Cariati, F. and Salvatori, T. (1973), *J. Organomet. Chem.* **59**, 379–394.
21. Albano, V.G., Ciani, G. and Martinengo, S. (1974), *J. Organomet. Chem.* **78**, 265–272.
22. Horváth, I. (1986), *Organometallics* **5**, 2333–2340.
23. Johnson, B.F.G., Lewis, J. and Matheson, T.W. (1974), *J. Chem. Soc., Chem. Comm.*, 441–442.
24. Bondietti, G., Ros, R., Roulet, R., Grepioni, F. and Braga, D. (1994), *J. Organomet. Chem.* **464**, C45–C48.

25. Bondietti, G., Suardi, G., Ros, R., Roulet, R., Grepioni, F. and Braga, D. (1993), *Helv. Chim. Acta* **76**, 2913–2925.

26. Suardi, G., Strawczynski, A., Ros, R., Grepioni, F. and Braga, D. (1990), *Helv. Chim. Acta* **73**, 154–160.

27. Orlandi, A., Frey, U., Suardi, G., Merbach, A.E. and Roulet, R. (1992), *Inorg. Chem.* **31**, 1304–1306.

28. Mann, B.E., Pickup, B.T. and Smith, A.K. (1989), *J. Chem. Soc., Dalton Trans.*, 889–893.

29. Stuntz, G.F. and Shapley, J.R. (1977), *J. Am. Chem. Soc.* **99**, 607–609.

30. Mann, B.E., Vargas, M.D. and Khadar, R. (1992), *J. Chem. Soc., Dalton Trans.*, 1725–1728.

31. Karel, K.J. and Norton, J.R. (1974), *J. Am. Chem. Soc.* **96**, 6812–6813; Stuntz, G.F. and Shapley, J.R. (1976), *Inorg. Chem.* **15**, 1994–1996.

32. Strawczynski, A., Suardi, G., Ros, R., and Roulet, R. (1993), *Helv. Chim. Acta* **76**, 2210–2226.

33. Besançon, K., Laurenczy, G., Lumini, T., Roulet, R. and Gervasio, G. (1993), *Helv. Chim. Acta* **76**, 2926–2935.

34. Stuntz, G.F. and Shapley, J.R. (1981), *J. Organomet. Chem.* **213**, 389–403.

35. Orlandi, A., Ros, R. and Roulet, R. (1991), *Helv. Chim. Acta* **74**, 1464–1470.

36. Bondietti, G., Ros, R., Roulet, R., Musso, F. and Gervasio, G. (1993), *Inorg. Chim. Acta* **213**, 301–309.

37. Strawczynski, A., Ros, R., and Roulet, R. (1988), *Helv. Chim. Acta* **71**, 867–871.

38. Churchill, M.R. and Hutchinson, J.P. (1980), *Inorg. Chem.* **19**, 2765–2769.

39. Shapley, J.R., Stuntz, G.F., Churchill, M.R. and Hutchinson, J.P. (1979), *J. Am. Chem. Soc.* **101**, 7425–7428.

40. Strawczynski, A., Ros, R., Roulet, R., Grepioni, F. and Braga, D. (1988), *Helv. Chim. Acta* **71**, 1885–1894.

41. Braga, D., Ros, R. and Roulet, R. (1985), *J. Organomet. Chem.* **286**, C8–C12.

42. Strawczynski, A., Hall, C., Bondietti, G., Ros, R. and Roulet, R. (1994), *Helv. Chim. Acta* **77**, 754–770.

43. Crowte, R.J., Evans, J. and Webster, M. (1984), *J. Chem. Soc., Chem. Comm.*, 1344–1345.

44. Malatesta, L. and Caglio, G. (1967), *Chem. Comm.*, 420–421.

45. Ciani, G., Manassero, M., Albano, V.G., Canziani, F., Giordano, G., Martinengo, S. and Chini, P. (1978), *J. Organomet. Chem.* **150**, C17–C19.

46. Davis, M.J. and Roulet, R. (1992), *Inorg. Chim. Acta* **197**, 15–20.

47. Ros, R., Scrivanti, A., Albano, V.G., Braga, D. and Garlaschelli, L. (1986), *J. Chem. Soc., Dalton Trans.*, 2411–2421.

48. Laurenczy, G., Ros. R. and Roulet, R., unpublished results.

49. Laurenczy, G., Bondietti, G., Merbach, A.E., Moullet, B. and Roulet, R. (1994), *Helv. Chim. Acta* **77**, 547–553.

50. Bondietti, G., Laurenczy, G., Ros, R., and Roulet, R. (1994), *Helv. Chim. Acta* **77**, in press.

STATIC AND DYNAMIC STEREOCHEMISTRY OF THE ORGANO-METALLIC CLUSTER COMPLEXES [(CpCo)$_3$(μ_3-ARENE)]

H. WADEPOHL

Anorganisch-Chemisches Institut
der Ruprecht-Karls-Universität
Im Neuenheimer Feld 270
D-69120 Heidelberg, Germany

Dedicated to the memory of Earl L. Muetterties

1. Introduction

Nearly twenty years ago, and based on even earlier propositions [1], Earl L. Muetterties stated that "discrete, molecular metal clusters may be reasonable models of metal surfaces in the processes of chemisorption and catalysis" [2]. This "cluster surface analogy" appeared immediately obvious when the first molecular cluster complexes with face capping benzene ligands, [Os$_3$(CO)$_9$(μ_3-C$_6$H$_6$)] **1** and [Ru$_6$C-(CO)$_{11}$(η^6-C$_6$H$_6$)(μ_3-C$_6$H$_6$)] **2**, were reported in 1985 [3]. In **1** and **2**, a benzene ligand is coordinated in the μ_3-η^2:η^2:η^2-bonding mode to a triangular array of metal atoms, very much like the so-called hollow [C$_{3v}$ (σ_d)] adsorption site of benzene on a close packed metal surface [4]. In one of the best studied systems, Rh(111)/C$_6$H$_6$/CO, benzene is adsorbed in such a hollow site and kept in place by the coadsorbed CO molecules. For this system, a strong Kekulé-type distortion of the C$_6$ ring was derived from an electron diffraction study, with alternating short (1.33(15) Å, "on top" of a metal atom) and long (1.81(15) Å, "between" two metal atoms) carbon carbon bonds [5]. In contrast, photoelectron spectroscopy pointed to an essentially sixfold symmetry for the adsorbed benzene, and a large distortion was ruled out [6]. This dis-

L. J. Farrugia (ed.), The Synergy Between Dynamics and Reactivity at Clusters and Surfaces, 175–191.
© 1995 Kluwer Academic Publishers.

crepancy still remains unsolved, although an interesting suggestion involving frustrated translations of the benzene parallel to the surface was recently made [7].

Both molecular cluster complexes **1** and **2** were structurally characterised in the solid state by X-ray structure analyses. An apparent carbon carbon bond length alternation was noted within the μ_3-benzene ligands [3]. However, the high standard deviations precluded a detailed analysis.

For some years we have been using cyclopentadienylcobalt fragments [CpCo] to construct hydrocarbon bridged oligonuclear complexes in a rational way [8]. Somewhat fortuitously, but not quite by accident, we came across a reaction which allowed us the synthesis of a large number of cluster complexes [(CpCo)$_3$(μ_3-alkenylarene)] **3** in one-pot reactions. Subsequently we set out to investigate the class of μ-arene metal cluster complexes more rigorously than had been possible before. Our synthesis gave access to μ_3-arene cluster complexes with a variety of substituted arene ligands. This was a prerequisite for the dynamic and equilibrium studies described below, all of which depended on the availability of quite specific derivatives. Cobalt as a relatively light atom also allowed us to obtain the geometric structures of our clusters in the solid state with much more precision from X-ray crystallography than has been possible in the case of the Os and Ru clusters.

2. Synthesis

The tricobalt cluster complexes [(CpCo)$_3$(μ_3-alkenylbenzene)] are obtained when alkenylbenzenes are treated with reactive sources of the CpCo complex fragment [8-10]. The Jonas reagent [CpCo(C$_2$H$_4$)$_2$] and [CpCo(C$_6$Me$_6$)] are particularly suited for this job. In these reactions, the trimetal clusters are assembled on the arene; the carbon carbon double bond in the side chain plays an essential role, serving as a "landing strip" for the first CpCo fragment [11]. Allylarenes and other phenyl substituted olefins, which have the double bond separated from the phenyl group by a number of methylene groups, can also be used. In these arenes, the double bond of the side chain is catalytically shifted into the α position to the arene ring in a pre-equilibrium [10].

3

Very significantly, benzene and alkyl substituted arenes do not react. Some $(CpCo)_3$ derivatives **4** with μ_3-alkylbenzene ligands can however be obtained via an indirect route which involves the synthesis of a $[(CpCo)_3(\mu_3$-alkenylarene)]$ **3** precursor followed by catalytic hydrogenation of the olefinic side chain. This very nicely shows that the carbon carbon bond in the side chain, although crucial for the formation of the metal cluster, does not contribute to the stability of the cluster complex and can be done away with easily without destruction of the $[(CpCo)_3(\mu_3$-arene)]$ unit.

3   **4**

It should be noted here that assembly of a tricobalt cluster at the aromatic ring is only one of several possible reactions which an alkenyl substituted arene may undergo with CpCo fragments [8]. Most of these involve the formation of trinuclear clusters. One particular side reaction, which involves CH-activation of the olefin, can even become dominant under certain reaction conditions. Here complexes of the type $[(\mu_3$-H)$(\mu_2$-H)$(CpCo)_3(\mu_3$-η^2:η^2:η^2-alkyne)]$ are formed; the hydrogen atoms, which were cleaved off the olefin, are retained in the products as bridging hydrido ligands [12]. CH-activation can also occur at the arene ring, *ortho* to the alkenyl substituent [13].

3. Solid State Structures

A number of derivatives of **3** and **4** have been studied by X-ray crystallography (Table 1) [9, 10]. In all the cases the μ_3-η^2:η^2:η^2 bonding mode of the arene to the Co$_3$ unit was found. Generally Co$_3$ triangle and arene hexagon are parallel to each other and adopt a staggered orientation (Figure 1). The substituents on the μ_3-arenes [i.e. alkenyl or alkyl carbons and hydrogens (when located in the diffraction study)] are bent away from the metal cluster and are displaced by 0.4-0.5 Å (C) resp. 0.2-0.3 Å (H) from the best plane of the μ_3-arene C$_6$ ring (Figure 1).

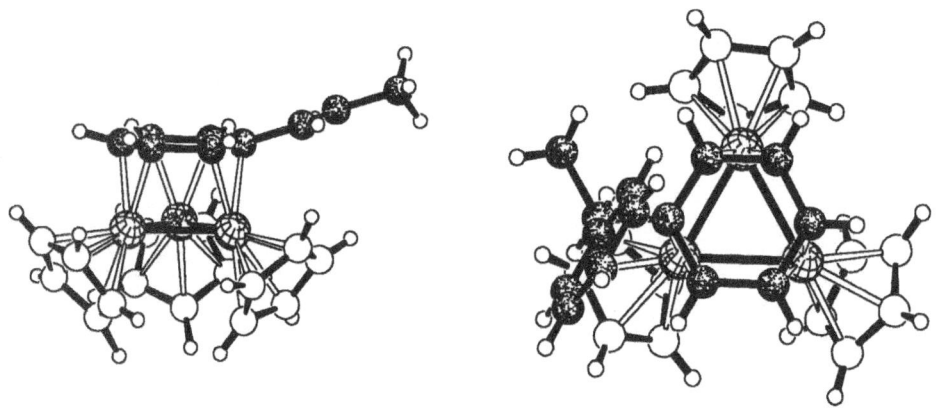

Figure 1. Graphical representations of the molecules **3a** (left) and **4a** (right).

There is some degree of alternation of the carbon carbon bond lengths in the μ_3-coordinated phenyl rings. If the 18 individual carbon carbon bonds, which are situated on top of a metal atom, and the 18 such bonds between two metal atoms in the six structures are considered to be independent observations (rigorously they are not), the weighted means given in Table 1 are statistically significantly different. However, the substituent on the arene induces asymmetry of the molecules, and therefore no two bonds are in the same environment.

TABLE 1. Carbon carbon bond lengths $d(CC)$ in some μ_3-η^2:η^2:η^2-arenes

	$d_{mean}(CC)^a$		$\sigma(d)^b$
[(CpCo)$_3$(μ_3-β-methylstyrene)] **3a**	1.446(14)	1.420(9)	0.005
[(CpCo)$_3$(μ_3-2-phenyl-2-butene)] **3b**	1.46(2)	1.41(2)	0.01
[(CpCo)$_3$(μ_3-1,1-diphenylethene)] **3c**	1.448(3)	1.416(6)	0.003
[(CpCo)$_3$(μ_3-1,1-diphenylethane)] **4a**	1.439(6)	1.414(8)	0.006
[(μ_3-H)(CpCo)$_3$(μ_3-1,1-diphenylethane)]$^+$ **5**	1.46(1)	1.41(1)	0.02
[(CpCo)$_3$(μ_3-1,2-diphenylethane)] **4b**	1.442(9)	1.417(7)	0.005
weighted mean of the above	1.447(3)	1.417(2)	

[a] in Å with standard deviation of the mean in parentheses; [b] standard deviation of the individual values in Å

The effect of a substituent on endocyclic carbon carbon bond lengths has been noted for the free arenes, but is difficult to quantify because of the effects of librational motion of the molecules in the crystals [14]. In some of our cluster complexes there are bulky substituents on C-α. Steric hindrance between those and the nearest Cp ring has an additional effect. Therefore, in some cases minor deviations from planarity of the μ_3-arene (the *ipso* carbon atom is slightly shifted away from the metal cluster) and even from the staggered orientation of Co$_3$ and C$_6$ rings are observed. This is illustrated in Figure 2 for **4a** and **5**.

5 is the protonation product of **4a**. The μ_3-hydrido ligand in **5** caps the Co$_3$ cluster face which is not occupied by the μ_3-arene. It has a stereochemical influence on the Cp ligands. These are pushed away from the hydride in the direction of the μ_3-arene, thereby restricting the space for the substituents on C-α.

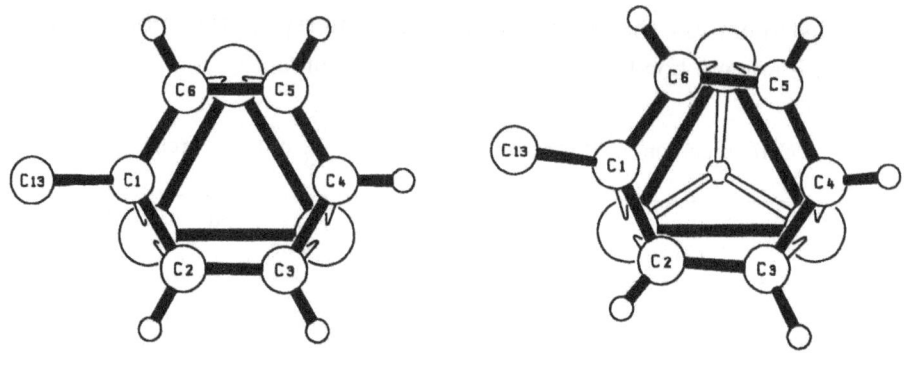

Figure 2. Views of the $Co_3(\mu_3$-arene) units in **4a** (left) and **5** (right). Cp-rings and substituents on the arenes (except C-α) are omitted for clarity.

The distribution of the individual endocyclic carbon carbon bond lengths in some derivatives is shown in Figure 3. If the experimental errors are taken into account there is considerable overlap of the regions of the "short" and "long" bonds.

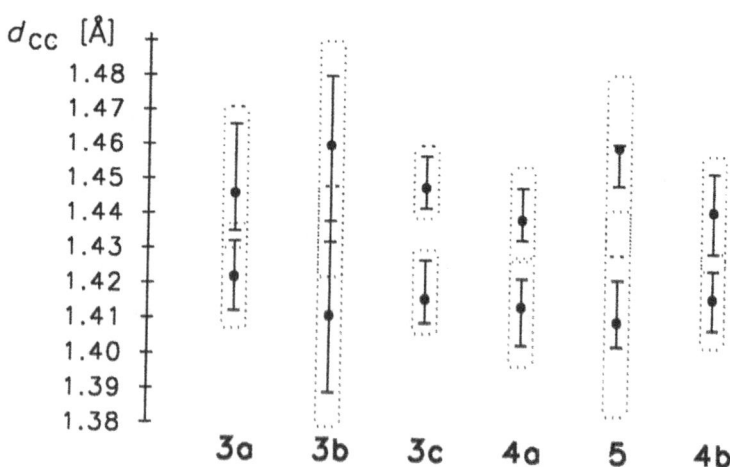

Figure 3. Distribution of the alternating short and long carbon carbon bonds in the μ_3-arene ligands of some cluster complexes $[(CpCo)_3(\mu_3\text{-}\eta^2\text{:}\eta^2\text{:}\eta^2\text{-arene})]$. The arithmetic means are indicated by filled circles, standard deviations by dotted lines.

4. Electronic Structure and the Reason for Distortion of the μ_3-Arenes

Molecular orbital calculations were carried out on the model complex $[(CpCo)_3(\mu_3\text{-}\eta^2\text{:}\eta^2\text{:}\eta^2\text{-}C_6H_6)]$ with the parameter-free but approximate Fenske-Hall SCF method [15]. The obvious partitioning of the molecule into a $(CpCo)_3$ cluster and the benzene ligand facilitates comparisons with both mononuclear complexes and benzene adsorbed on metal surfaces and slabs.

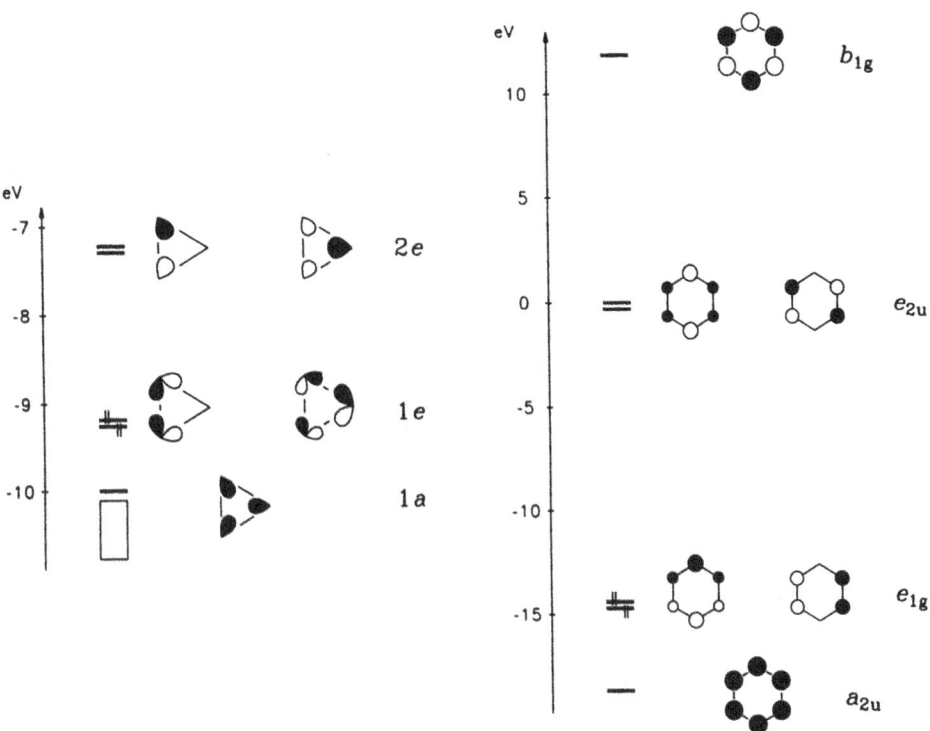

Figure 4. Frontier orbitals of $(CpCo)_3$ (left) and C_6H_6 (right). The orbitals are labelled according to their symmetry in the pointgroups C_3 [$(CpCo)_3$] and D_{6h} (C_6H_6). Note different energy scale for $(CpCo)_3$ and C_6H_6.

The frontier orbitals of $[(CpCo)_3]$ and C_6H_6 are depicted in Figure 4. The main bonding interactions in the complete molecule amounts to donation from the filled e_{1g} orbitals of benzene into the virtual $(CpCo)_3$ $2e$ set, accompanied by back donation from $(CpCo)_3$ $1e$ into benzene e_{2u}. However, under the C_3 symmetry of the cluster complex, both

HOMO (e_{1g}) and LUMO (e_{2u}) of benzene belong to the same irreducible representation (e), and therefore mixing becomes allowed. When in phase with the metal orbitals, this mixing is sufficient to cause a Kekulé distortion of the benzene ligand, as illustrated in Figure 5 for the combinations of the metal cluster $2e$ and benzene e_{1g} and e_{2u} orbitals. π-overlap is enhanced for carbon carbon bonds on top of a metal atom, while it is reduced for those bonds between two metal atoms.

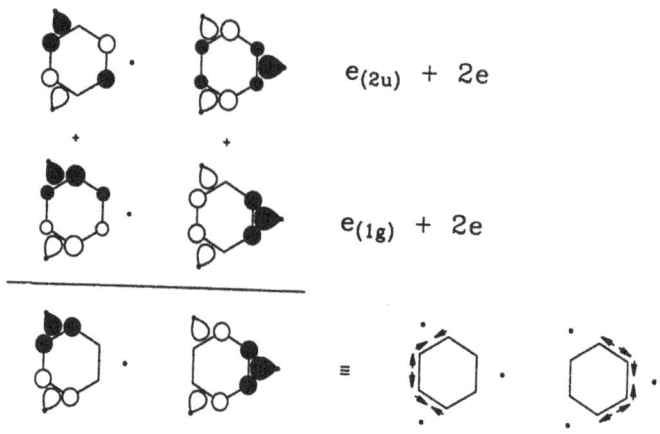

$$e_{(2u)} + 2e$$

$$e_{(1g)} + 2e$$

Figure 5. Schematic visualisation of how the mixing of benzene HOMO and LUMO can lead to a trigonal distortion. Only the combinations with the (CpCo)$_3$ $2e$ orbitals are shown.

There is a strong parallel between $[(CpCo)_3(\eta^2{:}\eta^2{:}\eta^2{-}C_6H_6)]$ and the mononuclear complex $[(CO)_3Cr(\eta^6{-}C_6H_6)]$. The frontier orbitals of $(CO)_3Cr$ are similar in shape and energy to the frontier orbitals of $(CpCo)_3$. Even if we take into account that in the latter fragment the orbitals are centered on three metal atoms in contrast to one in $(CO)_3Cr$, an isolobal relationship exists. It is therefore not surprising, that a similar small Kekulé distortion of the benzene ligand is observed in the chromium complex. This has been explained with an argument roughly equivalent to the one given here [16].

The experimentally observed bending of the benzene CH-bonds away from the metal cluster can also be satisfactorily explained by an increase of inter-fragment overlap populations. For geometrical reasons this distortion from planarity occurs in the opposite direction in the mononuclear $[(CO)_3Cr(\eta^6-C_6H_6)]$.

5. Dynamic Behaviour in Solution

As can be seen from Figure 1, the cluster complexes $[(CpCo)_3(\mu_3-\eta^2:\eta^2:\eta^2\text{-arene})]$ with monosubstituted arene ligands are asymmetric. Therefore, the three CpCo groups are inequivalent and should give rise to three separate resonances in the ^1H and ^{13}C NMR spectra. In solution three Cp resonances are however only observed at low temperature. On rising the temperature, the sharp singlets coalesce, and at room temperature and above only one resonance is usually found.

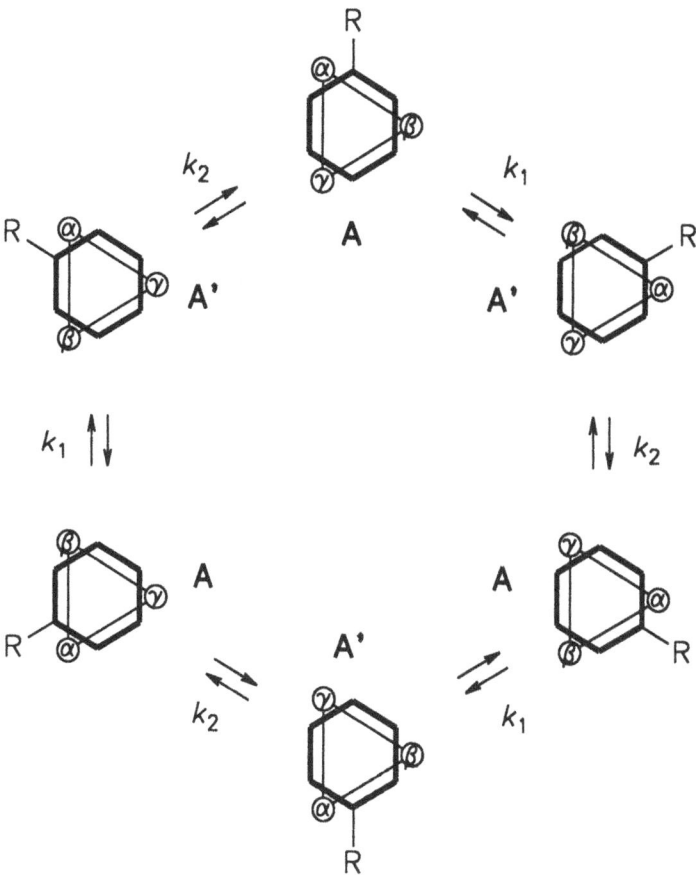

Scheme 1.

This behaviour can be explained by a hindered rotation of the Co_3 and C_6 rings relative to one another. If we for the moment assume as the elementary step successive [1,2] shifts of the cobalt atoms at the μ_3-arene (i.e. 60° rotation of the arene above the Co_3 cluster), six distinct chemical configurations are involved (Scheme 1).

Of the six molecules in Scheme 1, only two (**A, A'**) are geometrically different. **A** and **A'** are mirror images. At low temperature, interconversion of the enantiomers is slow on the NMR timescale. Since in isotropic solution NMR spectroscopy cannot distinguish between enantiomers (**A** and **A'** have identical spectra), only three CpCo resonances are observed for the racemic mixture in the slow exchange regime.

The mechanism of the arene rotation in Scheme 1 generates a complicated exchange matrix for the CpCo groups. With each elementary step only two of the three different chemical shifts α, β, γ are exchanged with one another (no simple cyclic permutation). This is shown in Scheme 2.

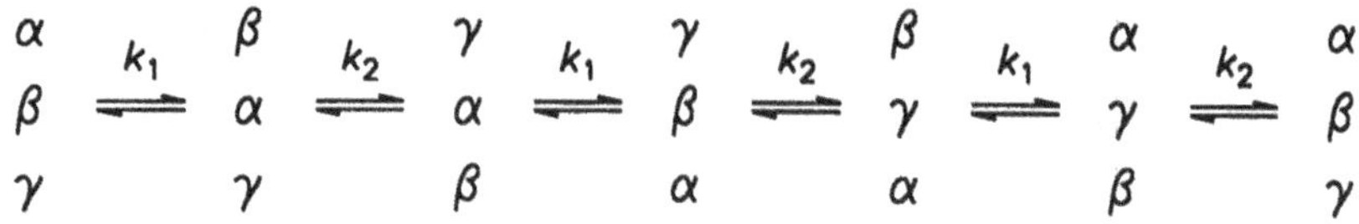

$$
\begin{matrix}
\alpha \\ \beta \\ \gamma
\end{matrix}
\underset{}{\overset{k_1}{\rightleftharpoons}}
\begin{matrix}
\beta \\ \alpha \\ \gamma
\end{matrix}
\underset{}{\overset{k_2}{\rightleftharpoons}}
\begin{matrix}
\gamma \\ \alpha \\ \beta
\end{matrix}
\underset{}{\overset{k_1}{\rightleftharpoons}}
\begin{matrix}
\gamma \\ \beta \\ \alpha
\end{matrix}
\underset{}{\overset{k_2}{\rightleftharpoons}}
\begin{matrix}
\beta \\ \gamma \\ \alpha
\end{matrix}
\underset{}{\overset{k_1}{\rightleftharpoons}}
\begin{matrix}
\alpha \\ \gamma \\ \beta
\end{matrix}
\underset{}{\overset{k_2}{\rightleftharpoons}}
\begin{matrix}
\alpha \\ \beta \\ \gamma
\end{matrix}
$$

Scheme 2.

It is improtant to note that whereas there is mutual interchange of α and β and β and γ, respectively, α and γ are never directly (i.e. in one elementary step) interconverted. Furthermore, two different rotation barriers have to be considered *a priori*, depending on whether the substituent moves between two CpCo groups (rate constant k_1) or across one such group (k_2).

Two dimensional ${}^1H$ EXSY NMR spectra directly prove this exchange mechanism. With a short mixing time in the EXSY experiment, only the quicker exchange process is observed (corresponding to k_1). In the 2D matrix cross signals occur between only two of the three Cp resonances

(Figure 6a). With longer mixing times the slower process also becomes apparent; of the three possible pairwise permutations of the chemical shifts ($\hat{=}$ cross peaks) only two are actually observed (Figure 6b), in complete agreement with Scheme 1.

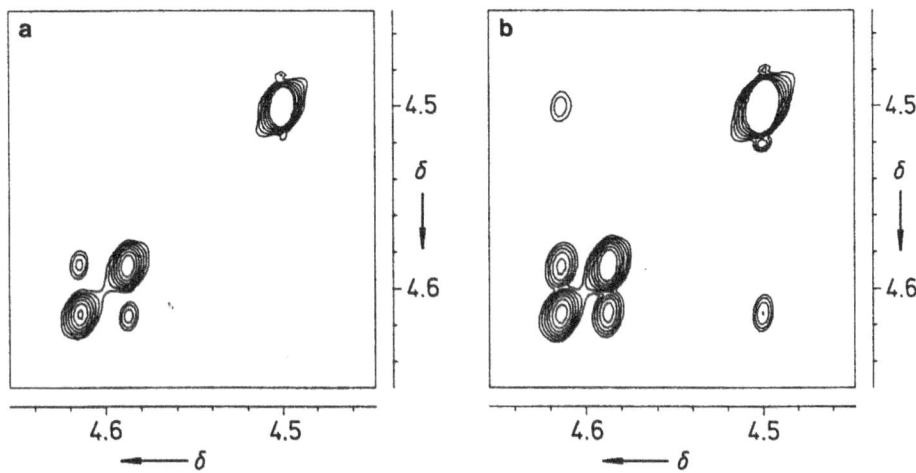

Figure 6. Section of the 200 MHz ^{1}H EXSY spectra of [(CpCo)$_3$(μ_3-β-methyl-styrene)] **3a** at 230 K (region of the Cp resonance signals). Mixing times τ: 60 ms (a) and 150 ms (b); echo detection.

If R in Scheme 1 is made prochiral, diastereotopy is observed in the case of slow arene rotation. An example of such a molecule is [(CpCo)$_3$(μ_3-*iso*-propylbenzene)] **4c** (R[1]= CH$_3$, R[2]= R[3]= H). At low temperature, there are three resonances for the CpCo groups in the ^{1}H and ^{13}C NMR spectra as usual. The methyl groups of the substituent appear as two distinct resonances which coalesce on rising the temperature.

Any one of the two rate processes already suffices to make the diastereotopic groups equivalent on the NMR timescale. This becomes obvious when a series of 2D EXSY spectra is recorded, each employing a different mixing time τ (the poor man's 3D spectrum). Exchange is detected between the methyl resonances as soon as two (and only two) of the three Cp signals interchange (Figure 7).

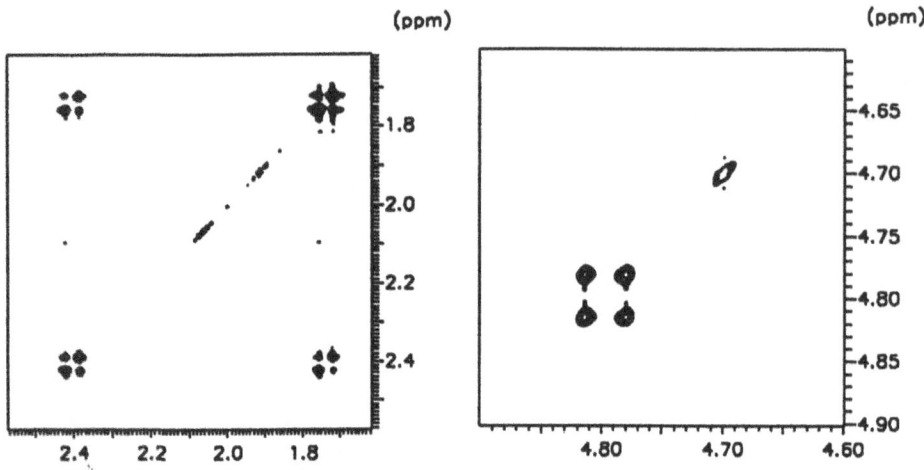

Figure 7. 200 MHz ^{1}H EXSY spectrum of [(CpCo)$_3$(μ_3-*iso*-propylbenzene)] **4c** at 250 K (τ= 150 ms). Left: CH$_3$ region; right: Cp region. Signals with 1.85 < δ < 2.2 are folded. The spectra were taken in the phase sensitive mode, all peaks are negative.

This situation corresponds to a 60° oscillation of the C$_6$ and Co$_3$ rings relative to each other, with the associated rate constant k_1 (Scheme 1). For a wide range of mixing times, the general features of the 2D EXSY spectra of **4c** remain the same, apart from a gain in intensity of the cross peaks when τ is increased.

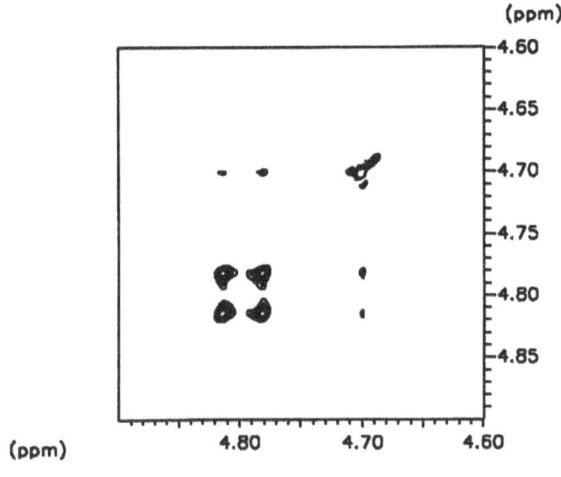

Figure 8. 200 MHz ^{1}H EXSY spectrum of [(CpCo)$_3$(μ_3-*iso*-propylbenzene)] **4c** (Cp region) at 250 K (τ= 850 ms). Other parameters were as in Figure 7.

A further increase of τ does not eventually generate a spectrum like the one in Figure 6b; instead cross peaks between all the Cp resonances appear (Figure 8). This can be explained by multi-step processes and is an indication of a large difference of the two potential energy barriers associated with k_1 and k_2.

When the symmetry of the face capping arene is lowered, chiral diastereomers may be obtained. This is illustrated in Scheme 3 and 4 for an unsymmetrically *meta*-disubstituted arene, and an arene with an asymmetric substituent, respectively. In both cases, the arene rotation interconverts the diastereomers but not the enantiomers. The latter process could only be accomplished when the tricobalt cluster is made to move from one to the opposite face of the μ_3-arene, a process which has never been observed so far.

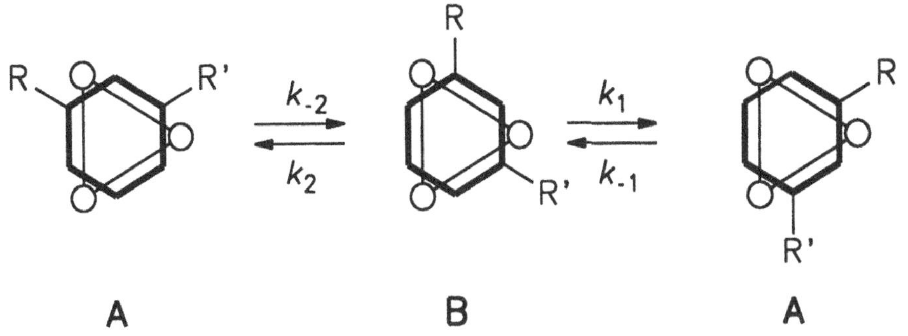

Scheme 3.

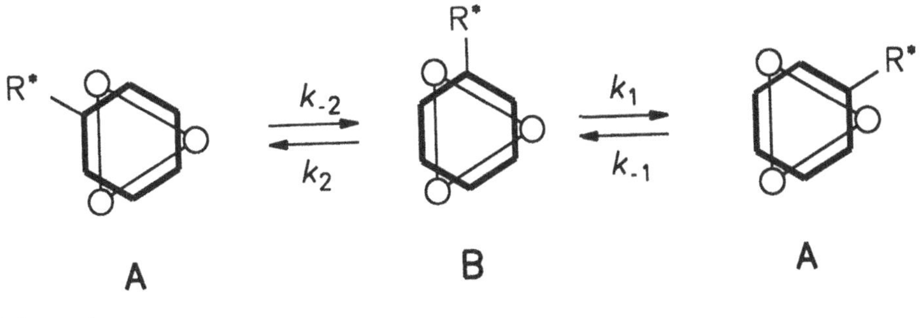

Scheme 4.

The two diastereomeric rotamers of $[(CpCo)_3(\mu_3$-*m*-methyl-β-methyl-styrene$)]$ **3d** (R^1= H, R^2= CH_3, R^3= *m*-CH_3) can be detected simultaneously at -90 °C in solution by NMR spectroscopy, where a total of six (two sets of three) singlets are observed for the Cp resonances). However, only three Cp resonances are present in the spectra of $[(CpCo)_3(\mu_3\text{-}1,1\text{-diphenylethane})]$ **4a** ($R^1 = C_6H_5$, $R^2 = R^3 = H$) at low temperature. In this case the difference in free enthalpy (ΔG^0) of the two diastereomers **A**, **B** is large enough to leave only one isomer popu-lated at low temperature. From the analysis of the lineshape of the methyl resonance, which is sharp at high and low temperature but broadens in between, the presence of an equilibrium between the two species can be deduced unambigously.

The quantitative determination of the energy barriers for the arene ro-tation is not trivial. The temperature dependent one-dimensional NMR spectra contain too little information for a good lineshape analysis, even more so because the Cp resonances can only indirectly be assigned to the three CpCo sites, and because of the presence of two barriers of different height. There are less problems with the methyl resonances in the *iso*-propyl derivative **4c**. The effective time constant for the exchange of the diastereotopic methyl sites in this complex is the weighted mean of k_1 and k_2, which should be close to k_1 (see above). Therefore we can get an approximation to the smaller barrier.

In order to obtain more precise results, the chemical system had to be made more simple, whereas at the same time the spectroscopic (spin-) system had to be made more complex. The first goal is met when a symmetrically disubstituted arene is used as the bridging ligand (Scheme 5).

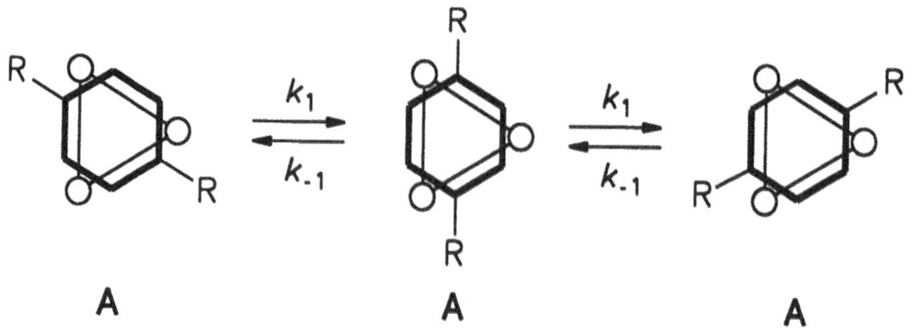

Scheme 5.

This introduces additional symmetry into the cluster complex. As a result the exchange process is degenerate (all participating molecules are the same), and there can only be one barrier.

The cluster complex [(CpCo)$_3$(μ_3-p-distyrylbenzene)] 3e (R^1= H, R^2= phenyl, R^3= p-styryl) is such a molecule. However, its ^{1}H NMR spectrumis quite simple, and therefore kinetic data cannot be obtained with high precision.

3e 4d

Both of the above conditions are met in [(CpCo)$_3$(μ_3-p-diethylbenzene)] 4d (R^1= H, R^2= CH$_3$, R^3= p-C$_2$H$_5$). The two methylene protons of each prochiral ethyl group are diastereotopic in the slow exchange regime, forming a nice exchanging spin system with differential lineshape changes as a function of temperature. Fairly precise activation parameters ($\Delta H^{\ddagger}$ and $\Delta S^{\ddagger}$) could be obtained from the iterative lineshape analysis of this system.

A summary of kinetic and thermodynamic data is given in Table 2. Both $\Delta G^{\ddagger}$ (free enthalpy of activation) and ΔG^0 (difference of the free enthalpies of the diastereomeric rotamers, if any) are characteristic for the energetics of the arene rotation process. With some caution (since the dataset is still relatively small) the cluster complexes in Table 2 can be grouped in two sets. The first set is characterised by smaller values of $\Delta G^{\ddagger}$ (around 50 kJ/mol, kinetics) and ΔG^0 (thermodynamics). In the second set both parameters are considerably higher.

TABLE 2. Kinetic [$\Delta G^{\ddagger}$ (in kJ/mol)] and thermodynamic [ΔG^0 (in kJ/mol)] data relating to the rotation of the arene ligands in [$(CpCo)_3(\mu_3\text{-}\eta^2{:}\eta^2{:}\eta^2\text{-}\{R^1R^2C_6H_4\})$]

	R^1	R^2	T [K]	$\Delta G^{\ddagger}$	ΔG^0
4d	p-C_2H_5	C_2H_5	250	49(1)	
			290	48(1)	
3e	p-CH=CHPh	CH=CHPh	250	50(1)	
4c	H	iso-C_3H_7	290	57(1)	
3d	m-CH_3	CH=CHCH$_3$	200		< 1
4a	H	C(H)(CH$_3$)Ph	250		> 9

The main stereochemical difference between the two sets is the degree of substitution on the α-carbon atom of the side chain on the μ_3-arene rings. Invariably a highly substituted C-α seems to go along with a much higher barrier for the rotation of the arene and, in the appropriate cases, a higher energy for the less favoured isomer. If we take into account our observations concerning the relative height of the two barriers in the isopropyl derivative 4c (see above), we can even conclude that μ_3-arene rotation is mainly governed by the steric hindrance caused by the bulky substituent passing "across" a CpCo group.

6. Acknowledgements

I am indebted to my coworkers and collaborators on this research. Their names appear in the list of references. Support by the Deutsche Forschungsgemeinschaft, the Fonds der chemischen Industrie and the Studienstiftung der Riedel-de Haen AG is gratefully acknowledged. International exchange of research students was made possible by the ARC and Vigoni programs, administered by the German Academic Exchange Service. Special thanks are due to the Deutsche Forschungsgemeinschaft for a Heisenberg Fellowship.

7. References

1. a) R.L. Burwell, J.B. Peri, *Ann. Rev. Phys. Chem.* **1964**, *15*, 131; b) P. Chini, *Inorg. Chim. Acta. Rev.* **1968**, *2*, 31; *J. Organomet. Chem.* **1968**, *200*, 37.

2. c) E.L. Muetterties, *Science (Washington)* **1977**, *196*, 839; E.L. Muetterties, T.N. Rhodin, E. Band, C.F. Brucker, W.R. Pretzer, *Chem. Rev.*, **1979**, *79*, 91 and references cited therein.

3. M.P. Gomez-Sal, B.F.G. Johnson, J. Lewis, P.R. Raithby, A.H. Wright, *J. Chem. Soc. Chem. Commun.* **1985**, 1682.

4. For recent reviews see: F.P. Netzer, *Langmuir* **1991**, 7, 2544; F.P. Netzer, *Crit. Rev. Solid State Mat. Sci.* **1992**, *17*, 397.

5. M.A. van Hove, R.F. Lin, G.A. Somorjai, *J. Am. Chem. Soc.* **1986**, *108*, 2532.

6. E. Bertel, G. Rosina, F.P. Netzer, *Surf. Sci.* **1986**, *172*, L515.

7. G. Witte, H. Range, J.P. Toennies, Ch. Wöll, *Phys. Rev. Lett.* **1993**, *71*, 1063.

8. H. Wadepohl, *Comments Inorg. Chem.* **1994**, *15*, 369 and references cited therein.

9. H. Wadepohl, *Angew. Chem.* **1992**, *104*, 253 and references cited therein.

10. H. Wadepohl, K. Büchner, H. Pritzkow, *Angew. Chem.* **1987**, *99*, 1294; H. Wadepohl, K. Büchner, M. Herrmann, H. Pritzkow, *Organometallics* **1991**, *10*, 861;

11. H. Wadepohl, K. Büchner, H. Pritzkow, *Organometallics* **1989**, *8*, 2745.

12. H. Wadepohl, T. Borchert, *Chem. Ber.* **1993**, *126*, 1615.

13. T. Borchert, K. Büchner, H. Pritzkow, H. Wadepohl, *XVI. Internatl. Conference on Organometallic Chemistry*, paper P70, Brighton, UK., **1994.**.

14. A. Domenicano, A. Vaciago, C.A. Coulson, *Acta. Cryst.* **1975**, *B31*, 221.

15. H. Wadepohl, L. Zhu, *J. Organomet. Chem.* **1989**, *376*, 115.

16. J.W. Chinn, Jr., M.B. Hall, *J. Am. Chem. Soc.* **1983**, *105*, 4930 and references cited therein.

NMR STUDIES ON THE DYNAMIC BEHAVIOUR IN SOLUTION OF RHENIUM-PLATINUM MIXED METAL CLUSTERS CONTAINING P-DONOR LIGANDS.

T. Beringhelli[a], G.D. Alfonso[a], A.P. Minoja[a] and R. Mynott[b]

[a] *Dipartimento di Chimica Inorganica, Metallorganica e Analitica, Università degli Studi dui Milano, Via Venezian 21, I-20133 Milano, Italy.*
[b] *Max-Plank Institut fur Kohlenforshung, Kaiser-Wilhelm-Platz 1, D-4330 Mulheim a.d. Ruhr, Germany.*

In the last few years we have been interested in the synthesis of Re-Pt mixed metal clusters as models or precursors of catalytic species and, following the approach proposed by Stone and coworkers, species of different nuclearity have been prepared [1-4]. These compounds show a manifold dynamic behaviour that comprises (i) ligand dynamics among different sites of the metal cluster; (ii) metal fragment mobility; (iii) metal frame rearrangements and (iv) metal fragment lability.

We have previously reported the dynamic behaviour in solution of the isomers of $[Re_2Pt(\mu-H)_2(PPh_3)_2(CO)_8]$ [5]. On treating a solution of $[Re_2(\mu-H)_2(CO)_8]$ with $[Pt(PPh_3)_2(C_2H_4)]$ at 273 K [1], a triangular cluster is obtained with the two phosphines still bound to Pt (**1a**) and the two hydrides bridging a Re-Pt and the Re-Re interaction respectively. This species exhibits two mutual exchange processes: the first equalizes the phosphines and the second, observed above 230 K, exchanges the two hydrides. At T>273 K an irreversible isomerization occurs, with the exchange of a carbonyl and a phosphine between a rhenium and the platinum atom, affording an equilibrium mixture of two interconverting species (**2a** and **3a**), where one phosphine is still bound to Pt and the other one is differently bound to one of the rhenium atoms. These observations prompted us to investigate the effects on the dynamic behaviour of the electronic and steric properties of the phosphine ligands and we have studied through multinuclear NMR spectroscopy the series of derivatives of general formula $[Re_2Pt(\mu-H)_2(CO)_8L_2]$, where L = PMe_2Ph, $P(OMe)_3$, $L_2 = (Ph_3)_2P(CH_2)_nP(Ph_3)_2$ (n = 1, dppm; 2, dppe; 3, dppp), and $Me_2P(CH_2)_2PMe_2$ (dmpe).

All these new products were prepared using as a precursor $[Re_2Pt(\mu-H)_2(CO)_8(COD)]$ (COD = 1,5-C_8H_{12}), a triangular cluster obtained by the reaction of $[Re_2(\mu-H)_2(CO)_8]$ and $[Pt(COD)_2]$, [2]. (Scheme I). The ^1H and ^{31}P NMR spectra at

193

L. J. Farrugia (ed.), The Synergy Between Dynamics and Reactivity at Clusters and Surfaces, 193–202.
© 1995 *Kluwer Academic Publishers.*

183 K of all these species are very similar proving their common structure. Tables 1 and 2 report their relevant chemical shifts and couplings.

On raising the temperature, all these compounds exhibit the same two types of fluxionality observed in the PPh$_3$ derivative. Indeed in some cases the low temperature process that equalizes the phosphines on the platinum atom is so fast that a limiting slow-exchange spectrum cannot be obtained even at the lowest temperature achievable at relatively medium-high field, 7 T. When possible, the ^{31}P{^{1}H}, ^{1}H{^{31}P} and ^{1}H variable temperature spectra of these species have been simulated in order to obtain the rate constants and the activation parameters of the observed processes and the overall results (Tables 3 and 4) allow the following observations.

TABLE 1. ^{1}H NMR parameters at 193 K (chemical shifts in ppm and coupling constants in Hz)

Ligand	$\delta(H_a)$	$J(H_a\text{-}P_b)$	$J(H_a\text{-}P_b)$	$J(H_a\text{-}Pt)$	$\delta(H_b)$
PPh$_3$	-9.22	-12.8	82.3	502	-15.62
PMe$_2$Ph	-10.15	-9.7	77.7	511	-15.67
P(OMe)$_3$	-8.80	115*		533	-16.32
dppm	-8.29	84*		593	-16.02
dppe	-7.73	-6	81	588	-15.88
dppp	-8.75	-9	81	522	-15.9
dmpe	-9.49	69*		541	-16.27

* The exchange rate did not allow the separate measure of the two coupling constants. The reported value is the absolute value of the difference of the coupling constants

TABLE 2. ^{31}P NMR parameters at 193 K (chemical shifts in ppm and coupling constants in Hz)

δP^*	Ligand	$\delta(P_a)$	$J(P_a\text{-}Pt)$	$J(P_a\text{-}P_b)$	$\delta(P_b)$	$J(P_b\text{-}Pt)$
-5.8	PPh$_3$	25.3	2167	16.5	15.2	4142
-44.8	PMe$_2$Ph	3.29	2194	18.3	-16.22	3890
140.6	P(OMe)$_3$	161.0	3480	n.d.	126.8	6266
-23.6	dppm	-15.4	1625	n.d.	-38.07	3386
-12.5	dppe	68.00	2116	5	48.07	3710
-17.3	dppp	8.28	2056	29	-0.46	3795
-49.4	dmpe	52.30	2065	n.d.	24.69	3467

* The first column reports the chemical shift of the free ligand

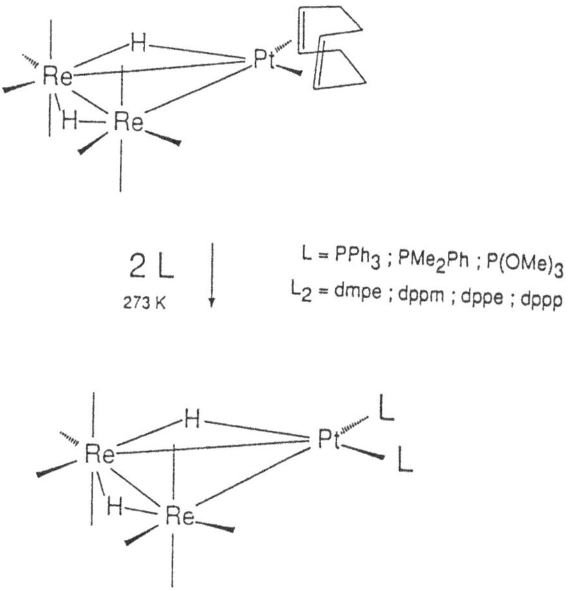

SCHEME I

The rates of the P-donor ligand exchange for the species with monodentate phosphines increase in the order PPh$_3$<<PMe$_2$Ph<<P(OMe)$_3$ and the corresponding activation energies are in the reverse order. For the bidentate ligands the exchange is generally easier and for two of them, dppm and dmpe, broadened ^{31}P spectra are observed even at 183 K. Combining simulation data and rough estimates of the rate constants at the temperature of coalescence, the overall order of the activation energies is PPh$_3$>PMe$_2$Ph$_3$,dppe,dppp>>P(OMe)$_3$,dppm>dmpe.

TABLE 3 Activation parameters for the exchange of the phosphines

Phosphine	E$_a$ (kJ/mol)	Θ(°)[a]	ν (cm^{-1})[a]
PPh$_3$	49 1	145	2068.9
PMe$_2$Ph	37 1	122	2065.3
P(OMe)$_3$	21 3	107	2079.5
dmpe	10	107	2063.1
dppm	20	121	2066.9
dppe	38 1	125	2066.5
dppp	38 2	127	2066.5

a) Tolman C. A. *Chem. Rev.*, 1977, 77 313.

The experimental results suggest therefore that the steric properties of the ligands are the major responsible factor in the activation barrier; the increase of the cone angle [6] and/or the increase of the P-Pt-P angle slows the exchange of the phosphines

This view is supported also by the preliminary results of the X-ray solid state structure of the PPh$_3$ derivative [7], which has the greatest activation energy. In this derivative, the steric requirements of the ligands (cone angle Θ = 145°) make the angle at Pt greater than 100° and moreover cause a remarkable tilt of the plane comprising the platinum and the two phosphorus atoms with respect to the plane of the metal cluster (dihedral angle *ca.* 38°) (Figure 1). The exchange of the two phosphines could be achieved either by the rotation of the PtL$_2$ fragment with respect to the [Re$_2$H$_2$(CO)$_8$] moiety or through the jump of the hydride bridging the Re-Pt edge from one side of the triangle to the other. This latter mechanism seems more likely since it requires less severe rearrangements in the transition state. A similar mechanism was also suggested for the less distorted COD precursor [3] on the basis of ^{13}C NMR evidence, and in order to explain its apparent C$_s$ symmetry.

TABLE 4. Activation parameters for the exchange of the hydrides

Phosphine	E$_a$(kJ/mol)	Θ(°)[a]	ν(cm^{-1})[a]
PPh$_3$	60 1	145	2068.9
PMe$_2$Ph	42 2	122	2065.3
P(OMe)$_3$	65 9	107	2079.5
dmpe	35 1	107	2063.1
dppm	82 2	121	2066.9
dppe	59 2	125	2066.5
dppp	62 6	127	2066.5

a) Tolman C. A. *Chem. Rev.*, 1977, 77, 313

The higher temperature mutual exchange process that equalizes the two hydrides is operative in all these derivatives and the estimated activation energies decreases in the order: dppm>P(OMe)$_3$>dppe, dppp, PPh$_3$>PMe$_2$Ph>dmpe. In this case it is apparent that the donor properties of the ligands play a major role even if the high activation energy calculated for the dppm derivative indicates that the decrease of P-Pt-P angle makes the system less amenable to exchange the hydrides. The concerted interchange of the hydride positions on the edges of the cluster (Scheme II) could be a possible mechanism of the observed behaviour and the observed trend could support a transition state with a significant decrease of the bonding interaction with Pt of the Re-μH-Pt hydride.

Dynamic processes concerning the motion of hydrides along cluster edges, alone or coupled with the motion of other ligands, are common in the cluster chemistry [5,8,9], and, conversely, it has also been observed that the presence of a bridging hydride ligand hinders the fluxional properties of the ligands on the adjacent metal sites [8,10]. Also in the present case, it is the vacancy of a ligand bridging one of the

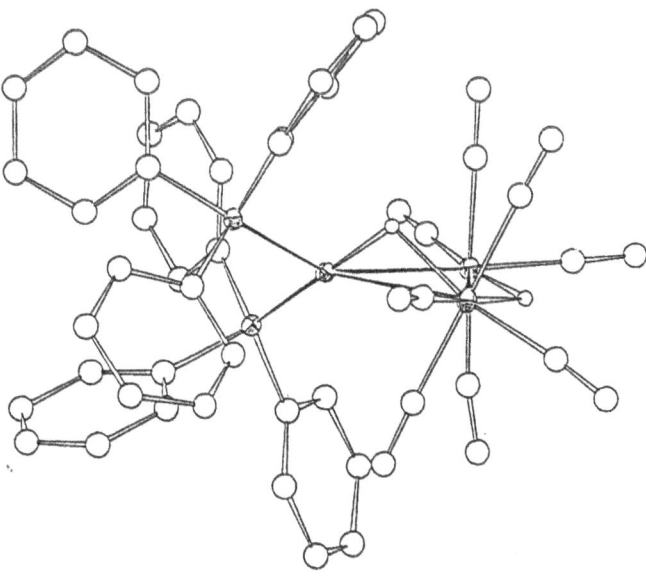

Figure 1. A View of the solid state structure of $[Re_2Pt(\mu-H)_2(CO)_8(PPh_3)_2$

rhenium-platinum bonds that allows both phosphine and hydride exchange processes. In order to prove this, compound **1a** has been protonated at low temperature and the new cationic species, $[Re_2Pt(\mu-H)_3(PPh_3)_2(CO)_8]^+$, (**4**), does not exhibit any fluxional behaviour in all the range of temperatures 183-298 K.

Figure 2 shows the hydridic region of the ^1H NMR spectrum of a solution of the protonation mixture and the inset shows that the multiplet due to the hydrides bridging the Pt-Re interaction is as required for an $[AX]_2$ spin system [^1H δ -11.62 ppm ($^2J_{HP}$ = 83 Hz, $^2J_{HP}$ = 17.0 Hz, $^1J_{HPt}$ = 591 Hz, $^2J_{HH}$ = 3.5 Hz, $^2J_{PP}$ = 3.5 Hz); -15.85 ppm ($^2J_{PPt}$ = 15 Hz)].

In contrast with the two mutual exchange processes described above, the irreversible transfer of one phosphine from Pt to Re and of one CO ligand from Re to Pt, observed above 273 K for the PPh_3 derivative, occurs only for the PMe_2Ph derivative (**1b**), affording two species (**2b** and **3b**), whose ^{31}P and ^1H NMR data indicate structural similarity with those of PPh_3. The $P(OMe)_3$ derivative at room temperature slowly decomposes and no formation of the new isomers was observed within two days. No evidence of phosphine transfer was found for the derivatives with the chelating ligands.

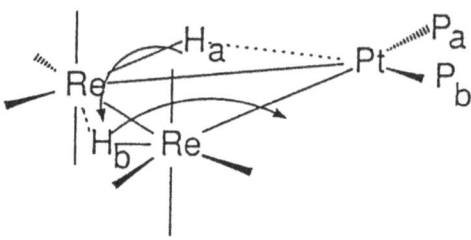

The isomerization of the PMe$_2$Ph derivative occurs at higher temperatures (above 280 K) and with a slower rate compared to the one of the PPh$_3$ derivative (apparent first order rate constant $k = 1.3 \ 10^{-4}s^{-1}$ at 298 K versus $k = 3.6 \ 10^{-4}s^{-1}$ at 294 K [1].

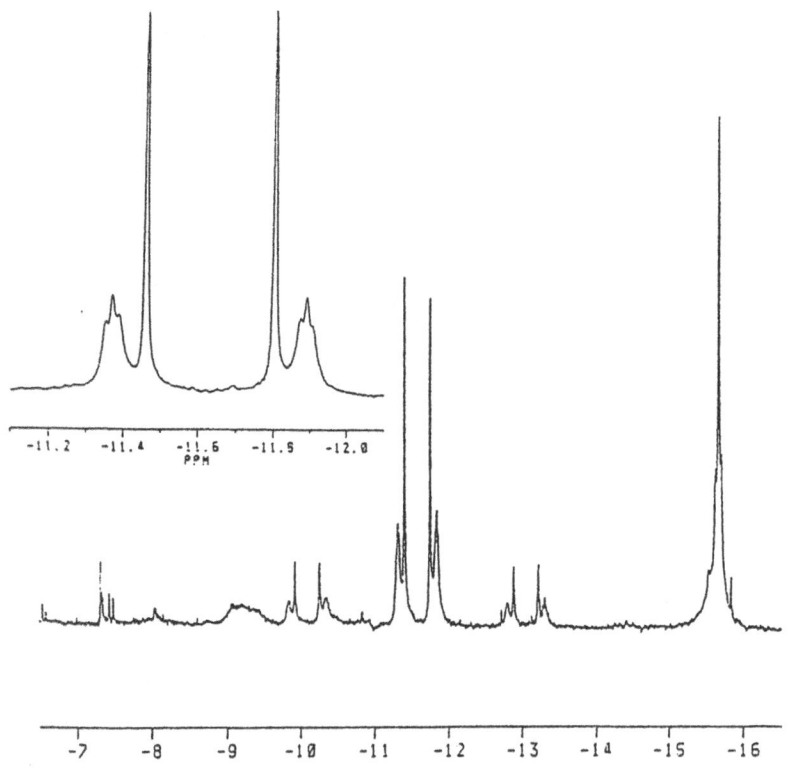

Figure 2. Hydridic region of the ^{1}H NMR spectrum of 4 (CD$_2$Cl$_2$, 200.13 Hz, 253K)

At present no definite conclusions concerning the mechanism of this irreversible isomerization are possible, but it is likely that the transfer of CO would require a μ-CO intermediate with the previous or concerted loss of one phosphine from Pt. Platinum assisted CO labilization on an adjacent metal has been reported for heterobimetallic compounds [11].   The authors demonstrated that, after the dissociation of one phosphine from Pt, one CO was transferred to Pt from the other metal site, via a μ-CO intermediate, and that the final loss of CO occurred at the same Pt site through its substitution by a phosphine molecule. The ease of dissociation of the phosphine from Pt appeared to be a key step of the reaction. Also in the present case the absence of isomerization of all the diphosphine derivatives suggest that the dissociation of a phosphine from Pt could be a crucial step for isomerization. The possible formation of a μ-CO intermediate is supported by the X-ray structure of the PPh$_3$ derivative in which one of the axial carbonyls on the rhenium atom of the μ-H bridged Re-Pt interaction leans significantly towards Pt(Pt --C contact 2.64 A). Similar hydrido-carbonyl double bridged M-Pt interactions have already been reported for M = Re [12] and M = Mn [13]

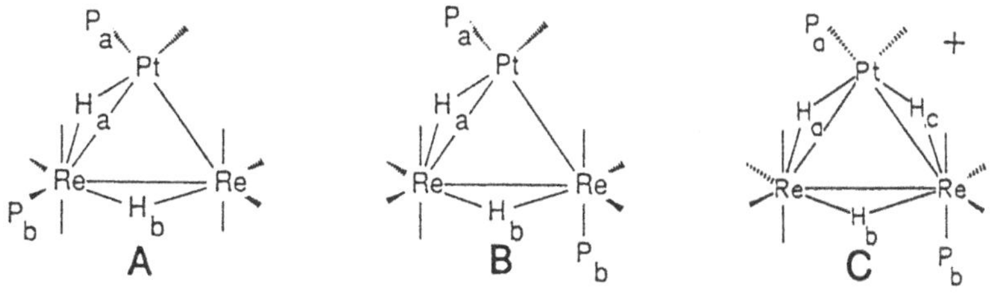

SCHEME III

It is particularly worth noting the absence of reaction of the dppm derivative since this ligand is known to stabilize heterometallic metal-metal bonds, being an efficient bridging ligand [14], that easily changes its coordination mode from η^2 to μ (the driving force of this reaction being the formation of a more stable five-membered ring compared to the former four membered ring [15]).

The ratio 2b/3b increases in polar solvents (3.5 in CD$_2$Cl$_2$ vs 2 in toluene-d_8) and decreases on increasing the temperature (ΔH° = -6.0 0.2 kJ/mol and ΔS° = -14 1J/molK), as in the case of the PPh$_3$ derivatives. The ^{31}P and ^{1}H NMR spectra allow the confident assignment of the structure of the major isomer 2b to the derivative with one phosphine bound axially to the rhenium adjacent to the μ-H bridged Re-Pt interaction, as shown also by the X-ray structure of the related species containing PPh$_3$ [1]. The main difference in the two isomers lies in the coupling pattern of the hydrides bridging the Pt-Re interaction, which indicates a different coordination mode of the phosphine bound to rhenium. The two most likely structures for the minor isomer **3b** are shown in the Scheme III.

The following results led us to reconsider the previous assignment (Structure A) made in the case of the PPh₃ derivative. Indeed, when the equilibrium solution of **2a** and **3a** is protonated, at 263 K, only one species is formed in relevant amounts. The ¹H NMR spectrum of the reaction mixture (Figure 3) shows that the coupling parameters of the three hydrides of this isomer of $[Re_2Pt(\mu\text{-}H)_3(PPh_3)_2(CO)_8]^+$ (**5**) are compatible only with the structure C [¹H: δ -8.36 ppm ($^2J_{HP}$ = 88.7 Hz, $^2J_{HP}$ = 13.0 Hz, $^1J_{HPt}$ = 646 Hz, $^2J_{HH}$ = 4.5 Hz); =11.53 ppm ($^2J_{HP}$ = 12 Hz, $^1J_{HPt}$ = 680 Hz, $^2J_{HH}$ = 4.5 Hz); -14.88 ppm ($^2J_{HP}$ = 10.7 Hz, $^2J_{HPt}$ = 20 Hz)].This strongly suggests that the minor isomer has the structure B and that, as expected, the protonation occurs preferentially on the isomer with a more electron rich unbridged Re-Pt interaction.

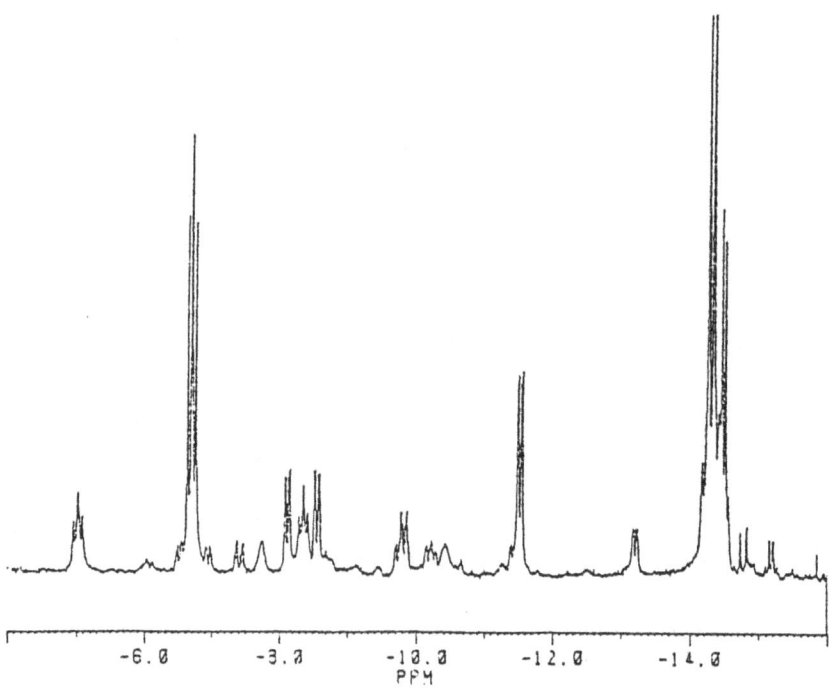

Figure 3. Hydridic region of the ¹H NMR spectrum of the mixture obtained by protonation of the equilibrium mixture 2a/3a.

As reported for the PPh$_3$ derivatives, complexes **2b** and **3b** also show a broadening of all the resonances on increasing the temperature, and ^{1}H and ^{31}P 2D EXSY experiments showed that the two isomers are in slow exchange at room temperature. With the new assignment of the structure of the minor isomer in mind, the mechanism of the exchange can be viewed as a rotation of an 'ethylene-like' [HRe$_2$(CO)$_7$P]$^-$ moiety with respect to [HPt(CO)P]$^+$ fragment (Scheme IV).

Similar dynamic processes have already been reported for the same or strictly related Pt fragments with respect to the 'ethylene-like' [Rh$_2$(CO)$_2$Cp*$_2$] [16] or other couples of isolobal fragments. It is worth noting the difference of *ca* three orders of magnitude of the rate constants for the dynamic process in these phosphine containing Re-Pt derivatives compared to the systems previously reported: $k = 0.185$ s^{-1} at 298 K for **2a** <==> **3a** [1], $k = 0.64$s^{-1} at 298 K [1] for **2b** == **3b** while $k = 147$ s^{-1} at 278 K for [Rh$_2$Pt(μ-H)(μ-CO)$_2$(CO)(PPh$_3$)(η^5-Cp*)$_2$]$^+$.

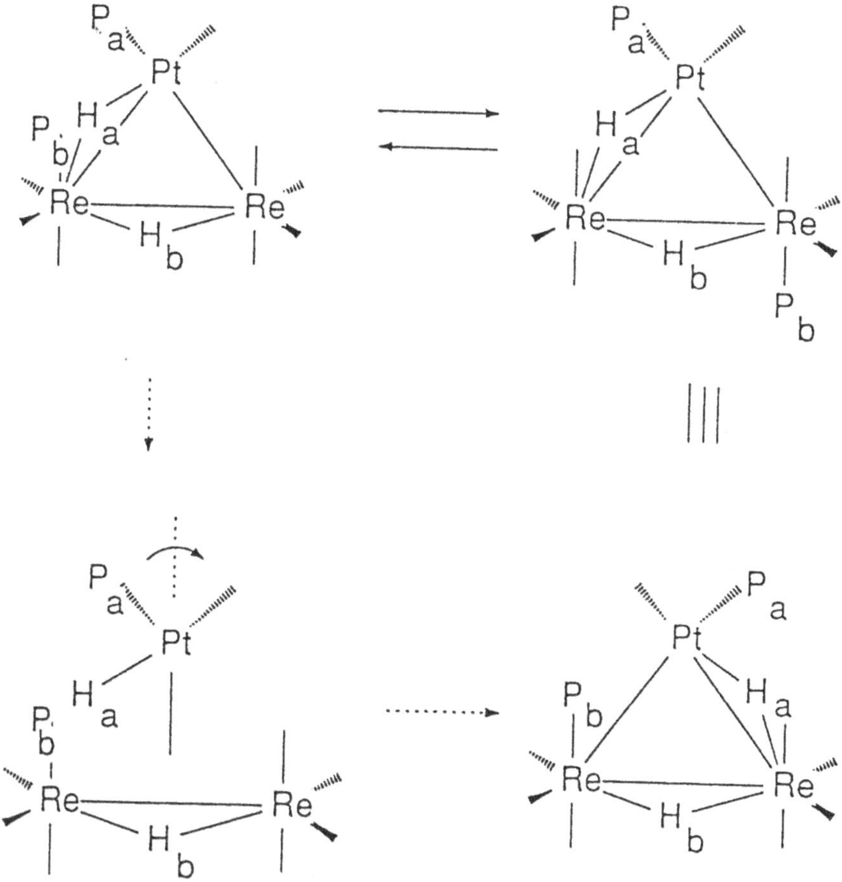

SCHEME IV

References

1. Beringhelli, T., Ceriotti, A., D'Alfonso, G., Della Pergola, R., Ciani, G., Moret, M., and Sironii, A. (1990) *Organometallics*, **9**, 1053-1059.

2. Antognazza, P., Beringhelli, T., D'Alfonso, G., and Minoja, A. (1992) . *Organometallics*, **11**, 1777-1784.

3. Ciani, G., Moret, M., Sironia, A., Berighelli, T., D'Alfonso, G., and Della Pergola, R. (1990) *J. Chem. Soc. Chem. Commun.*, 1668-1670.

4. Ciani, G., Moret, M., Sironi, A., Antognazza, P., Beringhelli, T., D'Alfonso, G., Della Pergola, R. and Minoja, A.P. (1991) *J. Chem. Soc. Chem. Commun.*, 1255-1257.

5. Beringhelli, T., D'Alfonso, G., and Minoja, A.P. (1991) *Organometallics*, **10**, 394-400.

6. Tolman, C.A. (1977) , *Chem. Rev.*, **77**, 313-348.

7. Ciani, G., Moret, M., Sironi, A., personal communication.

8. Ewing, P., Farrugia, L.J., Rycroft, D.S. (1988) *Organometallics*, **7**, 859-870

9. Farrugia, L.J. (1989) *Organometallics*, **8**, 2410-2417.

10. Beringhelli, T., D'Alfonso, G., Ciani, G., Sironi, A., Molinari, H. (1990), *J. Chem. Soc. Dalton Trans.*, 1901-1906.

11. (a) Powell, J., Gregg, M.R. and Sawyer, J.F. (1989) *Inorg. Chem.*, **28**, 4451-4460.(b) Powell, J., Sawyer, J.F., and Stainer, M.V.R. (1989) *,Inorg. Chem.*, **28**, 4461-4470.

12. Beringhelli, T., D'Alfonso, G., Minoja, A.P. and M. Freni (1992) *Gazz. Chim. Ital.*, **122**, 375-382.

13. Braunstein, P., Geoffroy, G.L., Metz, B. (1985) *Nouv. J. Chim.*, **9**, 221-223.

14. Chaudret, B., Delavaux, B., and Poilblanc, R. (1988) *Coord. Chem. Rev.*, vol. 86, 191-243. Elsevier Science Publishers B.V. Amsterdam printed in The Netherlands.

15. Braunstein, P., de Meric de Bellefon, C., and Ostwald, B. (1993) *Inorg. Chem.*, **32**, 1649-1655.

16. Green, M., Mills, R.M., Pain, G.N., Stone, F.G.A., Woodward, P. (1982) , *J. Chem. Soc. Dalton Trans.*, 1321-1326.

ORGANOMETALLIC MIGRATIONS OVER CLUSTERS AND SURFACES

Symmetry-allowed shifts via aromatic transition states

MICHAEL J. McGLINCHEY, LUC GIRARD
and ANDREAS DECKEN
Department of Chemistry
McMaster University
Hamilton, Ontario, Canada L8S 4M1

1. Introduction

Within the general theme of molecular dynamics of organometallic complexes, we have focussed our recent efforts on understanding how organic moieties migrate over cluster faces and, conversely, how organometallic fragments interact with polycyclic organic ligands. In the first case, the cluster face can be regarded as a tailor-made microsurface and the molecular dynamics are controlled primarily by the frontier orbital interactions between the organic unit and one or more metal centers. In the second case, an ML_n fragment is initially π-complexed to one ring of the polycyclic system but is able to undergo haptotropic shifts so as to bind to one or more different rings.

Our goals are three-fold: firstly, one must establish the overall reaction sequence by characterizing the initial and final molecular structures, for example by x-ray crystallography; secondly, the activation energy barrier for the process must be evaluated, normally by variable-temperature NMR or by 2D-EXCHANGE techniques. Thirdly, we try to elucidate the actual trajectory followed by the migrating fragment during the course of the rearrangement. This can sometimes be accomplished by a combination of molecular orbital calculations and application of the structure correlation approach of Bürgi and Dunitz.

We shall discuss two projects: we begin by looking in some detail at a fluxional process which has been observed for a variety of cluster types, and for which extensive NMR and x-ray crystallographic data are available. Secondly, we describe preliminary synthetic and mechanistic studies on the migration of metal tricarbonyl and (cyclopentadienyl)metal fragments over a tetracyclic framework and show that the mobility of the ML_n unit is crucially dependent on the π-electron count of the organic microsurface.

L. J. Farrugia (ed.), The Synergy Between Dynamics and Reactivity at Clusters and Surfaces, 203–216.
© 1995 *Kluwer Academic Publishers.*

2. Molecular Dynamics of Cluster Cations

The ability of transition-metal clusters to stabilize carbocations has been widely studied in recent years. The prototypical system was the tricobalt cluster, $[Co_3(CO)_9C\text{-}CH_2]^+$, **1**, first reported by Seyferth [1] These investigations encompassed not only the structural, but also the dynamic aspects of this chemistry. Extended Hückel molecular orbital calculations [2], complemented by a variable-temperature NMR study [3], painted a picture in which the vinylidene fragment bonds in a "side-on" manner (**1b** rather than **1a**) to a single cobalt vertex but is capable of undergoing an antarafacial shift (via **1c**) to a neighboring vertex.

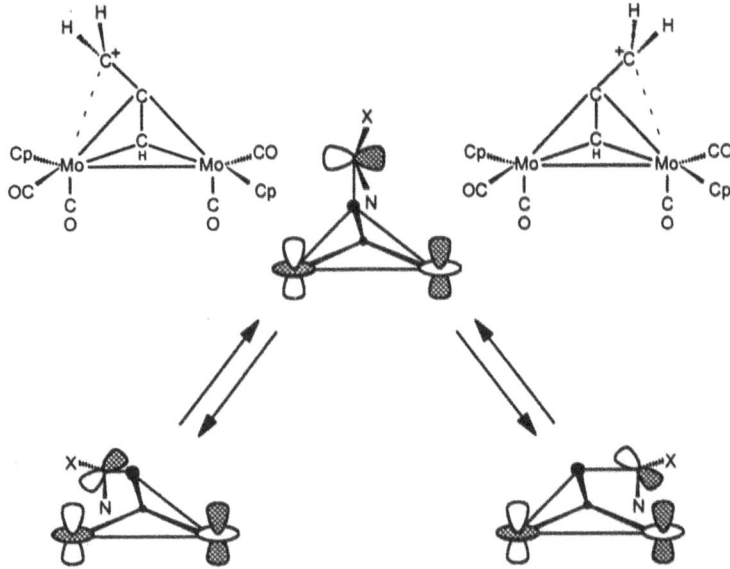

Analogous data for the propargyl cluster cations $[(HC\equiv C\text{-}CH_2)Co_2(CO)_6]^+$, **2**, or $[(HC\equiv C\text{-}CH_2)Mo_2(CO)_4(C_5H_5)_2]^+$, **3**, have been reported and similar mechanistic proposals have been advanced [4-6]. This process is illustrated in Scheme 1.

Scheme 1. Antarafacial migration of a methylene group between metal centers.

Studies on the molybdenum cluster cations **3** are particularly advantageous since, unlike their dicobalt analogues, they readily yield crystals of x-ray quality; a number of $[(RC\equiv C\text{-}CR'R'')Mo_2(CO)_4Cp_2$ structures are now available in which the R' and R'' substituents range from H, Me or ferrocenyl to steroidal and terpenoid moieties. The pathway for this antarafacial migration in **3** has been investigated both by EHMO calculations and also by a Bürgi-Dunitz trajectory analysis [7].

| 2 | 3 | 4 |

The molecular orbital calculations were carried out on the relatively rigid $[(\eta^5:\eta^5\text{-fulvalene})Mo_2(CO)_4(HC\equiv C\text{-}CH_2)]^+$ system, **4** [8], so as to avoid complications arising from semi-bridging carbonyl groups and cyclopentadienyl rotamers [9]. Figure 1 depicts the EHMO-calculated hypersurface in which the $[C\text{-}CH_2]^+$ capping group is free to swivel over the Mo_2C triangular base of **4**. The vinylidene fragment has three degrees of freedom: it can bend through an angle θ from a vertical position with respect to the centroid of the tetrahedron so as to lean towards the Mo-Mo bond; secondly, the $[C\text{-}CH_2]^+$ group can swivel through angle ϕ away from the molecular mirror plane towards a molybdenum atom. Moreover the vinylidene unit was allowed to rotate from $\omega = 0°$ to $\omega = 180°$ and the minimum energy orientation for each (θ, ϕ) position was computed. We see from Figure 1 that the most favorable position for the CH_2 fragment is when $\theta \approx 140°$ and $\phi \approx 50°\text{-}70°$, that is, leaning down directly towards a molybdenum atom.

The calculated minimum energy trajectory for the methylene carbon, C(1), is shown in Figure 2a and the steep nature of the descent from the transition state into the potential energy well is evident. Figure 2b depicts the changing orientation of the methylene fragment as the rearrangement proceeds.

This calculated migration pathway can be compared with the experimental trajectory obtained via a Bürgi-Dunitz analysis of the available x-ray crystallographic data. In this approach, one requires a series of structures of related molecules in which the differing electronic and steric requirements of the substituents cause minor geometric perturbations. When placed in an appropriate sequence these x-ray structures can provide a series of "snapshots" of the fluxional process as it unfolds [10].

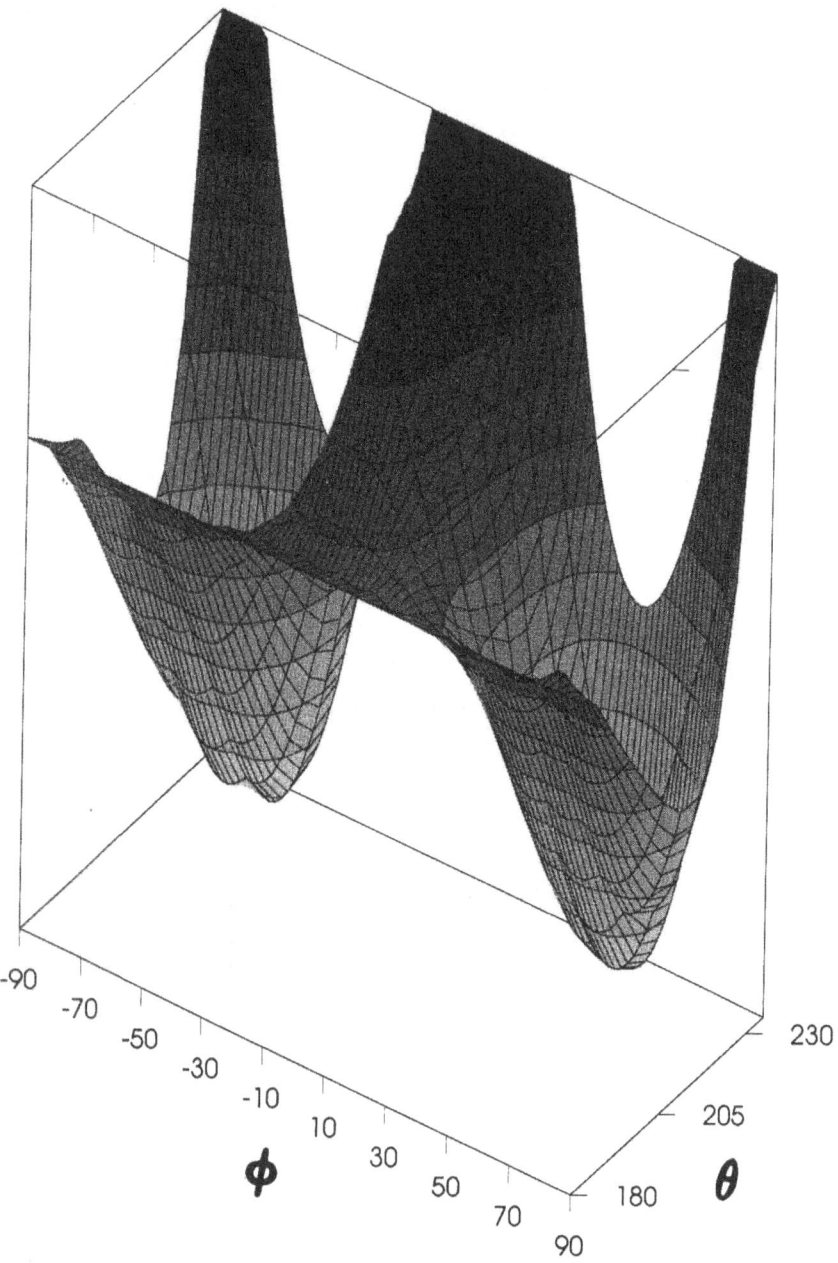

Figure 1. EHMO-calculated hypersurface for the migration of a methylene group in
$[(HC{\equiv}C\text{-}CH_2)Mo_2(CO)_4(C_5H_5)_2]^+$,**3**.

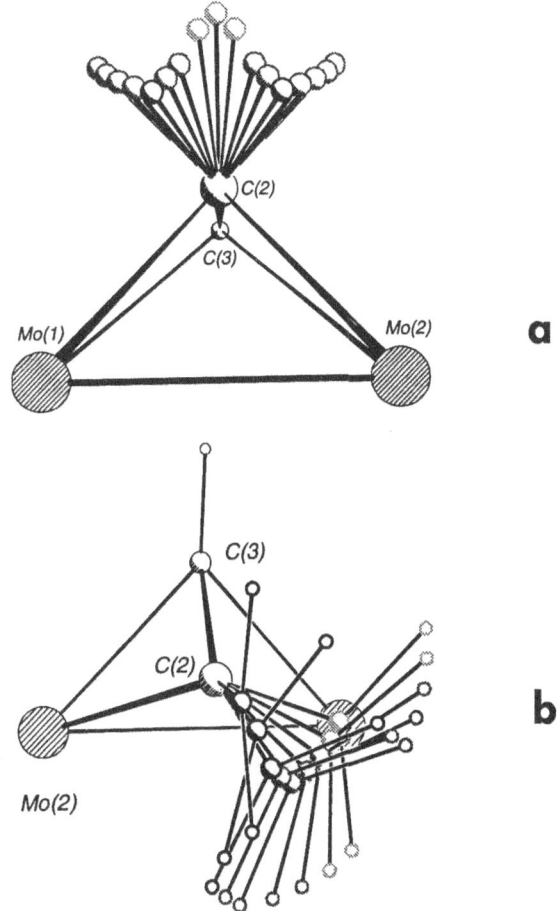

Figure 2.(a) View of the EHMO-calculated trajectory of C(1) during the migration process
(b) Bird's eye view of the twisting motion of the CH_2 fragment.

In Figure 3 we show a superposition of 11 crystal structures of $[(RC{\equiv}C-CR'R'')Mo_2(CO)_4Cp_2]^+$ cations, and again we see a range of C(1) positions which parallel almost exactly the EHMO-calculated trajectory. It is particularly noteworthy that all the structures where $R' = R'' = H$ lie at the bottom of the potential well; as the $CR'R''^+$ group becomes a secondary carbocation, and finally a tertiary carbocation, C(1) gradually climbs the wall of the potential energy well and approaches the calculated transition state geometry. These structural data are in complete accord with the NMR-derived barriers for rearrangement from one molybdenum to the other; in the CH_2^+ systems, $\Delta G^{\#}$ is found to be $\approx$ 18-19 kcal/mole, but this barrier falls to $\approx$ 10 kcal/mole for the CR_2^+ cations, which have less need of anchimeric assistance from the metal.

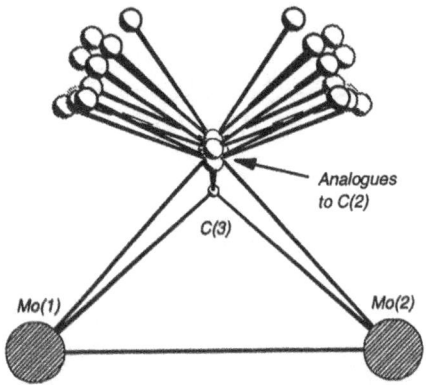

Figure 3. Superposition of the C(1) cationic centers in 11 different x-ray crystal structures.

To summarize, we now have a relatively complete picture of the structures, energies and molecular dynamics of a fluxional process which is widespread in cluster chemistry. This information can be gleaned from experimental data (obtained both in solution and in the solid state) complemented by molecular orbital calculations.

3. Haptotropic Shifts over Polycyclic Surfaces

We turn now to the more general problem of how organometallic fragments can move over polycyclic surfaces. This was originally considered by Albright and Hoffmann and their colleagues for the migration of a π-bonded ML_n unit between the two rings of an indenyl or of a naphthalene ligand [11]. They showed that the least-motion pathway across the center of the common bond between the two rings was strongly disfavored since most of the bonding interactions between the polycyclic π-system and the acceptor orbitals of the ML_n moiety are lost during the inter-ring transit. Instead they proposed a more circuitous route involving an exocyclic transition state, as shown below:

Our aim was to extend these predictions to larger polycyclic surfaces, and to use a combination of EHMO calculations and experimental measurements to probe the reaction trajectories. To this end, we prepared a series of organometallic derivatives of 4*H*-cyclopenta[*def*]phenanthrene, cppH **5**, in the expectation that, upon deprotonation, these cpp-ML_n systems would exhibit haptotropic behavior.

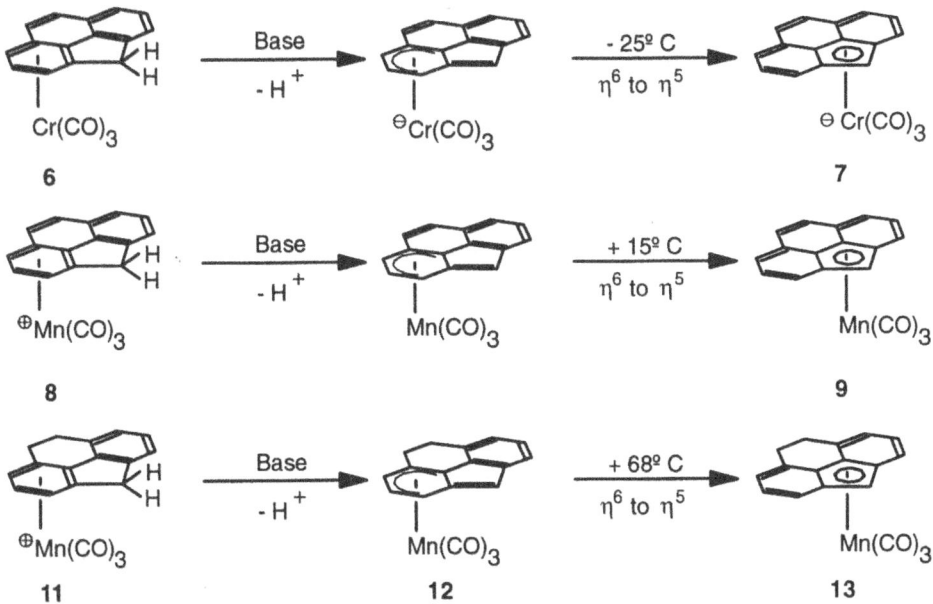

Scheme 2. Haptotropic shifts in cpp-metal complexes.

Indeed, as shown in Scheme 2, deprotonation of $(\eta^6\text{-cppH})Cr(CO)_3$, **6** to give $[(\eta^5\text{-cpp})Cr(CO)_3]^-$, **7**, proceeds rapidly even at -40 °C. Likewise, the reaction of $[(\eta^6\text{-cppH})Mn(CO)_3]^+$, **8** to yield $(\eta^5\text{-cpp})Mn(CO)_3$, **9**, is also a facile process [12]. As examples of "before" and "after" rearrangement, the molecules **6** and **9** have been characterized by x-ray crystallography.

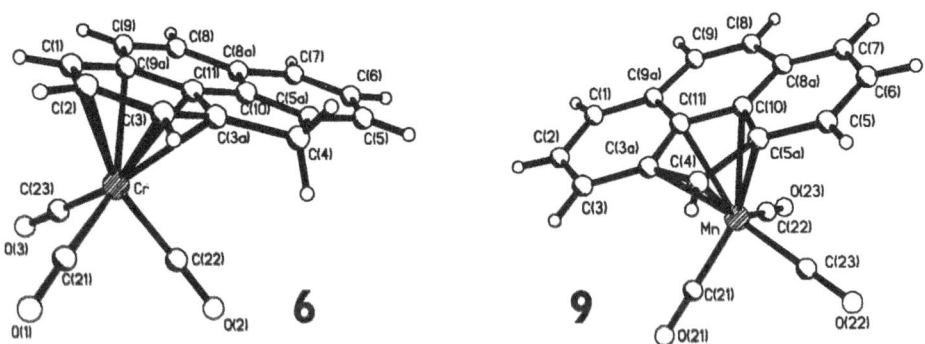

210

The most favored trajectory for an η^6- to η^5- haptropic shift from **8** to **9** has been computed at the Extended Hückel level and is shown in Figures 4a and 4b. The least-motion pathway (route **A**) across the common bond between 6- and 5-membered rings is strongly disfavored; route **B** which passes through transition state **10B** in the central 6-membered ring is better, but pathway **C** provides a low energy ($\approx$ 18-20 kcal/mole) barrier. This route not only involves an excursion into an exocyclic η^3-structure **10C** but also requires the Mn(CO)$_3$ tripod to rotate in such a way as to maximize its bonding interactions with the cpp ligand. This pathway is depicted in Figure 4b.

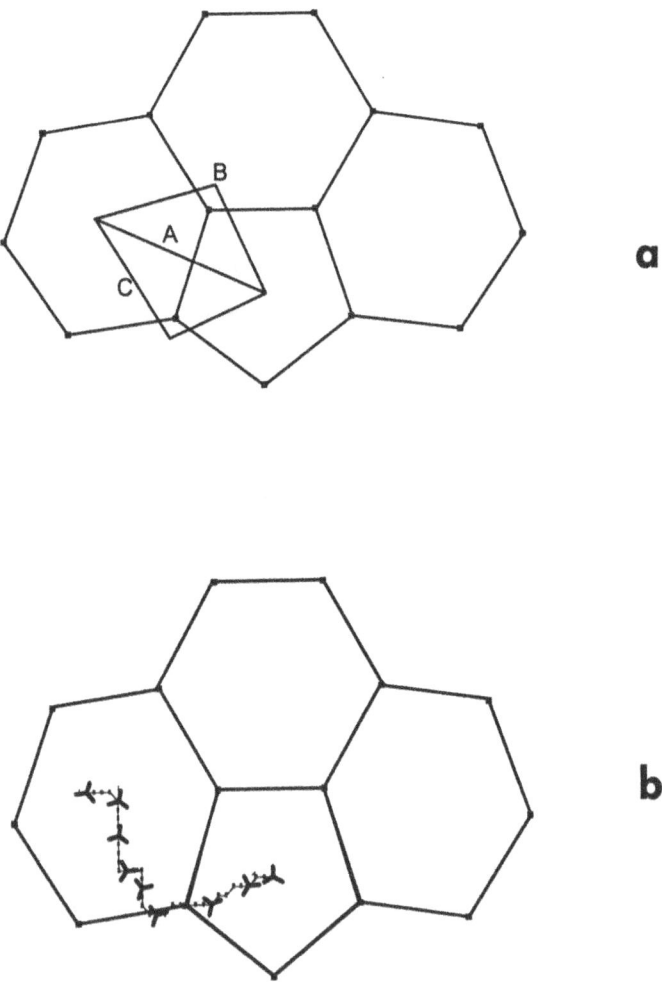

Figure 4. (a) Routes **A**, **B** and **C**. (b) Favored pathway for Mn(CO)$_3$ migration.

It has been suggested that a major factor in stabilizing a transition state such as **10C** is the development of a 10π-Hückel manifold (a naphthalene-type aromatic system) which greatly favors route **C**. As shown below, this aromatic stabilization is not available for a migration that proceeds through route **B**.

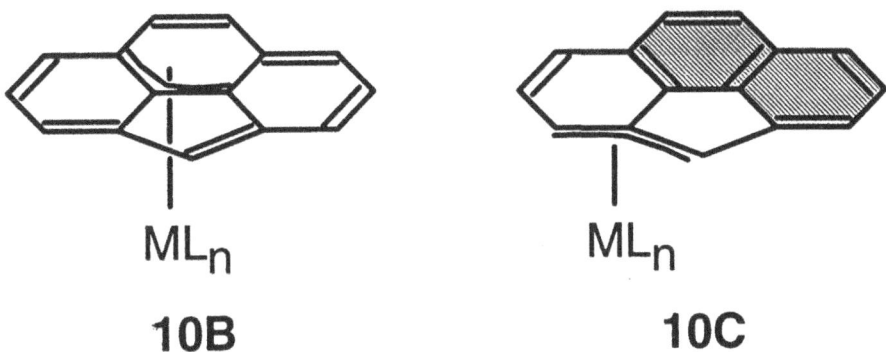

ML$_n$ ML$_n$

10B   **10C**

In support of this idea, hydrogenation of the 9,10 double bond in **8**, to give $[(\eta^6\text{-H}_2\text{-cppH})\text{Mn(CO)}_3]^+$, **11**, allows the ready observation of the intermediate $(\eta^6\text{-H}_2\text{-cpp})\text{Mn(CO)}_3$, **12**, which must be heated to bring about the haptotropic shift that generates $(\eta^5\text{-H}_2\text{-cpp})\text{Mn(CO)}_3$, **13** (see Scheme 2). Of course, in these dihydro-cpp complexes there is no possibility of accessing a 10π Hückel manifold and the H$_2$-cpp system behaves effectively as a fluorenyl ligand [13].

4. η^1-Cpp Complexes of Iron and Manganese

In an attempt to trap an η^3-exocyclic system, the complexes $(\eta^1\text{-cpp})\text{-Fe(CO)}_2(\text{C}_5\text{H}_5)$, **14**, and $(\eta^1\text{-cpp})\text{Mn(CO)}_3(\text{PEt}_3)_2$, **15**, were synthesized. It was hoped (see Figure 5) that decarbonylation of the iron complex **14** might yield either $(\eta^3\text{-cpp})\text{Fe(CO)}(\text{C}_5\text{H}_5)$, **16**, or even the ferrocene analogue $(\eta^5\text{-cpp})\text{-Fe(C}_5\text{H}_5)$, **17**. Interestingly, **17** is not observed when $[(\eta^6\text{-cppH})\text{Fe(C}_5\text{H}_5)]^+$ is deprotonated. This experimental result is in accord with the EHMO calculations on the (cpp)Fe(C$_5$H$_5$) hypersurface which predict a rather high barrier of ≈ 33 kcal/mole for the transformation of $(\eta^6\text{-cpp})\text{Fe(C}_5\text{H}_5)$ into $(\eta^5\text{-cpp})\text{Fe(C}_5\text{H}_5)$ [14].

Indeed, careful thermolysis of $(\eta^1\text{-cpp})\text{Fe(CO)}_2\text{Cp}$, **14**, does not yield products arising from loss of carbon monoxide, rather in an apparent free radical coupling process, one observes the iron dimer $[(\text{C}_5\text{H}_5)\text{Fe(CO)}_2]_2$ and the cpp-trimer **18**. This latter molecule exhibits a spectacular chemical shift range for its aromatic protons; presumably, this phenomenon is attributable to its stacked structure (Figure 6) which places several protons directly above other rings and in their ring current shielding regions.

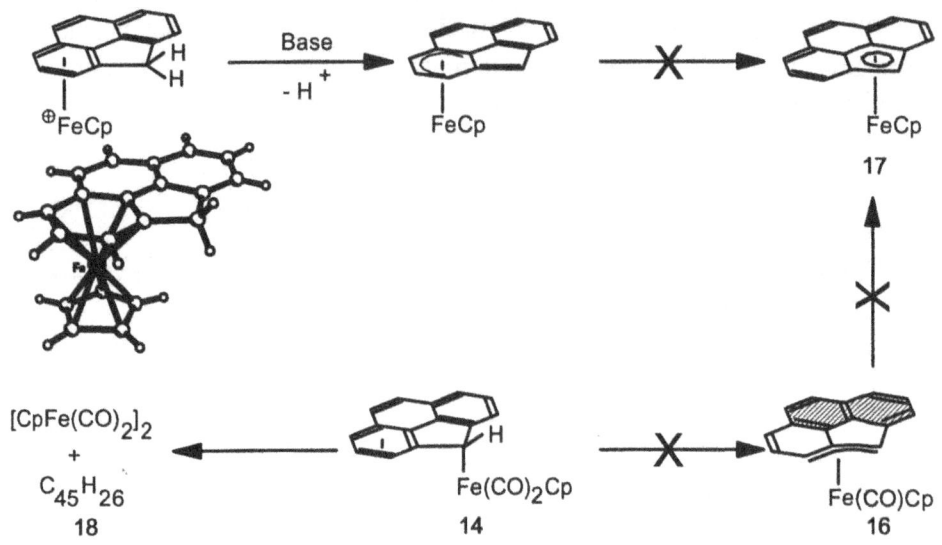

Figure 5. Reactions of (η^1-cpp)Fe(CO)$_2$Cp, **14**.

The manganese complex (η^1-cpp)Mn(CO)$_3$(PEt$_3$)$_2$, **15**, was also thermolyzed in an attempt to prepare (η^3-cpp)Mn(CO)$_2$(PEt$_3$)$_2$, but again the cpp-trimer **18** was formed. Nevertheless, **15** does exhibit an interesting NMR spectrum: the ^{1}H, ^{13}C and ^{31}P spectra all indicate that the two triethylphosphine ligands are inequivalent. The first thought was that the Et$_3$P ligands were disposed one axially and the other equatorially. However, there is a literature report by Biagioni that in the analogous (η^1-fluorenyl)Mn(CO)$_2$(PEt$_3$)$_2$ complex, the NMR coupling constants indicate that the molecule adopts a trans-meridional geometry [15]. Such a structure would suggest that there is hindered rotation about the cpp—Mn σ-bond. Attempts to verify this hypothesis for (η^1-cpp)Mn(CO)$_3$-(PEt$_3$)$_2$, **15**, by high-temperature ^{31}P NMR with the intent of coalescing the two phosphorus peaks were thwarted by thermal decomposition. However, the 2D-EXCHANGE experiment could be used to evaluate this barrier. Figure 7 shows the results of a 2D-NOESY experiment for the ethyl protons of the Et$_3$P ligands in **15**.

With a short mixing time (τ_m = 25 ms, Figure 7a) no off-diagonal cross-peaks are detectable; however, when τ_m is 50 ms (Figure 7b), magnetization transfer between the two sets of methylene protons and between the two methyl environments is evident. This process can be quantified by means of a series of selective inversion experiments to determine exchange rates, at different

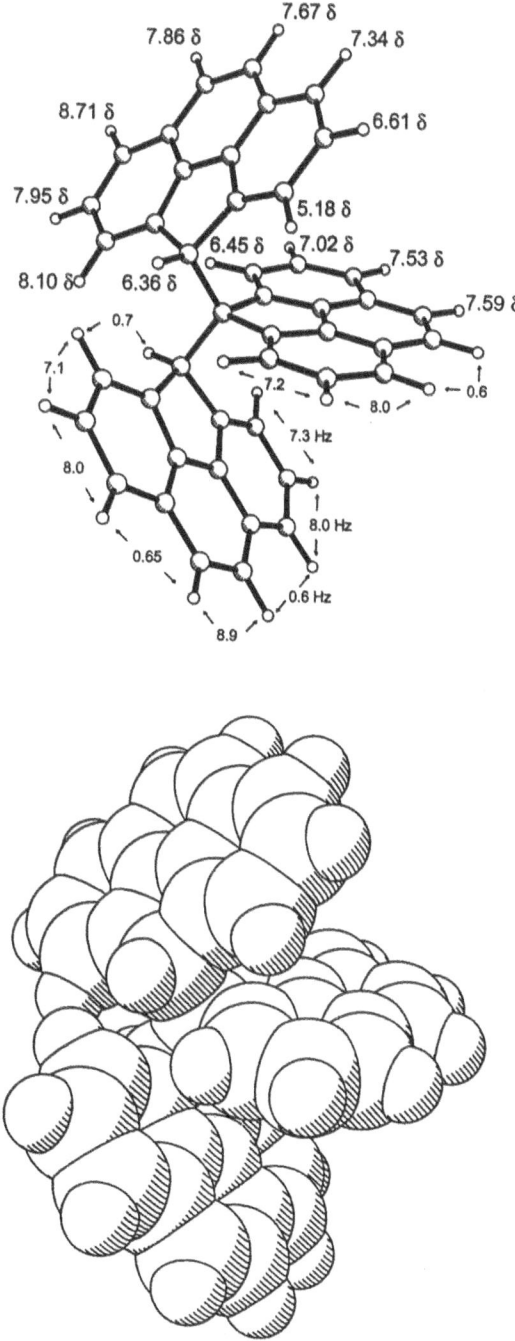

Figure 6. Structure and NMR data for the cpp-trimer, **18**.

214

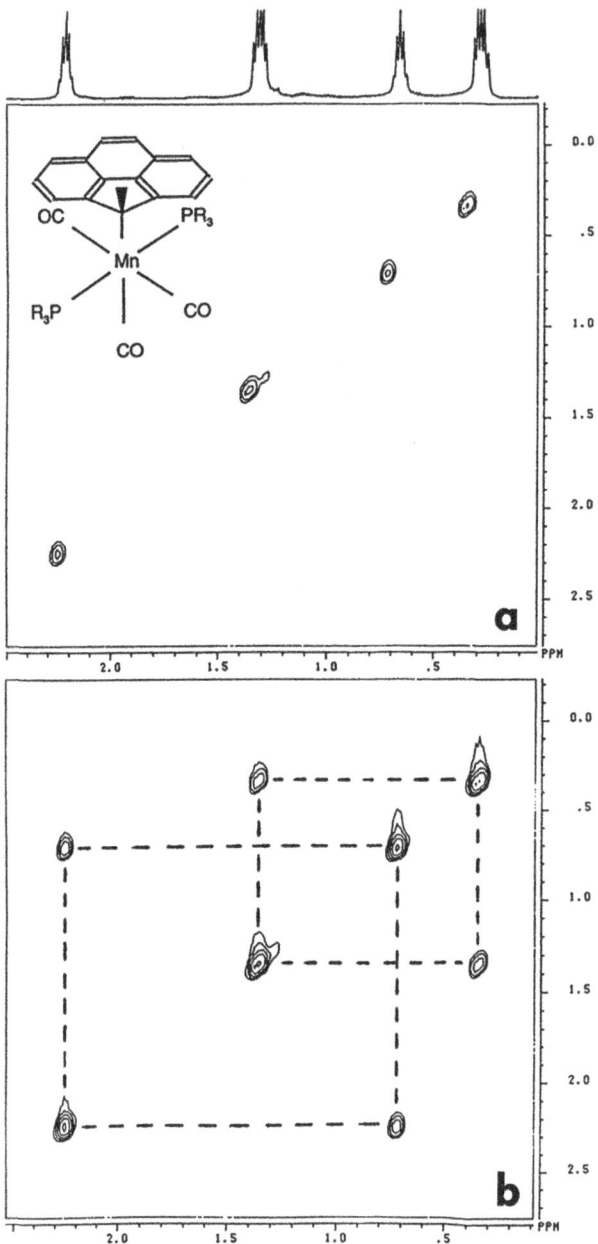

Figure 7. 2D-NOESY spectra for (η^1-cpp)Mn(CO)$_3$(PEt$_3$)$_2$, **15**.

temperatures, between the two methylene signals. These data yielded an Arrhenius plot from which a $\Delta G^{\#}$ value of 13.7 ± 0.2 kcal/mol was extracted.

The trans-meridional structure of **15** was confirmed by an x-ray crystallographic study and a view of the molecule appears as Figure 8. The rather high barrier to rotation about the Mn—cpp σ-bond is readily rationalized in terms of steric crowding in the symmetical structure when the hydrogens at positions C(3) and C(5) obtrude into the cone angles of the bulky Et_3P ligands.

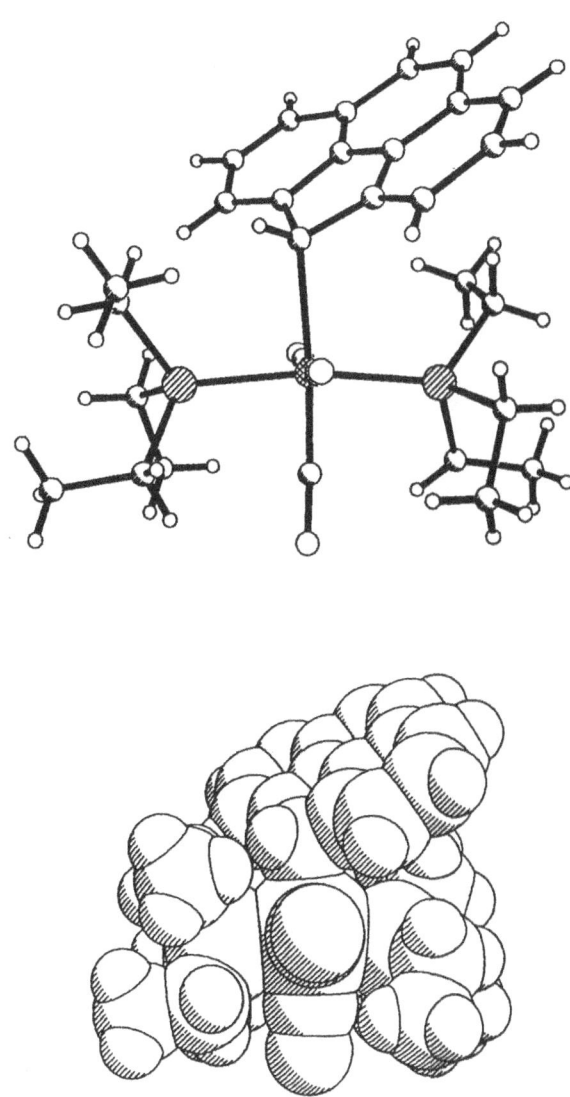

Figure 8. X-ray crystal structure of $(\eta^1\text{-cpp})Mn(CO)_3(PEt_3)_2$, **15**.

To conclude, the synthetic, crystallographic, spectroscopic and computational data on the series of cpp-ML_n complexes suggests that the trajectory for η^6- to η^5- haptotropic shifts involves an exocyclic η^3-cpp pathway. Attempts to trap such a π-allylic system have not yet been successful and other approaches towards their isolation are in progress.

5. Final Remarks

The continuing dialogue between experimentalists and theoreticians provides a necessary bridge between surface scientists and cluster chemists. It is evident that many parallels exist between these two domains, and it is our hope that we can continue to make contributions by our efforts both *in vitro* and *in silico*.

6. References

1. Seyferth, D. (1976) *Adv. Organometal. Chem.* **14**, 97.
2. Schilling, B.E.R. and Hoffmann, R. (1979) *J. Am. Chem. Soc.* **101**, 3456.
3. Edidin, R.T, Norton, J.R.; Mislow, K. (1982) *Organometallics* **1**, 561.
4. Padmanabhan, S. and Nicholas, K.M. *J. Organometal. Chem.* **268**, C23.
5. Schreiber, S.L., Klimas, M.T., and Sammakia, S. (1987) *J. Am. Chem. Soc.* **109**, 5749.
6. Meyer, A., McCabe, D.J., and Curtis, M.D. (1987) *Organometallics* **6**, 1491.
7. Girard, L., Lock, P.E., El Amouri, H., and McGlinchey, M. J. (1994) *J. Organometal. Chem.* in press.
8. El Amouri, H., Vaissermann, J., Besace, Y., Vollhardt, K.P.C., and Ball, G.E. (1993) *Organometallics* **12**, 605.
9. Tondu, S., Jaouen, G., D'Agostino, M.F., Malisza, K.L., and McGlinchey, M.J. (1992) *Can. J. Chem.* **70**, 1743.
10. Bürgi, H.B. and Dunitz, J.D. (1983) *Acc. Chem. Res.* **16**, 153.
11. Albright, T.A., Hofmann, P., Hoffmann, R., Lillya, C.P., and Dobosh, P.A. (1983) *J. Am. Chem. Soc.* **105**, 3396.
12. Decken, A., Britten, J.F. and McGlinchey, M.J. (1993) *J. Am. Chem. Soc.* **115**, 7275.
13. Johnson, J.W. and Treichel, P.M. (1977) *J. Am. Chem. Soc.* **99**, 1427.
14. Decken, A., Girard, L., Britten, J.F., Rigby, S., Bain, A.D., and McGlinchey, M.J. unpublished results.
15. Biagioni, R.N., Lorkovic, I, Skelton, J., and Hartung, J.B. (1990) *Organometallics* **9**, 547.

STRUCTURE, MELTING, AND REACTIVITY OF NICKEL CLUSTERS FROM NUMERICAL SIMULATIONS*

J. JELLINEK and Z. B. GÜVENÇ †
Chemistry Division
Argonne National Laboratory
Argonne, IL 60439, USA

1. Introduction

Structure and reactivity are aspects common to surface science, organo- and inorganometallic chemistry, and to the relatively new field of cluster research. Here we use the term *cluster* to designate bare ("neet", "naked") clusters of like atoms (molecules). A broader usage of the term also includes organo- and inorganometallic compounds, i.e., ligated clusters. Although cluster research grew out of studies of bulk and surface properties of materials (clusters were first utilized to inquire about the microscopic origins of these properties), it soon acquired the status of an independent field with its own subjects, objectives, and methodologies. As the field matured, it became clear that the relation between bulk and surface properties, on the one hand, and those of clusters of atoms (molecules) of a given material, on the other, may not be as straightforward as originally thought. One of the reasons is that the finiteness of clusters, especially those of smaller and intermediate sizes, has a major effect on their electronic and energetically favorable geometric structure(s), as well as other properties. Many of these properties turn out to be quite different from the corresponding macroscopic properties of the materials. Another intriguing and complicating issue is that the properties of clusters change and eventually converge to bulk and surface properties, as the clusters grow in size, not necessarily in a monotonic or even unique way.

We shall discuss here properties of metal clusters. These are especially close to organometallic and inorganometallic molecular compounds as they form the skeletal framework of these compounds. Issues such as stability, geometric forms, fluxionality, isomerization transitions, and others are common for both bare and ligated metal clusters. The precise meaning and the details of these issues, as dictated by the nature of the objects studied and as emphasized by the practitioners of the different fields, are not necessarily identical. Thus, for

---
*Work performed under the auspices of the Office of Basic Energy Sciences, US-DOE under contract number W-31-109-ENG-38.
†Present address: Department of Physics, University of California, Riverside, CA 90521.

L. J. Farrugia (ed.), The Synergy Between Dynamics and Reactivity at Clusters and Surfaces, 217–240.
© 1995 *Kluwer Academic Publishers.*

example, stable and metastable structures and their possible interconversions are subjects of primary importance and concern in studies of bare metal clusters. One can even talk about a meltinglike transition (loss of structure) in these clusters as their energy is increased. Notions of fluxionality and isomerization as applied to metal-containing molecular compounds, however, often refer to the mobility and bonding rearrangement of the ligands rather than that of the metal framework to which these ligands are attached. Rearrangement of the metal framework may require energies that would cleave the intraligand and metal-ligand bonds and thereby alter the very object of the study. The relationship between the fields of surface science, chemistry of metal-containing molecular compounds, and physics and chemistry of clusters is that of complementarity. Studies of metal clusters and of their interactions with molecules furnish a link between the traditional areas of surface science and the chemistry of metal-based molecular compounds. A comprehensive picture of metal systems and their interactions with ligands that covers the range from a few atoms to the bulk material will emerge as a result of continuing progress in each of the three research areas.

The subject of our discussion is structural and dynamical properties of nickel clusters as derived from numerical simulations. A brief description of the potential energy functions used to represent the intracluster interactions and the interaction of a Ni_{13} cluster with a D_2 molecule is given in Section 2, which also contains a sketch of the theoretical and computational methodologies utilized. A sample of results on structures of nickel clusters of different sizes is presented in Section 3. The evolution of the patterns and peculiarities of the intracluster dynamics with the cluster energy, which leads to structural isomerization and solid-to-liquid-like transitions, is discussed in Section 4. In Section 5 we present and discuss results on the $Ni_{13} + D_2$ collision system. In particular, we analyze the dynamics of the reactive channel of the interaction (i.e., dissociative adsorption of the molecule on the cluster) as defined by the initial rovibrational state of D_2, the system collision energy, and the structure and temperature of Ni_{13}. The calculated data are compared with the available experimental results and a "tuning" of the cluster-molecule interaction potential based on the interplay between theory and experiment is performed. A brief summary is given in Section 6.

2. Theory and Methodology

Ideally one would like to obtain the energies of the interaction and the forces between the atoms in a cluster or in a cluster-molecule system from first principle calculations. Despite the immense recent progress in computational methodologies and computer technology, such calculations for systems containing more than a few transition metal atoms remain beyond the limits of practical feasibility, especially if, as in dynamical simulations, they must be repeated many times. The alternative is to use semiempirical potentials. We employ Voter and Chen's form [1] of an embedded-atom potential V_{EA} to describe the interatomic interactions in nickel clusters. This potential is fitted to seven properties of bulk nickel (equilibrium lattice constant and cohesive energy, bulk modulus, three cubic elastic constants, and vacancy formation energy), as well as the binding energy and equilibrium bond length of Ni_2. It is expected to be more adequate for clusters than those potentials that incorporate only properties of bulk matter.

Globally or locally stable structures (isomers) of clusters of different sizes are obtained by simulated thermal quenching from configurations generated along phase space trajectories. These trajectories are produced through constant-energy molecular dynamics simulations. The initial coordinates and momenta of the constituent atoms are defined in such a way as to supply no total linear and angular momenta to the cluster. The forces experienced by the individual atoms are calculated from V_{EA}. The time evolution of the cluster is obtained by solving numerically Hamilton's equations of motion for all atoms. Hamming's modified predictor-corrector propagator is used with a step size of 10^{-15} s. To assure an adequate sampling of the configuration space the trajectories should correspond to sufficiently high values of the internal energy and they have to be run sufficiently long. Even in the longest runs performed ($3 \cdot 10^6$ steps) the total energy is conserved within 0.03%.

The evolution of the dynamics with the cluster energy is analyzed in terms of the following characteristic quantities:

1) Short-time averaged (internal) kinetic energy of the cluster per atom $\overline{E}'_k$ as a function of time. The goal of the averaging is to suppress the changes in the kinetic energy due to vibrational motion of the atoms. The averaging is performed over intervals of $5 \cdot 10^{-13}$ s. (Note for calibration that the vibrational period of Ni_2 is $1.01 \cdot 10^{-13}$ s, and the period corresponding to the Debye frequency of bulk nickel is $1.18 \cdot 10^{-13}$ s);

2) Relative root-mean-square (rms) bond length fluctuation δ,

$$\delta = \frac{2}{n(n-1)} \sum_{i<j}^{n} \frac{(<r_{ij}^2> - <r_{ij}>^2)^{1/2}}{<r_{ij}>}, \quad (1)$$

where n is the number of atoms in a cluster, r_{ij} is the distance between atoms i and j, and $<>$ denotes time-averaging over the entire trajectory;

3) Mean square displacement $\Delta \vec{r}^2$,

$$\Delta \vec{r}^2(t) = \frac{1}{n \cdot n_t} \sum_{j=1}^{n_t} \sum_{i=1}^{n} \left[\vec{r}_i(t_{oj}+t) - \vec{r}_i(t_{oj}) \right]^2, \quad (2)$$

as a function of time t. In Eq. (2), $\vec{r}_i$ is the position vector of atom i and n_t is the number of time origins t_{oj};

4) Diffusion coefficient D, which is related to the mean square displacement $\Delta \vec{r}^2$ by the equation

$$D= \frac{1}{6} \lim_{t \to \infty} \frac{d(\Delta \vec{r}^2 (t))}{dt} . \tag{3}$$

The total interaction energy of the $Ni_{13}+D_2$ collision system is described by the potential V,

$$V = V_{EA} + V_{LEPS} , \tag{4}$$

where V_{LEPS} is a London-Eyring-Polanyi-Sato representation of the D-D and D_2-Ni_{13} interactions. We employ the form of V_{LEPS} suggested by Raghavan et al. [2] and modified by a smoothing function of the form used by Truong et al. [3]. The values of the parameters of V_{LEPS} are those of the potential energy surface (PES) II of Raghavan et al. [2]. These values are chosen to reproduce the binding energy and the binding height of an H atom over a (triangular) face of an icosahedral Ni_{13} as calculated using the so-called corrected effective-medium approach. We examine the adequacy of the PES II by comparing the calculated data on the dissociative adsorption of the molecule on the cluster with the available measured data and suggest a modification of V_{LEPS} that brings the theoretical and experimental results into agreement.

The quasiclassical trajectory approach is used to mimic collisions of D_2 with Ni_{13}. The initial coordinates and momenta of the D atoms are chosen in such a way as to place the center of mass (COM) of the D_2 molecule at a distance 8.5 Å from the COM of the cluster (the energy of interaction of the molecule with the cluster is zero at this distance) and to supply to the molecule amounts of vibrational and rotational energy that correspond to specified vibrational v and rotational j quantum numbers, respectively. The quasiclassical quantization procedure used is described in Ref. [4]. The molecule is propelled toward the cluster with a specified collision (i.e., relative translational) energy E_{tr} and impact parameter b. The initial state of the nontranslating and nonrotating cluster is specified by its geometry and temperature T,

$$T = \frac{2<E_k>}{(3n-6)k} , \tag{5}$$

where E_k is the total internal kinetic energy of the cluster and k is the Boltzmann constant. The time evolution of the collision system is mimicked by solving Hamilton's equations of motion for all fifteen atoms as described above. Upon collision, the molecule either adsorbs on the cluster (dissociatively or nondissociatively) or is scattered by it. The state of molecular adsorption, which we refer to as a *resonance*, is a transient state with a finite lifetime. The cluster-molecule complex is a *reactive* resonance if the adsorbed molecule eventually dissociates on the cluster. The resonance is *nonreactive* if the molecule desorbs from the cluster without ever dissociating. Thus the two ultimate channels of the cluster-molecule interaction (both may proceed via direct and resonance-mediated pathways) are the reactive one (dissociative adsorption of the molecule) and the nonreactive one (scattering of the molecule). We ran N=500 to 10000 trajectories for each fixed set of initial conditions to calculate the probabilities of the two channels. The trajectories for a given set of initial conditions differ by the initial phase of the D_2 oscillator and by the initial relative orientation of the molecule and of the cluster. Each trajectory is propagated until either the D-D separation reaches the value of 2.223 Å (three times the equilibrium bond length of the D_2 molecule), in which case the trajectory is counted as reactive, or the molecule after collision with the cluster returns into the asymptotic region, in which case the trajectory is qualified as nonreactive. The state-specific probability $P_{v,j,T,...}(E_{tr},b)$ of a channel is defined as

$$P_{v,j,T,...}(E_{tr},b) = \frac{\tilde{N}_{v,j,T,...}(E_{tr},b)}{N} , \tag{6}$$

where $\tilde{N}_{v,j,T,...}(E_{tr},b)$ is the number of trajectories resulting in the channel of interest. The nonreactive channel probabilities are resolved also with respect to the final states of the molecule and of the cluster (for details see Ref. [5]). Here we consider only the reactive channel. The detailed probabilities are calculated on a grid of energies and impact parameters with increments of $\Delta E_{tr}=0.05$ eV and $\Delta b=0.25$ Å, respectively. The state-resolved reaction cross sections $\sigma_{v,j,T,...}(E_{tr})$ are obtained as

$$\sigma_{v,j,T,...}(E_{tr}) = 2\pi \int_0^{b_{max}} bP_{v,j,T,...}(E_{tr},b)db , \tag{7}$$

where b_{max} is the largest impact parameter contributing to the dissociative adsorption of the molecule. The translational-temperature T_{tr} dependent state-specific reaction rate constants $\kappa_{v,j,T,...}(T_{tr})$ are calculated using the formula

$$\kappa_{v,j,T,...}(T_{tr}) = \pi\mu \left(\frac{2}{\pi\mu kT_{tr}}\right)^{3/2} \int_0^\infty \sigma_{v,j,T,...}(E_{tr}) E_{tr} \exp\left(-\frac{E_{tr}}{kT_{tr}}\right)dE_{tr} , \tag{8}$$

where μ is the reduced mass of the cluster-molecule collision system.

3. Structures

In Fig. 1 we present the lowest energy structures, as obtained from V_{EA}, for a variety of small nickel clusters [6,7]. The major feature to notice is that these structures, for the cluster sizes shown, are either complete icosahedra (Ni_{13}, Ni_{55}) or icosahedra-based geometries with missing (Ni_9, Ni_{12}, Ni_{18}) or added (Ni_{14}, Ni_{15}, Ni_{20}) atoms. Ni_{19} is a so-called double icosahedron. Most of these

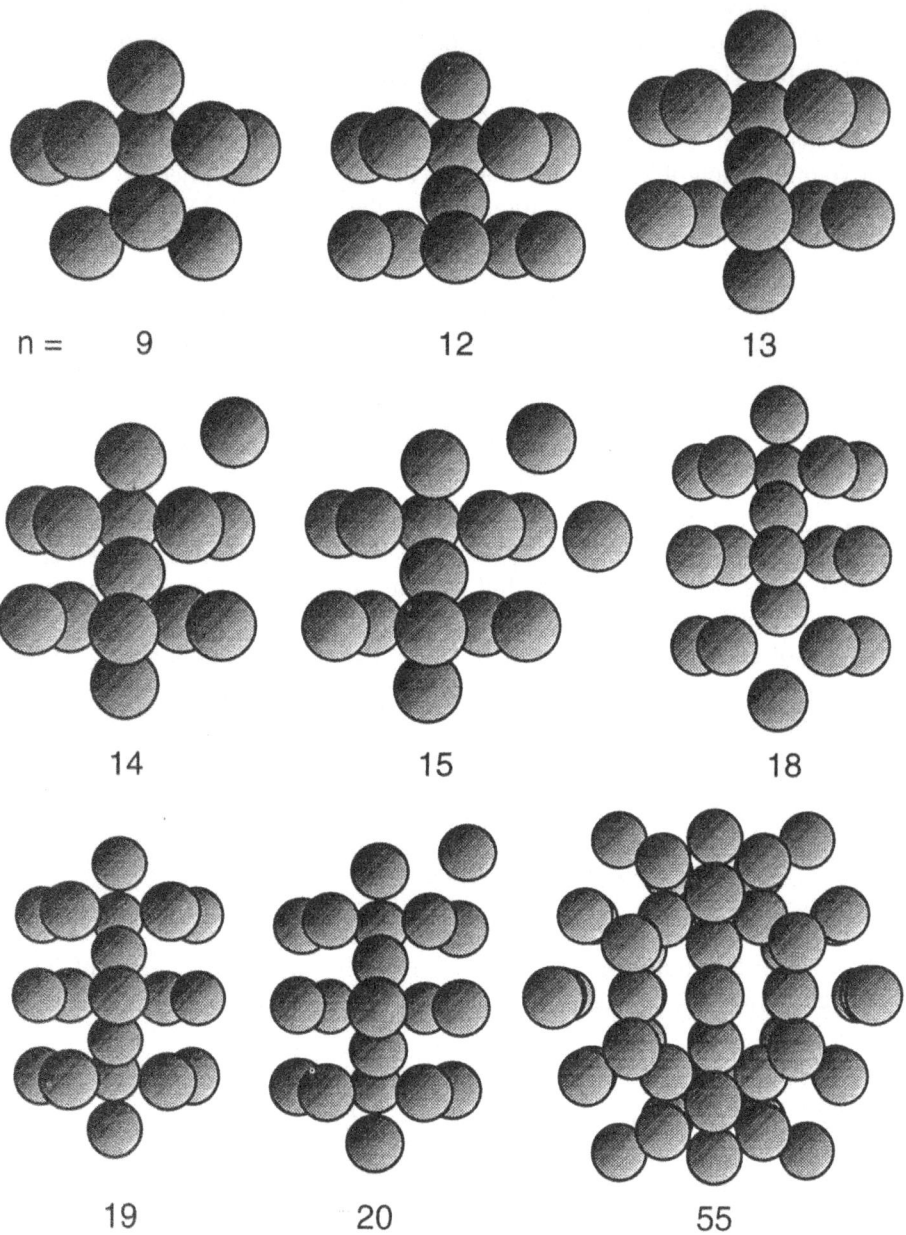

Figure 1. Most stable structures of Ni$_n$ clusters as obtained from V$_{FA}$. Their energies in eV are: n=9: -25.29; n=12: -36.48; n=13: -41.12; n=14: -44.09; n=15: -47.70; n=18: -58.52; n=19: -63.01; n=20: -65.98; n=55: -203.02.

structures are in agreement with those calculated using alternative approaches [2,8-11], as well as with the ones inferred from experiments on the reactivity of nickel clusters [12]. Under normal conditions bulk nickel assumes the face - centered cubic (fcc) lattice structure. It emerges as a general rule that the energetically most favorable geometries of small [tens of atoms (molecules)] and intermediate [hundreds to a few thousands of atoms (molecules)] size clusters are different from the corresponding bulk structures of the materials, irrespective of the fact that the nature of the interatomic forces in these materials may be very different. One of the tantalizing questions is what are the critical sizes for clusters of different materials at which they prefer to assume the bulk structure, and how do these sizes depend on the nature (e.g., anisotropy and range) of the interatomic forces [13-15]. A related issue is the pathways of the evolution of the preferred geometries with the cluster size. There are indications that these pathways may be quite intricate in that they may involve polycrystalline states [16].

Another morphological property of clusters is that for fixed sizes they form hierarchies of structural isomers. These isomers correspond to different minima of the (3n-6)-dimensional potential energy surfaces (we abstract ourselves from minima related by permutational symmetry). In Fig. 2 five isomers of Ni_9 obtained from V_{EA} are shown. The energies of some of these isomers are indeed very close.

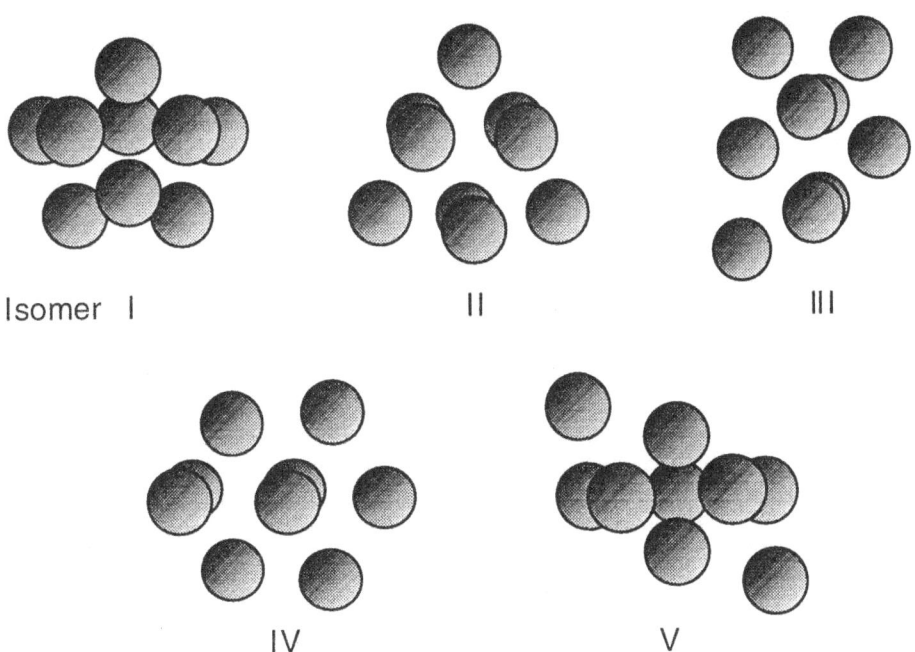

Isomer I   II   III

IV   V

Figure 2. Isomers of Ni_9. Their energies in eV are: I: -25.29; II: -25.20; III: -24.86; IV: -24.84; V: -24.65.

Figure 3 displays three highly symmetrical structures of Ni_{13}. The icosahedral (ico) isomer of the cluster is separated from its hexagonal close-packed (hcp) and cuboctahedral (cubo) forms by a relatively large energy gap, whereas the energies of the hcp and the cubo structures are again very close. The cubo structure can be viewed as a fragment of the fcc lattice. The two lowest energy structures of Ni_{14}, as defined by V_{EA}, are obtained by placing an extra atom over a face (threefold hollow site) and over an edge (twofold site), respectively, of the ico Ni_{13} and allowing for a small relaxation. The energy difference between the thus obtained two isomers of Ni_{14} is 0.19 eV. The fact that both have an ico-based structure has a profound effect on the way the dynamics of Ni_{14} responds to an increase of the internal energy of this cluster (see section 4).

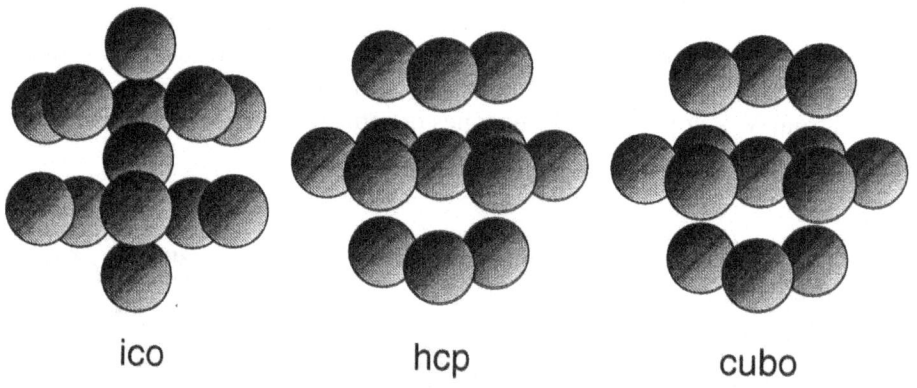

ico   hcp   cubo

Figure 3. Highly symmetrical structures of Ni_{13}.
Their energies in eV are: ico: -41.12; hcp: -39.53; cubo: -39.51.

Figure 4 shows three isomers of Ni_{55}. A 55-atom cluster is the smallest one in which a complete closure of two spherical atomic shells can be achieved. The lowest energy ico form of Ni_{55} is just such a structure. The next isomer in the figure, labeled as "defico", is especially interesting in that its structure is that of an ico with a defect: one of the edge atoms is missing and an extra atom appears on the opposite side of the cluster causing a small relaxation away from the ideal ico geometry. The cubo structure - a fragment of the fcc lattice - is a metastable form of the cluster. A more complete description of the structures of nickel clusters of various sizes, as determined from V_{EA}, will be given elsewhere.

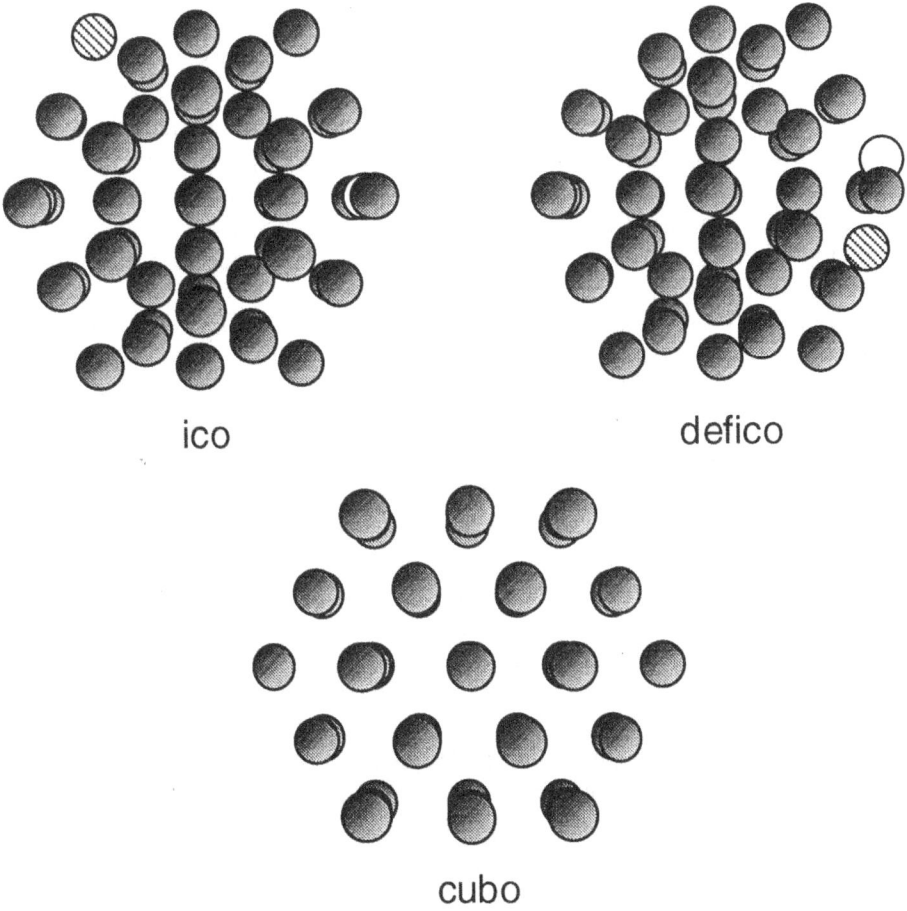

Figure 4. Isomers of Ni$_{55}$. Their energies in eV are: ico: -203.02; defico: -201.68; cubo: -199.63. The shade-coding is used to establish a correspondence between the structures rather than between the individual atoms of the ico and defico isomers.

4. Isomerization and Melting

In this section we present evidence for isomerization and solid-to-liquid-like transitions in small nickel clusters caused by an increase of their internal energy. In particular, we illustrate the important role of the cluster size and the associated with it structures in determining the particular mechanisms and stages of the meltinglike transition. Bulk melting is usually viewed and treated as a thermodynamic phenomenon. The essence of this phenomenon can, however, be expressed and understood in dynamical terms as a transition from a limited and correlated (periodic) motion of atoms, which is typical for solids, to an essentially uncorrelated motion, which is typical for liquids. More precisely, melting is characterized by a drastic reduction in the characteristic correlation time and length. Clusters, as systems with a relatively small number of atoms (molecules), afford us the opportunity to carry out detailed dynamical investigations of the analog of the bulk melting in finite systems.

In earlier studies of the meltinglike phenomenon in van der Waals and ionic clusters [17-22] the solid-to-liquid-like transition has been characterized as a change in the cluster dynamics from being confined to a part of configuration space ("catchment area", "basin of attraction") associated with a single, e.g., lowest energy, structure of a cluster to exploring all its possible configurations. We exclude from consideration here those parts of configuration space that correspond to fragmentation of the cluster (cf. Ref. 11). The change in the pattern of dynamics is caused by an increase of the cluster energy. (As specified above, we consider nontranslating and nonrotating clusters, and therefore the total energy of a cluster is its internal energy. In general, the energy may be partitioned between the vibrational modes and the overall rotation; the translational motion is of no consequence for phenomena considered here. For a discussion of the effects of the rotational motion in clusters see Refs. [23-25]). As the cluster energy increases, it reaches the value of the lowest barrier separating the catchment area of the initial geometry from the rest of the configuration space. Then, unless the initial conditions specify a so-called trapped trajectory, the cluster dynamics finds its way out of the basin of attraction of the initial structure, i.e., the cluster undergoes an isomerization transition. As the cluster energy continues to increase so does the number of isomers sampled. Extensive numerical simulations [17-22] confirm the prediction derived from an analytical model [26] that there is a range of energy (or temperature, if temperature is the control parameter) of finite size that separates the solidlike and the liquidlike forms of a cluster. This range represents the isomerization stage in the meltinglike transition. The dynamics carries the cluster through the basins of its different isomeric forms, but between successive isomerization transitions it gets trapped in the catchment areas of the individual isomers for relatively long times (as measured by the characteristic vibrational period of the cluster) so that these isomers can be identified. This stage is called also the "coexistence" stage, and the energy (temperature) range corresponding to it is called the "coexistence" range. Different isomers of the cluster sequentially appear, and in this sense coexist, along the same phase space trajectory. The term acquires a more direct meaning when one considers the ensemble counterparts of the dynamics - microcanonical for constant energy dynamics and canonical for constant temperature dynamics (cf., e.g., Refs. [17,18]). At any energy (temperature) from the coexistence range solidlike and liquidlike clusters coexist within the same corresponding ensemble. With an increase of the energy, not only the number of isomeric forms visited, but also the frequency of isomerization transitions increases. The dwell times between these transitions decrease. A cluster is qualified as liquidlike when its dwell times in the basins of the individual isomers become comparable to or less than its characteristic vibrational period. Numerical simulations [6-9,27-29] confirm that this general picture applies to metal clusters as well. However, these simulations also reveal that metal clusters may possess new elements in the mechanisms of their meltinglike transition.

In Fig. 5 the time dependence of the short-time averaged kinetic energy (per atom) $\bar{E}_k$ is shown for Ni_{13} at different values of the total energy (per atom) [6]. Graphs a) and b) display an essentially constant $\bar{E}'_k$, which, taking into account the constancy of the total energy of the cluster, means constancy of its short-time averaged potential energy. The reason for this constancy is that at energies corresponding to graphs a) and b) the cluster dynamics explores the catchment area of a single isomer - in this case of the lowest energy ico isomer. Thermal quenches from configurations traced out by the trajectories that generate graphs a) and b) confirm this assertion. Graph c), which corresponds to a higher energy of the cluster, displays well-identifiable branches that persist for appreciable times and that are intermittently revisited by the dynamics. These branches represent different averaged potential energies, i.e., different structures of the cluster. The highest-$\bar{E}'_k$ branch corresponds to dynamics trapped in the basin(s) of the ico isomer. The lower-$\bar{E}'_k$ branches represent sampling of the catchment areas of the higher energy metastable isomers. The pattern of graph c) is typical of the coexistence stage. As the energy of the cluster increases further, the pattern of $\bar{E}'_k$ becomes that of graph d). Large fluctuations and the almost complete disappearance of resolvable individual branches indicate frequent interbasin transitions - the cluster becomes liquidlike.

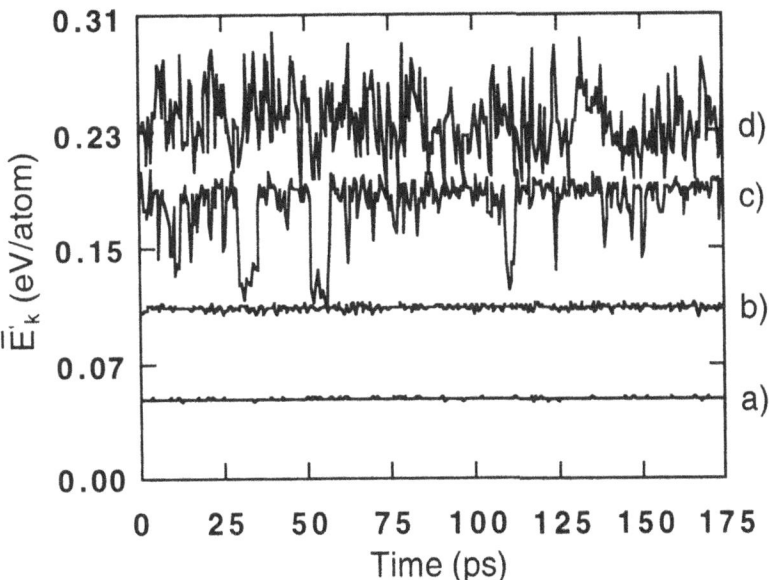

Figure 5. Short-time averaged kinetic energy per atom as a function of time for Ni_{13}. The graphs correspond to the following total energies (in eV) per atom: a) -3.05; b) -2.92; c) -2.73; d) -2.49.

The above assignment of the solidlike and liquidlike states of N_{13} is corroborated by Fig. 6. It shows the graphs of the mean square displacement for two energies considered in Fig. 5 (graphs corresponding to the same energy are labeled by the same letter in the two figures). The long-time tale of graph b) has a zero slope indicating no diffusion in the cluster at the energy corresponding to this graph [cf. Eq. (3)]. The cluster is solidlike despite the fact that its temperature exceeds 1000 K (cf. Fig. 7). In contrast, the large slope of graph d), Fig. 6, signifies a well-developed diffusive motion, which is a signature of a liquidlike state. Further corroboration of the meltinglike transition in the cluster is given by Fig. 7. The abrupt increase in the rms bond length fluctuation with the temperature is analogous to the Lindemann criterion of bulk melting [30]. The transition starts only when the temperature of Ni_{13} reaches the value of approximately 1300 K, which indicates a high degree of stability of this cluster. (The melting temperature of bulk nickel is 1726 K). The general qualitative features of the meltinglike transition in Ni_{13}, with a possible exception of its extra stability, are shared by nickel clusters of other sizes, such, for example, as n=12, 15, 18, 19, etc. The double-icosahedral Ni_{19} also shows a considerable resistivity to melting. The mechanism of the solid-to-liquid-like transition in these clusters is similar to that in van der Waals and ionic clusters [17-22].

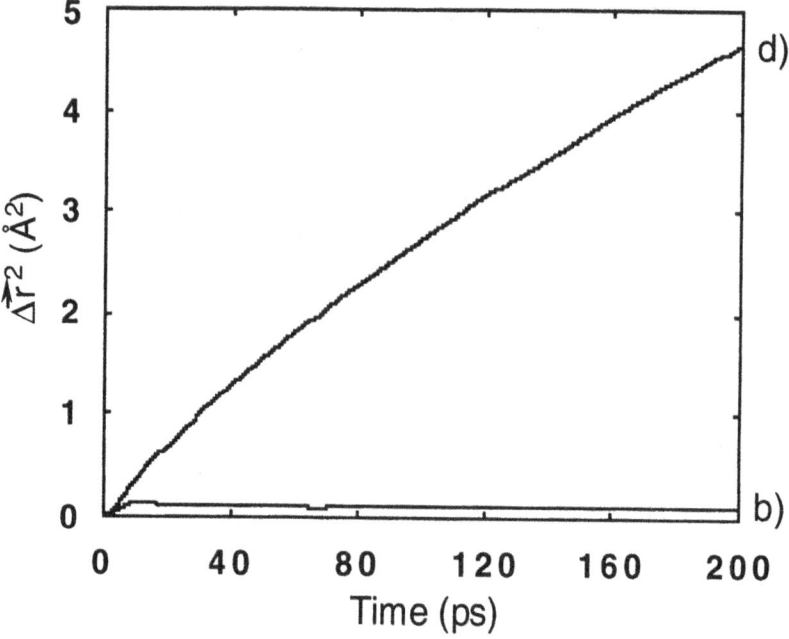

Figure 6. Mean square displacement as a function of time for N_{i13}. Graphs b) and d) correspond respectively to the same total energies as graphs b) and d) of Fig. 5.

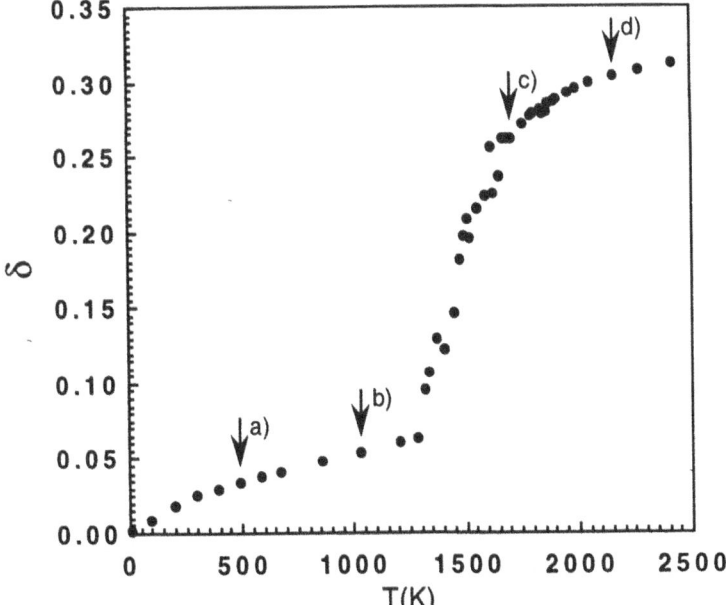

Figure 7. Relative rms bond length fluctuation as a function of cluster temperature for Ni_{13}. Arrows indicate points of the graph that correspond to the total energies considered in Fig. 5.

The picture is quite different for Ni_{14}. The rms bond length fluctuation as a function of temperature is shown for this cluster in Fig. 8. The graph clearly displays two abrupt increases. Such a pattern of δ has not been obtained for nonmetallic clusters. The first increase starts at $T \approx 300$ K and is associated with the so-called premelting phenomenon [6,8,9]. This phenomenon involves only the two lowest energy structures of Ni_{14} mentioned above. It includes a coexistence (isomerization) stage, when the 14-th atom intermittently changes its position from a threefold to a twofold site and back, followed, as the energy increases, by a local melting of the cluster. The local melting involves frequent migrations of the 14-th atom between its threefold and twofold sites and occasional insertion of this atom into the framework of the icosahedral 13-mer. The insertion is accompanied by one of the neighboring atoms of the 13-mer moving outward and becoming the "extra" 14-th atom. The relatively large amplitude motions are localized to the vicinity of the 14-th atom. The rest of the cluster preserves its overall icosahedral symmetry. Premelting is a consequence of the local destabilizing effect of the 14-th atom on the otherwise stable structure of the underlying icosahedral 13-mer. The second abrupt increase of δ takes place at considerably higher temperatures. It signifies complete melting of Ni_{14}. This complete melting requires melting of the underlying icosahedral 13-mer and it includes all the stages described above for Ni_{13} [6]. Inspection of Figs. 7 and 8 shows that, indeed, the second abrupt increase of δ for Ni_{14} takes place over a temperature range that is part of the transition range of Ni_{13}. We have used visualization techniques to disentangle the different elements in the quite intricate melting mechanism of Ni_{14}. The Ni_{20} cluster, which is similar to

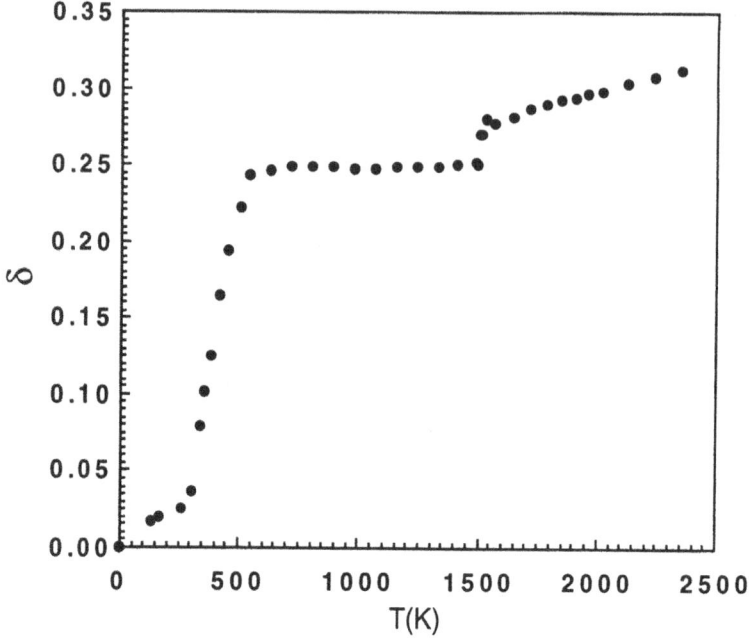

Figure 8. Relative rms bond length fluctuation as a function of cluster temperature for Ni$_{14}$.

Ni$_{14}$ in that its lowest energy structure is obtained by adding an extra atom to an especially stable network - the 19-atom double icosahedron - also exhibits the premelting stage in its meltinglike transition. We conjecture that the premelting phenomenon is a feature of nickel clusters of those sizes that form complete icosahedra with one (or, possibly, more than one for clusters of larger sizes) extra atom. The n-body character of metallic cohesion certainly plays a role in emergence of the premelting phenomenon (cf. Refs. [8,9]). Not all 14-atom metal clusters, however, exhibit it. The reason is that the lowest energy structure of these clusters may be different from that of Ni$_{14}$. Au$_{14}$ is one example, and it does not exhibit premelting [29]. An important factor in defining the lowest energy structures of metal clusters is the range of the attractive and repulsive components in the n-body interactions.

Yet another possible melting scenario is shown in Fig. 9, which displays the rms bond length fluctuation for Ni$_{55}$. The ico isomer of the cluster is chosen as its solid form. As mentioned, this isomer contains two complete spherical shells of atoms, and its core is a 13-atom icosahedron. The graphs show the temperature dependence of δ calculated for the surface, core, and all atoms of the cluster. Whereas the network of the 42 surface atoms that form the external shell begins to undergo a meltinglike transition at T$\approx$850 K, the core remains solidlike up to T$\approx$1100 K. Thus, there is a finite temperature range for which Ni$_{55}$ can be viewed as a liquidlike mantle encapsulating a solidlike core. At T$\approx$1200 K the entire cluster becomes liquidlike. Analysis of other physical quantities, for example of the diffusion coefficients, confirms the phenomenon of surface melting in Ni_{55} (the details are given in Ref. [7]). Surface melting has also been predicted for Cu_{55} and noble gas clusters of a variety of sizes [31], as well as gold clusters of larger sizes [32].

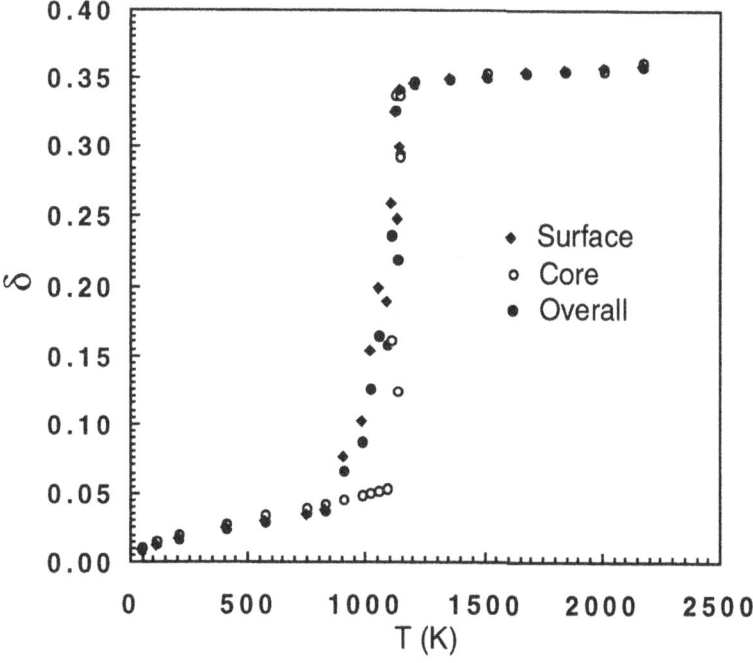

Figure 9. Relative rms bond length fluctuation as a function of cluster temperature for Ni_{55} (see text).

5. Reactivity

In this section we present results on the reaction $Ni_{13} + D_2 \rightarrow Ni_{13} \cdot 2D$ [5,33-36]. First we consider results obtained with the PES II (see section 2) parameters of V_{LEPS}. The cross sections of the dissociative adsorption as a function of the collision energy are shown for the room-temperature ico Ni_{13} and different initial rovibrational states of D_2 in Fig. 10. The (v=0, j=10) and (v=1, j=0) states of the molecule are energetically almost degenerate. They are chosen to test the mode-selectivity of the reaction, i.e., its sensitivity to the initial partitioning of the energy between the vibration and rotation of the molecule. The major features displayed by Fig. 10 are: 1) the reactivity of the cluster increases monotonically with an increase of the system collision energy; 2) in order to initiate the reaction the collision energy has to be larger than a (small) (v, j)-dependent threshold value; 3) larger initial rovibrational energy of the molecule results in larger reaction cross sections; 4) the reaction is mode-selective - an initial vibrational excitation of the molecule leads to considerably larger reaction cross sections than an energetically equal initial rotational

232

excitation. We have determined that change of the initial cluster temperature in the range T=0-300 K has no effect on the graphs of Fig. 10 (cf. also below).

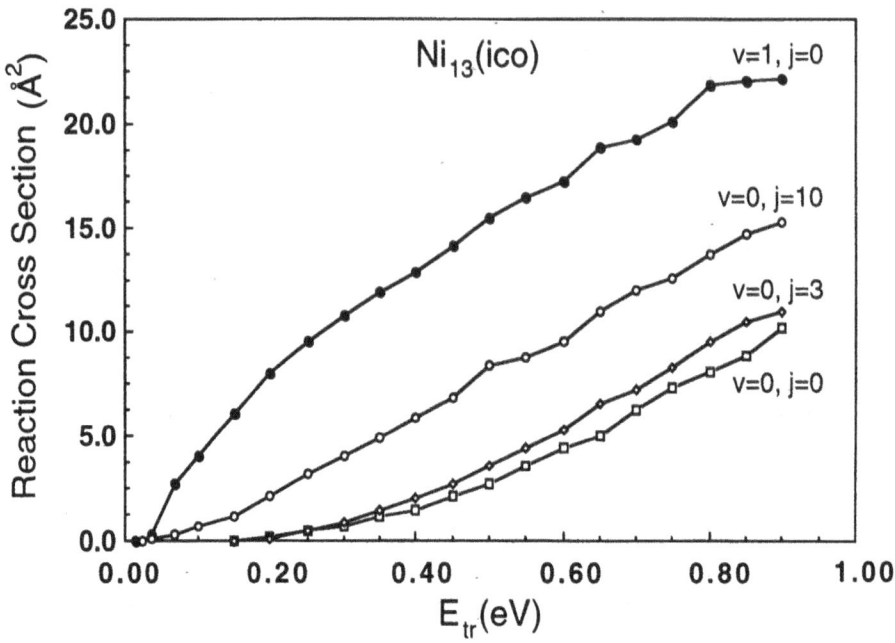

Figure 10. Reaction cross sections as functions of the collision energy for a room-temperature ico Ni_{13}. The different graphs correspond to different initial rovibrational states of D_2.

Because the cubo and hcp structures of Ni_{13} are metastable, the reactivity of these forms is studied at zero initial temperature of the cluster. The reaction cross sections for the cubo and hcp geometries are shown in separate panels of Fig. 11. The qualitative features of the corresponding graphs in the two panels are strikingly similar, with the cubo structure being slightly more reactive. As in the case of the ico isomer, higher initial rovibrational energies of D_2 result in larger cross sections, and the reaction remains mode-selective. In all other respects the features of the graphs in Fig. 11 are different from those in Fig. 10. Overall, the cubo and hcp forms of Ni_{13} are more reactive than its ico isomer. The molecule dissociates on the cubo and hcp clusters at any collision energy. But more importantly, the graphs of Fig. 11 show a nonmonotonic change of the reaction cross section with the collision energy. Especially intriguing are the peaks in the graphs at very low collision energies. We have used time-dependent reaction probabilities and cross sections to differentiate between the direct and the precursor-mediated reaction pathways [33,34,36] and determined that these peaks are almost entirely due to formation of reactive resonances. At very low collision energies the probability of molecular adsorption ("sticking") on the cluster is high. This is true for all three structures of Ni_{13} considered here. A large fraction of the adsorbed D_2 molecules then proceeds to dissociate on the cubo and hcp forms of the cluster. In contrast, the precursor states

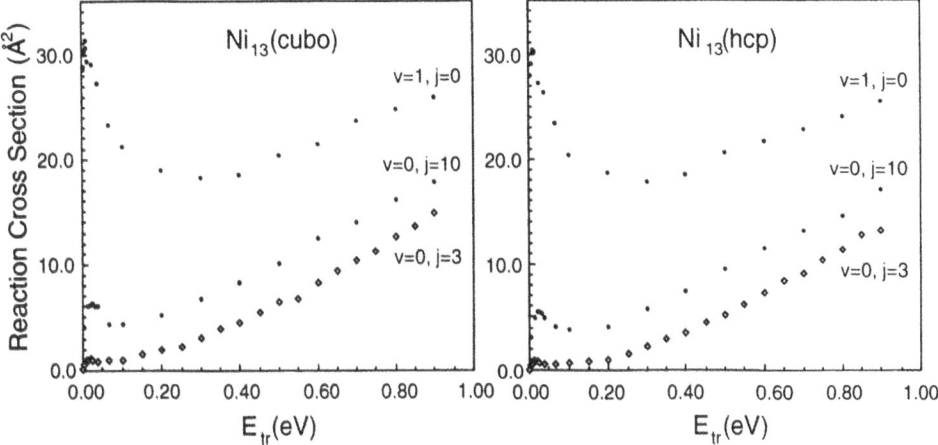

Figure 11. The same as Fig. 10 but for zero-temperature cubo and hcp forms of Ni_{13}.

formed on the ico Ni_{13} at low collision energies turn out to be nonreactive resonances, i.e., the adsorbed D_2 molecules eventually desorb from the cluster without ever dissociating. As a consequence, the graphs of Fig. 10 do not possess the low-E_{tr} peaks. Upon increase of the collision energy the probability of molecular adsorption and, consequently, of the indirect reaction, decreases. This is the reason for the decrease of the cross sections over a finite range of E_{tr} exhibited by the graphs of Fig. 11. As the collision energy continues to increase, the direct reaction begins to play a more important role. Its probability increases with E_{tr}, and eventually the cross sections become increasing functions with values that are less sensitive to the cluster structure. It is clearly the difference in the surface topology of the ico isomer, on the one hand, and of the cubo and hcp forms, on the other, that is responsible for the difference in the low-E_{tr} parts of the graphs in Figs. 10 and 11. The ico isomer has only threefold faces whereas the cubo and hcp structures possess threefold and fourfold faces. (An energy argument regarding the difference in the reactivity of the different structures of Ni_{13} at low collision energies is given below). Characteristic results on the reactive resonance formation probabilities and lifetimes, as well as the branching ratio between the direct and indirect reaction pathways are given in Table 1 (for details see Refs. [33,34,36]). With increase of the collision energy, the lifetime of the reactive resonances decreases and the relative contribution of the direct reaction into the process of dissociative adsorption of the molecule increases. Experimental evidence corroborating the important role of reactive resonances in reaction of D_2 with nickel clusters has been obtained recently [37,38].

For the ico isomer one can examine the effect of the cluster temperature on its reactivity. Results covering a broad temperature range are shown in Fig. 12.

TABLE 1. Probabilities $P_{...}^{DR}$ and $P_{...}^{IR}$ of the direct and indirect reaction, respectively, fractional contribution $f_{...}^{DR}$ of the direct reaction, and reactive resonance lifetime $\tau_{...}$ for a cubo Ni_{13} (T=0K), D_2(v=1, j=0), b=0.25 Å and different values of E_{tr}.

E_{tr} (eV)	0.038	0.30	0.50	0.90
$P_{...}^{DR}$	0.06	0.11	0.13	0.20
$P_{...}^{IR}$	0.44	0.39	0.52	0.57
$f_{...}^{DR}$ (%)	12	22	19	26
$\tau_{...}$(ps)	0.07	0.03	0.02	0.01

As mentioned, the reactivity of the cluster does not change in the range T=0-300 K. This has been confirmed experimentally [37]. Because of the relatively large mass of nickel atoms, heating from 0 to 300 K causes only slow motions in the cluster and only small departures in its structure from the ideal

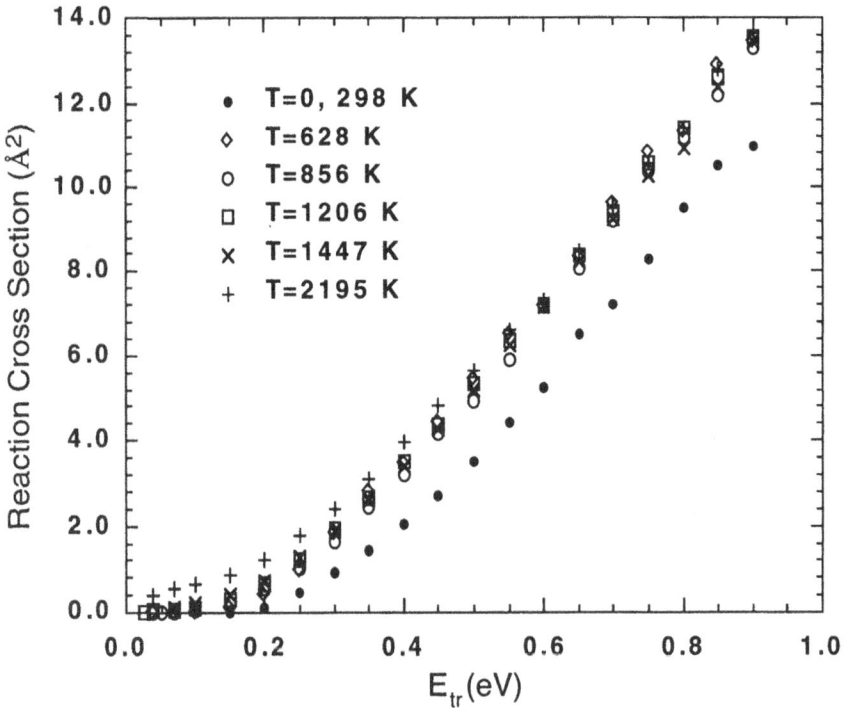

Figure 12. Reaction cross sections as functions of the collision energy for dissociative adsorption of D_2(v=0, j=3) on an ico Ni_{13} at different initial temperatures T of the cluster.

icosahedral geometry. It is somewhat surprising, however, that further heating and even melting of the cluster has only a minor effect on its reactivity.

Whereas a number of predictions [5,33-36] derived using the PES II parame-terization of V_{LEPS} has been confirmed experimentally, the computed value of the reaction rate constant at room temperature turns out to be orders of magni-tude lower than that inferred from the experiments [37]. This indicates that the original PES II parameterization has to be modified. As a guidance for "tuning" of V_{LEPS} we used the following considerations. The binding energy of a D atom in a threefold hollow site of an ico Ni_{13} calculated using the original V_{LEPS} is 2.43 eV. This value is about 10% lower than the measured energy of binding of a D atom to the (111) and (100) faces of bulk nickel. No experimental or ab initio results on the D-Ni_{13} binding energy are available at present. Arguments have, however, been put forward [39] that this energy may be close to or even exceed the corresponding value for the atom-surface binding. We modified the PES II parameters [35,36] so that the "tuned" V_{LEPS} gives a 10% larger D-Ni_{13}(ico) binding energy than the original one. The reaction cross section cal-culated with the modified V_{LEPS} for the case of a room-temperature ico isomer of the cluster and (v=0, j=3) initial rovibrational state of D_2 is shown in Fig. 13.

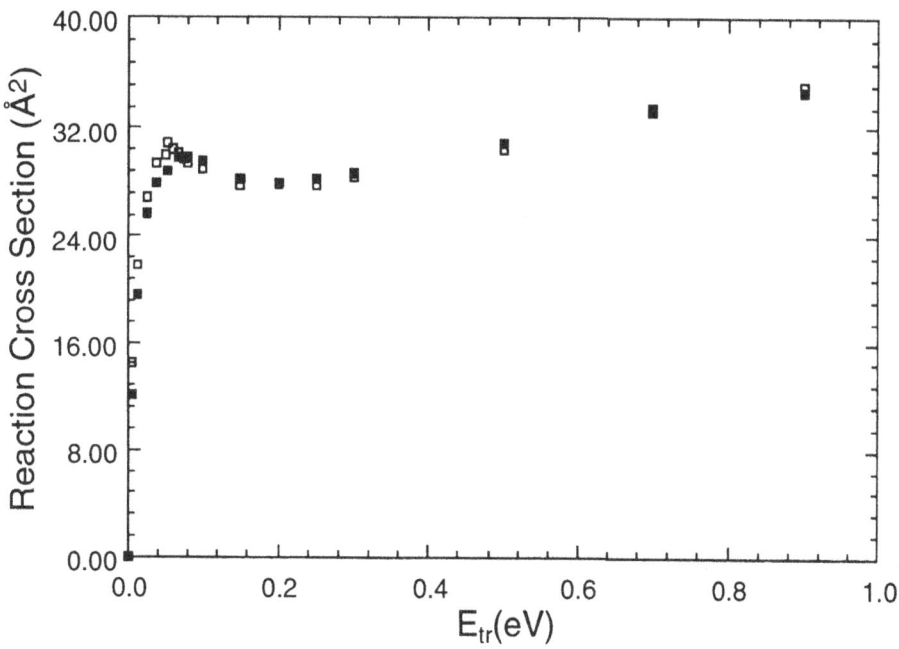

Figure 13. Reaction cross sections for room-temperature ico (full squares) and zero-temperature cubo (empty squares) forms of Ni_{13} obtained using the "tuned" V_{LEPS}. The graphs correspond to $D_2(v=0, j=3)$.

Comparison with the corresponding graph of Fig. 10 shows that the modified V_{LEPS} results in a much higher reactivity of the ico cluster. The reaction rate constant at room temperature computed using the cross sections of Fig. 13 is $3 \cdot 10^{-10}$ cm^3/s. This value is in good agreement with the experimentally derived estimate of $(8 \pm 4) \cdot 10^{-10}$ cm^3/s [37].

Tuning of V_{LEPS} also has a profound effect on the qualitative characteristics of the reactivity of ico Ni_{13}. The low-E_{tr} peak in the graph of Fig. 13 corresponding to this isomer indicates that with the modified V_{LEPS} the ico form of the cluster also supports reactive resonances. The figure displays the reaction cross section obtained with the modified V_{LEPS} for the cubo isomer as well. The ico and cubo forms of the cluster show now almost identical reactivities, except at very low collision energies where the cubo isomer is slightly more reactive. Regarding the temperature dependence of the reactivity, the modified V_{LEPS} gives essentially the same result as the original one. Except at low E_{tr}, the reactivity of the ico isomer is practically independent of its temperature. The only noticeable effect is the gradual disappearance of the low-E_{tr} peak in the graph as the cluster temperature increases [40]. The reason for this is that the resonance formation probabilities (cross sections) are decreasing functions of the cluster temperature.

In order to understand the profound effects of a relatively minor modification of the cluster-molecule interaction potential (the modification performed is within the limits of uncertainty of the original PES II parameters) we have mapped out the original and the tuned V_{LEPS} functions (for details see Refs. [35,36]). The minimum energy paths (MEPs) defined by these functions for ico Ni_{13} are shown together with the Morse potential (and the zero-point energy) of D_2 as functions of the D-D distance in Fig. 14. The minima in the MEPs with the smallest D-D separation are the molecular adsorption wells. The next five minima that correspond to larger D-D distances are the dissociative adsorption wells. They represent the five topologically different possible pairs of threefold faces of the ico isomer as preferred sites for adsorption of two D atoms. The maxima between the dissociative adsorption wells are the diffusion barriers. They correspond to one of the D atoms being over a threefold site (face) and the other over a twofold site (edge) of the cluster. The maxima separating the molecular adsorption wells are the dissociation barriers. The zero-point energy of D_2 is below the dissociation barrier defined by the original V_{LEPS}. This explains the collision energy thresholds in Fig. 10 and the fact that the low-E_{tr} resonances are nonreactive. The dissociation barriers defined by the original V_{LEPS} for the cubo and hcp forms of the cluster are lower than the zero-point energy of D_2 (not shown) [40], which explains the qualitative and quantitative differences between the graphs of Figs. 10 and 11. As seen in Fig. 14, apart from a 10% downward shift, the performed modification of V_{LEPS} does not change the "topology" of the exit channel part of the original MEP. It does, however, have a considerable effect on the transition state. The dissociation barrier gets reduced quite substantially, and its value becomes lower than the zero-point energy of D_2. Figure 13 reflects the consequences of this change. Further discussion on tuning of V_{LEPS} and its implications can be found in

Refs. [35,36]. Here we remark only that use of parameterized potentials appears to be at present (and will, probably, remain for quite some time) the only feasible way for carrying out comprehensive simulation studies of systems as complex as clusters interacting with molecules. An adequate representation of the cluster-molecule interaction is of paramount importance. It can be achieved only through an intimate interplay between theoretical dynamics, ab initio calculations, and experiment.

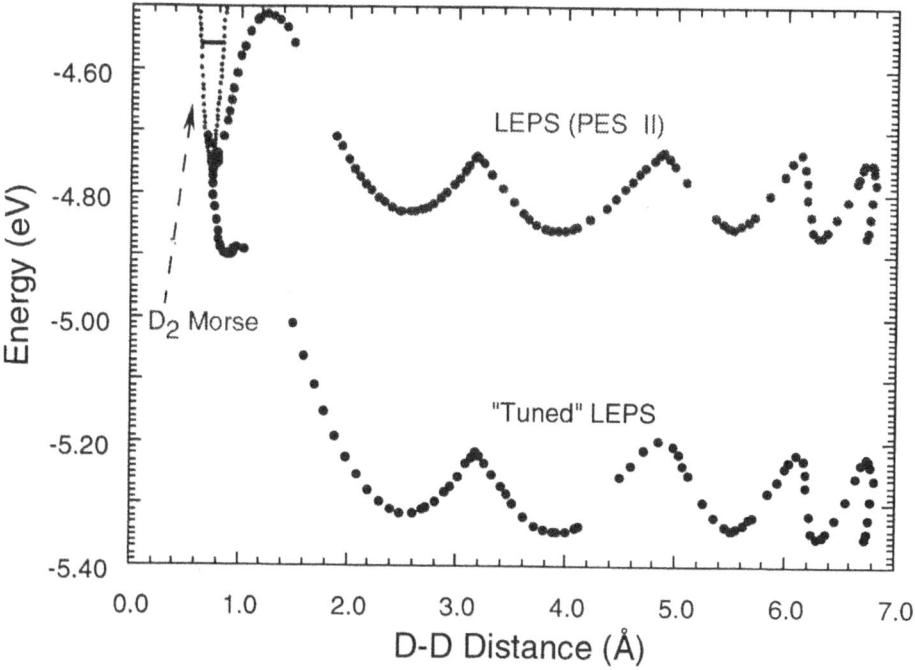

Figure 14. The minimum energy paths corresponding to the PES II and the "tuned" parameterizations of V_{LEPS} as functions of the D-D separation for the case when the cluster is frozen in its ico geometry. The Morse potential and the zero-point energy of D_2 are also shown.

6. Summary

We have presented here a sample of results on the structural and dynamical properties of nickel clusters derived from numerical simulations. These properties are exhibited through the response of the clusters to an increase of their internal energy and through their chemical reactivity.

The interatomic interactions in the clusters are mimicked by a many-body embedded-atom potential. The most stable isomers of small nickel clusters obtained from this potential tend to favor icosahedra-based structures. The clusters exhibit a characteristic meltinglike transition as their internal energy is increased. Distinct stages in this transition can be identified, which depend on

238

the cluster size. A novel feature exhibited by Ni_{14} (Ni_{20}, etc.) is premelting or, alternatively, local melting. It is quite remarkable that in a system containing only 14-atoms a liquidlike part can coexist with a solidlike. This coexistence within the same cluster is different from the "canonical" coexistence between liquidlike and solidlike clusters of the same energy (or temperature). The meltinglike transition in the Ni_{55} cluster includes the stage of surface melting - a phenomenon that has been found to be more generic and shared not only by metal clusters.

Cluster-molecule interactions, more specifically reactions, can be character-ized at the same level of detail as the more extensively studied atom-molecule, molecule-molecule, and gas-surface interactions. We have presented here results on the dissociative adsorption of D_2 on Ni_{13} and described the dependence of this reaction on the system collision energy, initial rovibrational state of the molecule and the structure and temperature of the cluster. The reaction may be direct or resonance-mediated. The resonances are characterized in terms of their formation probabilities and lifetimes, both of which depend on the collision energy. An important issue is the correlation and agreement (or lack of it) between the computed and measured data. We have used the experimental value of the D-nickel-surface binding energy to modify ("tune") the D_2-Ni_{13} interaction potential and have shown that this modification brings the computed and the measured rate constants of the dissociative adsorption of the molecule on the cluster into agreement.

7. References

1.   Voter, A.F. and Chen, S.P. (1987) Accurate interatomic potentials for nickel, aluminum and nickel-aluminum (Ni_3Al), *Mater. Res. Soc. Symp.* **82**, 175-180.
2.   Raghavan, K., Stave, M.S., and DePristo (1989) Ni clusters: structures and reactivity with D_2, *J. Chem. Phys.* **91**, 1904-1917.
3.   Truong, T.N., Truhlar, D.G., and Garrett, B.C. (1989) Embedded diatomics-in-molecules: a method to include delocalized electronic interactions in the treatment of covalent chemical reactions at metal surfaces, *J. Phys. Chem.* **93**, 8227-8239.
4.   Porter, R.N., Raff, L.M., and Miller, W.H. (1975) Quasiclassical selection of initial coordinates and momenta for a rotating Morse oscillator, *J. Chem. Phys.* **63**, 2214-2218.
5.   Jellinek, J. and Güvenç, Z.B. (1991) Mode selectivity in cluster-molecule interactions: $Ni_{13} + D_2$, in J. Jortner, R. D. Levine, and B. Pullmann (eds), *Mode Selective Chemistry*, Kluwer Academic Publishers, Dordrecht, pp. 153-164.
6   Güvenç, Z.B., Jellinek, J., and Voter, A.F. (1992) Phase changes in nickel clusters from an embedded-atom potential, in P. Jena, S. N. Khanna, and B. K. Rao (eds), *Physics and Chemistry of Finite Systems: From Clusters to Crystals*, Vol. 1, Kluwer Academic Publishers, Dordrecht, pp. 411-416.
7.   Güvenç, Z.B. and Jellinek, J. (1993) Surface melting in Ni_{55}, *Z. Phys. D* **26**, 304-306.
8.   Jellinek, J. and Garzón, I.L. (1991) Structural and dynamical properties of transition metal clusters, *Z. Phys. D* **20**, 239-242. Values of kinetic energies and of temperature in this paper and in Refs. [9,29] below have to be divided by two. This does not affect the validity of the qualitative conclusions.

9. Garzón, I.L., and Jellinek, J. (1992) Melting of nickel clusters, in P. Jena, S.N. Khanna, and B.K. Rao (eds), *Physics and Chemistry of Finite Systems: From Clusters to Crystals*, Vol. 1, Kluwer Academic Publishers, Dordrecht, pp. 405-410.

10. Stave, M.S. and DePristo, A.E. (1992) The structure of Ni_N and Pd_N clusters: $4 \leq N \leq 23$, *J. Chem. Phys.* **97**, 3386-3398.

11. López, M. J. and Jellinek, J. (1994) Fragmentation of atomic clusters: a theoretical study, *Phys. Rev. A* **50**, 1445-1458.

12. Parks, E. K., Zhu, L., Ho, J., and Riley, S. J. (1994) The structure of small nickel clusters. I. Ni_3-Ni_{15}, *J. Chem. Phys.* **100**, 7206-7222; and to be published.

13. Farges, J., de Feraudy, M.F., Raoult, B., and Torchet, G. (1987) From five-fold to crystalline symmetry in large clusters, in J. Jortner, A. Pullman, and B. Pullman (eds), *Large Finite Systems*, D. Reidel Publishing Company, Norwell, pp. 113-119.

14. Torchet, G., Farges, J., de Feraudy, M.F., and Raoult, B. (1989) Structural study of CH_4, CO_2 and H_2O clusters containing from several tens to several thousands of molecules, *Ann. Phys. Fr.* **14**, 245-260.

15. Cleveland, C.L. and Landman, U. (1991) The energetics and structure of nickel clusters, *J. Chem. Phys.* **94**, 7376-7396.

16. Raoult, B., Farges, J., de Feraudy, M.F., and Torchet, G. (1989) Stability of relaxed Lennard-Jones models made of 500 to 6000 atoms, *Z. Phys. D* **12**, 85-87.

17. Jellinek, J., Beck, T.L., Berry, R.S. (1986) Solid-liquid phase changes in simulated isoenergetic Ar_{13}, *J. Chem. Phys.* **84**, 2783-2794.

18. Davis, H.L., Jellinek, J., and Berry, R.S. (1986) Melting and freezing in isothermal Ar_{13} cluster, *J. Chem. Phys.* **84**, 6456-6464.

19. In: P. Jena, B.K. Rao, and S.N. Khanna (eds), (1987) *Physics and Chemistry of Small Clusters*, Plenum Press, New York.

20. In: I. Prigogine and S.A. Rice (eds), (1988) *Adv. Chem. Phys.* Vol. 10, Part 2, Wiley-Interscience, New York.

21. In: G. Benedek, T.P. Martin, and G. Pacchioni (eds), (1988) *Elemental and Molecular Clusters*, Springer, Berlin.

22. In: P. Jena, S.N. Khanna and B.K. Rao (eds), (1992) *Physics and Chemistry of Finite Systems: From Clusters to Crystals*, Vols. 1 and 2, Kluwer Academic Publishers, Dordrecht.

23. Jellinek, J. and Li, D.H. (1989) Separation of the energy of overall rotation in any N-body system, *Phys. Rev. Lett.* **62**, 241-244.

24. Li, D.H. and Jellinek, J. (1989) Rotating clusters: centrifugal distortion, isomerization, fragmentation, *Z. Phys. D* **12**, 177-180.

25. Jellinek, J. and Li, D.H. (1990) Vibrations of rapidly rotating N-body systems, *Chem. Phys. Lett.* **169**, 380-386.

26. Berry, R.S., Jellinek, J., and Natanson, G. (1984) Melting of clusters and melting, *Phys. Rev. A* **30**, 919-931.

27. Sawada, S. and Sugano, S. (1989) Structural fluctuation and atom-permutation in transition-metal clusters, *Z. Phys. D* **14**, 247-261.

28. Rey, C., Gallego, L.J., García-Rodeja, J., Alonso, J.A., and Iñiguez, M.P. (1993) Molecular-dynamics study of the binding energy and melting of transition-metal clusters, *Phys. Rev. B* **48**, 8253-8262.

29. Garzon, I.L. and Jellinek, J. (1993) Peculiarities of structures and meltinglike transition in gold clusters, *Z. Phys. D.* **26**, 316-318.

30. Fisher, I.Z. (1966) *Statistical Theory of Liquids*, University of Chicago Press, Chicago.

31. Cheng, H.P. and Berry, R.S. (1992) Surface melting of clusters and implications for bulk matter, *Phys. Rev. A* **45**, 7969-7980.

32. Ercolessi, F., Andreoni, W., and Tosatti, E. (1991) Melting of small gold particles: mechanism and size effects, *Phys. Rev. Lett.* **66**, 911-914.

240

33. Jellinek, J. and Güvenç, Z.B. (1991) Molecule-cluster collisions: reaction of D_2 with Ni_{13}, in R. Schmidt, H.O. Lutz, and R. Dreizler (eds), *Nuclear Physics Concepts in the Study of Atomic Cluster Physics, Lecture Notes in Physics*, Vol. 404, Springer-Verlag, pp. 169-177.

34. Jellinek, J. and Güvenç, Z.B. (1992) The $D_2 + Ni_{13}$ reaction: mode-specific and structure-specific features, in P. Jena, S. N. Khanna, and B. K. Rao (eds), *Physics and Chemistry of Finite Systems: From Clusters to Crystals*, Vol. 2, Kluwer Academic Publishers, Dordrecht, pp. 1047-1056.

35. Jellinek, J. and Güvenç, Z.B. (1993) Cluster-molecule systems: analysis and tuning of the interaction potential, *Z. Phys. D* **26**, 110-114.

36. Jellinek, J. and Güvenç, Z.B. (1994) Collisions of molecules with clusters: a quasiclassical study, in B. Remaud, A. Calboreanu and V. Zoran (eds), *Topics in Atomic and Nuclear Collisions*, Plenum Press, New York, pp. 243-256.

37. Zhu, L., Ho, J., Parks, E.K., and Riley, S.J. (1993) Temperature dependence of the reaction of nickel clusters with deuterium, *J. Chem. Phys.* **98**, 2798-2804.

38. Zhu, L., Ho, J., Parks, E.K., and Riley, S.J. (1993) Physisorption of deuterium on deuterated nickel clusters, *Z. Phys. D* **26**, 313-315.

39. Mlynarski, P. and Salahub, D.R. (1991) Local and nonlocal density functional study of Ni_4 and Ni_5 clusters. Models for the chemisorption of hydrogen on (111) and (100) nickel surfaces, *J. Chem. Phys.* **95**, 6050-6056.

40. Güvenç, Z.B. and Jellinek, J. to be published.

HETERODOX BONDING EFFECTS BETWEEN TRANSITION METAL ATOMS

Santiago Alvarez, Pere Alemany, Gabriel Aullon,
Ana A. Palacios, Juan J. Novoa
Departament de Química Inorgànica and Departament de Química-
Física, Universitat de Barcelona
Diagonal 647, 08028 Barcelona (Spain)

1. Abstract

Two center-two electron bonds can be described in terms of a Lewis structure and may be considered as orthodox. Heterodox metal-metal interactions comprise the short contacts between d^8 square planar or d^{10} linear complexes of transition metals, adduct formation of the same compounds with Lewis acids, through-ring bonds in diamond-shaped clusters or binuclear bridged complexes, or subtle effects superimposed on orthodox bonds due to, e.g., changes in the degree of pyramidalization around a metal atom.

A Molecular Orbital analysis of the bonding in a variety of coordination compounds and small clusters, facilitates an understanding of the differences in bond distances within families of compounds. The well established frontier orbitals of transition metal and main group-element fragments, ML_m and XR_n, allow us to predict structural trends in related families of compounds ranging from organic molecules to transition metal clusters in solid state compounds.

2. Metal-Metal Interaction in Dimers of Square-Planar d^8 Metal Ions

The existence of *short* intermolecular M⋯M contacts between d^8-ML_4 square planar complexes is well documented both in solution [1] and in the solid state [2] . In most cases though, the dimers or chains are held together by bridging ligands and some doubts persist as to whether there is really a metal-metal bonding interaction or not.

241

L. J. Farrugia (ed.), The Synergy Between Dynamics and Reactivity at Clusters and Surfaces, 241–255.
© 1995 *Kluwer Academic Publishers.*

Our theoretical studies in this field are aimed at (a) testing via *ab initio* calculations the bonding nature and energetics of such metal-metal interactions; (b) predicting the way in which chemical modifications of the metal coordination sphere affect the strength of the metal-metal interactions and (c) comparing the relative stability of different oligomers.

2.1. AB INITIO RESULTS FOR DIMERS OF Pt(II) AND Rh(I)

Ab initio calculations on model dimeric compounds of Pt(II) and Rh(I) were carried out, considering electron correlation at the MP2 level [3] . The studied compounds were $[PtCl_2(CO)_2]_2$ (in two different conformations), $[PtCl_2\{HNCH(OH)\}_2]_2$, and $[RhCl(CO)_3]_2$. The three dimers are computationally stable, with dissociation energies larger than 9 kcal/mol. The smaller dissociation energy corresponds to the former compound in its rotated conformation. The eclipsed conformation is sensibly more stable (29 kcal/mol), indicating the existence of stabilizing intermolecular ligand-ligand interactions. Still more stable is the amido complex (70 kcal/mol), probably due to intermolecular hydrogen bonding.

The calculated M$\cdots$M bond distances are 3.319 and 3.285 Å for $[PtCl_2(CO)_2]_2$ in its rotated and eclipsed conformations, respectively. For $[RhCl(CO)_3]_2$ a distance of 3.551 Å is calculated, in fair agreement with the 3.519 Å reported for the dimer $[RhCl(CO)_2(Imidazole)]_2$ [4] . Finally, a distance of 3.268 Å is calculated for the Pt(II) amido complex, to be compared with experimental values reported for the unsupported dimers $[PtCl_2\{HNCBu^t(OH)\}_2]_2$ (3.165Å) [5] and $[Pd_2Cl_4(H[9]\text{-}aneN_3)_2]^{2+}$ (3.311 Å) [6]

2.2. QUALITATIVE BONDING MODEL AND THE EFFECT OF AXIAL GROUPS ON THE M-M INTERACTION

Once a good agreement between calculations and the experimental data has been found, we describe a qualitative model of the M$\cdots$M bonding, which will be useful to explore the possibility of enhancing such bonding interactions. The simple model proposed early by Gray [7] is based on the hybridization of the metal d_{z^2} orbitals, through mixing with p_z orbitals. A slightly different version of the same model is presented in the Figure 1: the d_{z^2} orbital of each metal acts as a donor to the p_z orbital of the other (interaction 2), accounting for two bonding contributions; simultaneously, a repulsion between the two d_{z^2} orbitals (interaction 1) arises. Depending on the relative strength of the attractive and repulsive interactions a net bonding situation may result.

The analysis of the bonding capabilities of the d^8-ML$_4$ complexes[8] indicates that the addition of a Lewis base to an axial position enhances the basicity of the d_{z^2} electrons, whereas the addition of a Lewis acid enhances the acidity of the p_z acceptor

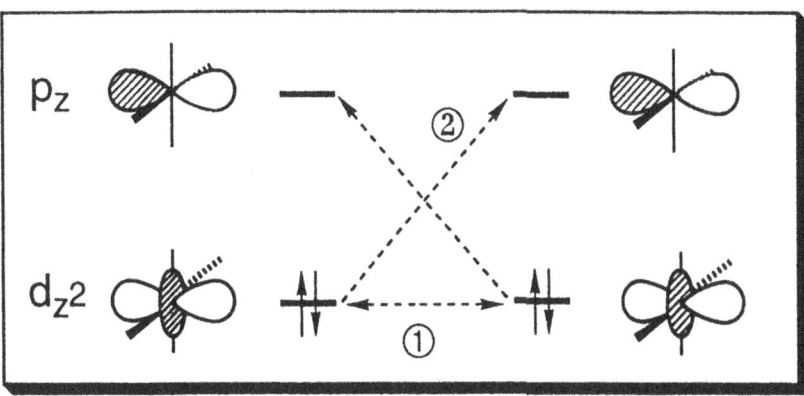

Figure 1. Schematic diagram for the orbital interaction between two d8 square-planar complexes.

orbital. Since the M···M bonding contributions 2 are acid-base interactions, the best way to strengthen the intermolecular bonding is to make one metal a better Lewis base, the other metal a better Lewis acid, that is, to add a base to one metal and an acid to the other one. Addition of one acid to each metal center is expected to enhance only slightly the M···M bonding. Finally, addition of two bases increases the d_{z^2}/d_{z^2} repulsion and is expected to weaken the interaction. All this qualitative reasoning based on the simple orbital model of Figure 1 can be summarized in the following stability series:

$$L_4BM \cdots MBL_4 < L_4BM \cdots ML_4 \approx L_4AM \cdots MAL_4 \approx$$
$$L_4M \cdots ML_4 < L_4M \cdots MAL_4 < L_4BM \cdots MAL_4 \tag{1}$$

where B represents a Lewis base and A a Lewis acid. This stability series is confirmed by extended Hückel and ab initio MP2 calculations [3] on $[YRh^ICl_4\cdots Rh^ICl_4Z]$ and $[YPtCl_2(CO)_2\cdots PtCl_2(CO)_2Z]$, respectively (Y, Z = H^+, AuCl, Cl^-, or CO). The metal-metal distances in known structures of Ni(II) or Pt(II) dimers and of their adducts, shown in Table 1, are also in agreement with the predictions of eq. 1. It must be noted that binuclear oxidative addition products, such as our model compound $[HPtCl_2(CO)_2\cdots PtCl_2(CO)_2Cl]$, are just a particular case of the general class of $BL_4M\cdots ML_4A$ adducts.

TABLE 1. $M\cdots M$ distances (Å) for dimers of d^8-ML_4 complexes and for their adducts $[Y$-ML_4-ML_4-$Z]$ (3).

M	Y	Z	M⋯M (dimer)	M⋯M (adduct)	ref
Pt	Ag(I)	Ag(I)	2.90-2.98[a]	2.892 (1)	[9]
Pt	ROH	ROH	3.165 (1)	3.399 (1)	[5]
Ni	aniline	aniline	3.208 (1)	3.654 (2)	[10, 11]
Ni	4,4'-bipy	4,4'-bipy	3.208 (1)	3.909 (3)	[11, 12]

a) Data for bare dimer comes from six Pt compounds found in a structural database search [13].

3. The Effect of the Pyramidalization of the Metal Atom

3.1. PYRAMIDALITY AND BOND STRENGTH IN M-M BONDS

In metal-metal bonded binuclear complexes (1), of the types M_2X_8 or $M_2X_8L_2$, the pyramidality angle, α, is defined as the average of the MMX bond angles. The larger α is, the larger the separation of the M atom from the X_4 plane (that is, the pyramidality) is. The degree of pyramidalization turns out to be very important for metal-metal bonding. From the orbital point of view, it is easy to understand that in square planar MX_4 complexes ($\alpha = 90°$) the metal d orbitals cannot mix with the s or p orbitals. Increased pyramidalization ($\alpha > 90°$) allows mixing of s and p orbitals with d_{z^2}, d_{xz} and d_{yz}, hybridizing them toward the other metal atom, whereas a decreased pyramidalization ($\alpha < 90°$) favors hybridization away from the other metal atom (**2**).

The outcome is that, other things being equal, a metal-metal bond becomes stronger as the pyramidality angle increases [14, 15] , as shown by CASSCF calculations on Os compounds (Figure 2). This trend is nicely reproduced by the experimental data for 15 families of multiply-bonded binuclear complexes, comprising more than 240 structures. In general, the experimental trend conforms to a least-squares linear equation,

$$d(M\text{-}M) = b + c \cos \alpha \qquad (2)$$

(see Table 2 for the least-squares fitting parameters). The independent parameter of eq. 2 corresponds to the M-M distance for a pyramidality angle of 90° and is referred to as the *intrinsic* metal-metal distance, whereas the slope c gives a measure of the bond *susceptibility* to pyramidalization.

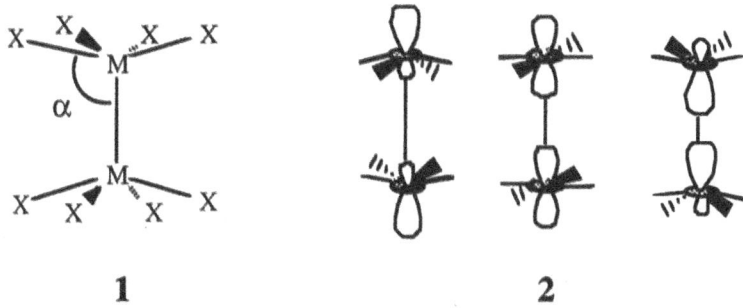

1   **2**

Since most of the studied structures correspond to bridged compounds, one might suspect that eq. 2 is merely the mathemathical expression of a geometrical constraint. If this were the case, eq. 3 should apply. Therefore, the parameter b would be the non-bonded X···X distance, that is, the bite of the bidentate ligand, assuming that it is practically constant for analogous ligands in a family of compounds. Also, the parameter c should then correspond to twice the M-X bond distance. The analysis of the least-squares data (Table 2) indicates that the pyramidality-dependence of the M-M bond length is not only due to the bridging nature of the ligands. This is consistent with the fact that the same type of dependence on α is found in extended Hückel and CASSCF calculations of compounds with non-bridging ligands, as well as with experimental data for unsupported dimers.

$$d(M\text{-}M) = (X\cdots X) + 2(M\text{-}X) \cos \alpha \qquad (3)$$

This simple bond angle-bond length correlation explains some of the unsolved enigmas in the field of metal-metal bonds. For instance, the large variation of the Cr-Cr bond distances found in the quadruply-bonded Cr(II) carboxylates and analogous compounds[16] , spanning all the range of Cr-Cr distances between 1.8 and 2.6 Å is found to correspond to differences in the pyramidality angle. Another case is that of the complexes $[Re_2Cl_4(PMe_2Ph)_4]^{n+}$ (n = 0-2). In these compounds of Re-Re bond orders 3, 3.5 and 4, respectively, the unconventional relationship between bond order and bond distance is explained by taking into account the different pyramidality angles.

An interesting feature of the pyramidality effect is that the existence of axial ligands L has a very small influence on the M-M bond length: the same dependence on α is found for compounds with two, one or no axial ligand. The presence of axial ligands is only indirectly related to the M-M bond length, since they favor smaller values of α which, in turn, result in longer M-M bonds. It is worth stressing also that the bond length is found to depend on the *average* angle α within a family of analogous compounds, even when the four ligands around a metal atom show clearly different bond angles. This is found both in the structural data and in the results of molecular orbital calculations. Only in a few cases, when the different M-X bonds have markedly different covalent character, the bond distance seems to depend only on those angles involving the less electronegative ligand.

The relationship between pyramidality angle and bond length applies not only to molecules, but also to multiple metal-metal bonds in extended structures in which M_2X_8 fragments are found. This is the case of the isostructural compounds $Ca_xNb_3O_6$ (x = 0.95, 0.75) and $NaNb_3O_5F$, in which Nb_2O_6 or Nb_2O_5F clusters show shorter Nb-Nb bonds as the Nb-Nb-O angle increases [17, 18]

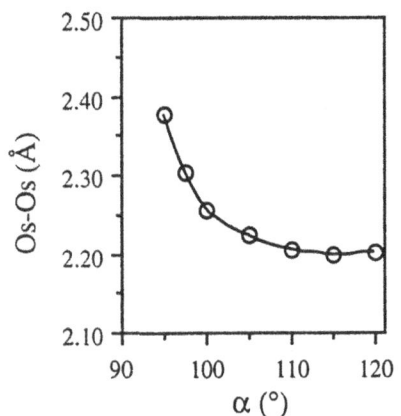 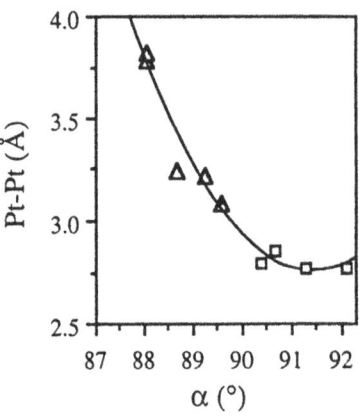

Figure 2. Calculated Os-Os bond distance (CASSCF) in the triply-bonded compound $[Os_2Cl_8]^{2-}$.

Figure 3. Intra- (squares) and intermolecular (triangles) experimental Pt-Pt distances in Pt(II) dithio-carboxylates.

TABLE 2. Least-squares parameters of equation 2 for several families of dinuclear complexes with different M-M bond orders.

M	ligands	bond order	b	c	d_{min}	d_{max}	σ (Å)	N
Cr	chelates	4	2.241	3.740	1.83	2.60	0.020	52
Cr	unsupported	4	3.138	3.847	1.98	3.63	0.071	5
Mo	chelates	4	2.158	1.774	2.06	2.14	0.009	62
Mo	carboxylate + phosphine	4	2.131	0.189	2.09	2.12	0.003	7
Mo	halide + phosphine	4	2.191	0.197	2.12	2.16	0.004	13
Mo	propyl-diphosphines	4	1.997	-0.568	2.13	2.16	0.005	6
W	chelates	4	2.222	1.873	2.16	2.24	0.013	21
W	amides	3	2.007	-1.536	2.29	2.33	0.001	5
W	chelate + alkyl	3	2.188	-0.466	2.19	2.30	0.008	8
Re	halides	4	2.337	0.455	2.20	2.30	0.015	34
Re	diphosphines	3	2.379	0.562	2.23	2.31	0.012	8
Rh	chelates	1	2.299	2.934	2.36	2.49	0.014	103
Rh	metallated phosphines	1	2.474	2.481	2.47	2.56	0.019	6
Co	chelates	1	2.141	4.605	2.26	2.86	0.043	6
Ni	dimethylglyoximes	0	3.252	5.993	3.21	3.91	0.018	5
Ni	carboxylates	0	2.291	3.437	2.48	2.77	0.010	8
Pd	carboxylates	0	2.444	1.948	2.54	2.62	0.009	10
Au	halo-ylides	0	3.093	0.196	3.07	3.09	0.004	4
Au	R-ylides	0	3.119	0.976	3.07	3.10	0.008	3

b is the *intrinsic* metal-metal bond distance, c the susceptibility to pyramidalization, σ the standard error of the estimate, and N is the independent number of structural data sets. All experimental distances are comprised in the range $d_{min} < d < d_{max}$. All values in Å.

The existence of a pyramidality effect seems to be widespread in structural chemistry. We have found such correlation for more than 100 singly-bonded Rh(II) binuclear complexes [19] , as well as for the much smaller family of Co(II) chelates [20] . Similar effects have been detected in organic compounds in which steric strain can induce unusual angles and a neat variation of C-C bond distances [21] .

3.2. PYRAMIDALITY IN d^8-d^8 INTERACTIONS

The analysis of the evolution of the molecular orbitals of a square planar unit MX_4 with a indicates that the p_z orbital is sensitive to pyramidalization, whereas d_{z^2} is not. Therefore, an increased pyramidality angle is expected to enhance the attractive interaction 2(Figure 1) without affecting the repulsion 1. Such simple qualitative argument indicates that a pyramidality effect should also be expected for the metal-metal contacts in dimers and chains of d^8-ML_4 complexes [20] .

Extended Hückel calculations on the model compound $[Rh_2O_8]^{14-}$ gives a neat dependence of the M-M overlap population as a function of the pyramidality angle α. Structural data for a few families of such dimers, grouped by metal and type of ligands, have been analyzed and show the expected pyramidality effect. The least-squares fitting parameters are shown in Table 2. The Pt···Pt distances and pyramidality angles for the family formed by chains of dimeric platinum(II) dithiocarboxylates [22-26] are plotted in Figure 3. Both the intra- and inter-molecular Pt···Pt distances are aligned along a parabola reminiscent of that found for Os-Os multiple bonds (see Figure 2).

3.3. GENERAL TRENDS OF THE PYRAMIDALITY EFFECT

Probably the most remarkable fact about the data presented in Table 2 is that pyramidality effects are observed for a large variety of M-M interactions, ranging from the quadruple bonds to the non-bonding contacts. Some trends that can also be observed in Table 2 are: (a) The first-row transition metals show larger susceptibility to pyramidalization and a wider range of M-M distances than the analogue compounds of second- and third-row metals. (b) For the first-row transition metals, compounds of the same metal and oxidation state with chelating bridges present clearly shorter intrinsic distances. The chelating nature of the ligand seems to have little influence on the intrinsic M-M distance for second- and third-row transition metals. (c) The intrinsic M-M distances in d^8···d^8 dimers are sensibly longer than those of bonded M-M pairs.

4. Through-ring M-M bonding in rhombic M_2X_2 frameworks

The rhombus is a highly common structural element found in bridged dinuclear complexes as well as in small clusters. In many instances, a regular rhombus appears, with long interatomic distances across the diagonals. However, in a large number of compounds a short through-ring distance is suggestive of a bonding interaction, typically between two metal atoms, but sometimes between main group-element atoms. In some cases, the bonding in the resulting squeezed diamonds can be unambiguously described by considering both metal-metal and metal-bridge bonds, according to the 18-electron rule. Even in those cases, both types of bonds share the same region of space and some fragment orbitals, and it is neither rigorous nor practical to attempt such a description. Alternatively, a delocalized description of bonding in the rhombic (e.g., M_2X_2) framework as a whole should account for the structural differences found in such compounds.

The aim of our research in this field has been to provide simple rules which might enable one to predict the existence of through-ring bonding in any A_2B_2 diamond, taking into account only the molecular topology, the vertices (i.e., A, B = ML_n or XR_m) and the number of electrons available to the rhombic framework (*framework electron count*, abbreviated FEC). Here we present first a general analysis of the framework bonding based on the topology of the molecular orbitals. Then, we study how ML_2 and ML_4 fragments fit into the general picture.

4.1. GENERAL MOLECULAR ORBITAL TOPOLOGY AND FRAMEWORK ELECTRON COUNT

In order to make an orbital model as general as possible while keeping a high framework symmetry, we consider diamonds with identical antipodal vertices (3). In a first approximation, we focus on the in-plane bonding and disregard the π out-of-plane interactions. Every corner in the diamond is therefore considered to contribute, at most, two orbitals to framework bonding (4a and 4b, where 4a can also be a d orbital with the same symmetry). Therefore, only the framework molecular orbitals will be considered, and only the electrons occupying such orbitals need to be counted.

As an example, in the case of a thiolato bridging ligand, RS⁻, the sulfur atom has two electrons involved in the R-S bond, a nonbonding electron pair occupying an out-of-plane orbital, and only two electron pairs are considered for framework bonding.

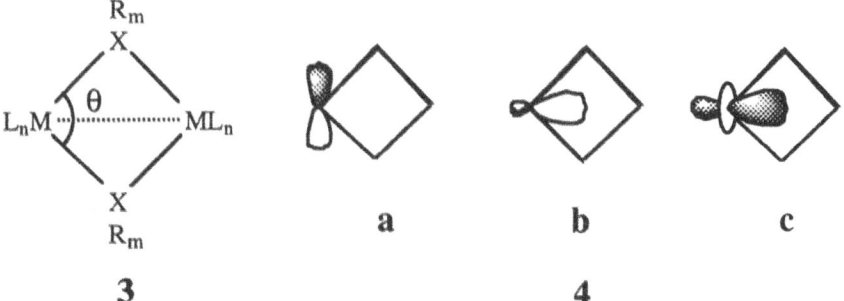

3   4

A general diagram for the orbital interaction between the two antipodal ML_n groups and the two XR_m bridges is presented in Figure 4. The energies of the resulting molecular orbitals must be considered only in a qualitative sense, since they are strongly dependent on the degree of squeezing of the diamond, given by the angle θ. What is essential is that four bonding and four antibonding framework molecular orbitals are formed. In the cases studied so far [27-29] , when the four bonding

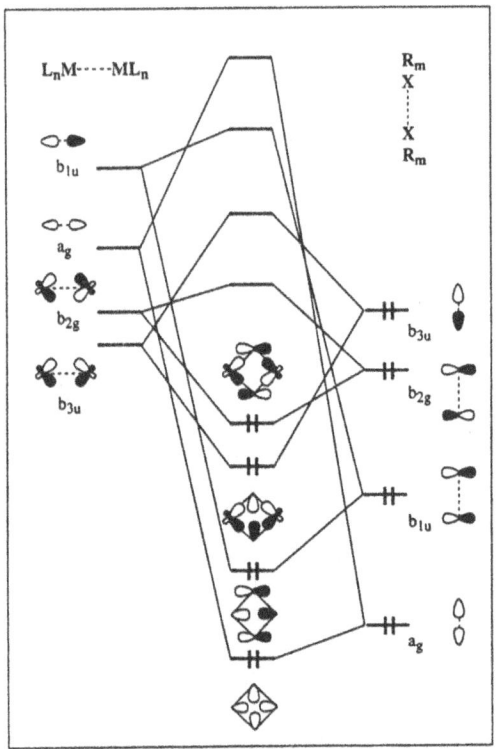

Figure 4. General interaction diagram for the framework orbitals of a $[(ML_n)_2(\mu\text{-}XR_m)_2]$ diamond. The occupation shown corresponds to an FEC of 8, and the energy ordering is that of a regular ring with $\theta = 90°$.

orbitals are occupied, the most stable geometry is that of a *regular* ring ($\theta \approx 90°$), in which every pair of antipodal atoms is kept at nonbonding distance. Hence, an FEC of 8 is expected to favor a structure without through-ring bonding.

If four electrons are removed, the two alternative squeezed structures (small or large q) are more stable than the regular ring. With a short M-M through-ring distance, the metal-metal antibonding orbitals b_{1u} and b_{2g} are strongly destabilized and become empty, while the occupied bonding counterparts, a_g and b_{3u}, are more stable. In a similar way, a short X-X through-ring distance stabilizes a_g and b_{1u} and destabilizes b_{2g} and b_{3u}. This topological analysis leads to the conclusion that for FEC = 4 a squeezed structure is preferred. For the intermediate case, i.e., FEC = 6, also the short through-ring distances are favored over the regular ring. In the latter case, however, an open structure XM-MX is energetically affordable.

The diamonds with monoatomic bridges having square planar ML_2 fragments generally present an FEC of 8 and conform to the general rule, having long through-ring M$\cdots$M distances (M = Ni^{II}, Pd^{II}, Pt^{II}, Rh^I and Au^{III}) [30] . In contrast, the Cu^I diamonds with tetrahedral coordination spheres and bridging halides show a shallow potential for the ring squeezing. Accordingly, even if the FEC is 8, a wide variety of through-ring Cu$\cdots$Cu distances are found in this family, the shorter ones being those observed for the less electronegative and bulkier halides [27] .

Some bridging groups may have only one orbital available for framework bonding. This is the case of CH_3, PR_3, H$^-$, pyridine or C_6H_5. In such cases, the general interaction diagram of Figure 4 must be slightly modified. The b_{1u} and b_{2g} combinations of the metals cannot interact with orbitals of the same symmetry from the bridges and only two framework bonding orbitals (FBO) remain, a_g and b_{3u}. Obviously, with such orbitals occupied, the squeezed structure with a short M-M distance is the most stable one. Generally speaking, it can be stated that in those cases in which two antipodal groups have only one orbital available, up to four framework bonding electrons can be accomodated (FEC = 4), and a short through-ring distance is expected between the other two vertices.

4.2. DIAMONDS OF ML_2 FRAGMENTS

For those M_2X_2 diamonds with tetrahedral ML_2 fragments, all the framework bonding involves the s and p metal orbitals, with the d orbitals being essentially nonbonding. Therefore, one must disregard 10 d electrons per metal atom for framework electron-counting purposes. Similarly, for square planar ML_2 fragments, one must account for eight non bonding d electrons. Consequently, the rotation of the terminal ligands leading from a square-planar to a tetrahedral coordination of the metal atom, allocates two framework electrons into nonbonding d electrons. Experimental evidence can be found in the two isomers of $[Rh_2(CO)_4(\mu\text{-}P^tBu_2)_2]$: the isomer with square-planar metal atoms (FEC = 8) has a long Rh-Rh distance (3.717

Å); the isomer containing one tetrahedral Rh atom (FEC = 6) has a short Rh-Rh distance (2.761 Å) [31] . A similar compound with two tetrahedral metal atoms, $[Rh_2(CO)_2(PMe_3)_2(\mu\text{-}P^tBu_2)_2$, (FEC = 4) has a still shorter distance (2.550 Å) [32]

Other examples of diamonds with square planar ML_2 fragments (Table 3) nicely illustrate the simplicity and predictive ability of the FEC rules: The first three complexes in that Table are formally d^9, d^8 and d^7 compounds. Having FEC's of 6 and 4, they all have a short through-ring distance. Many analogous Pt compounds with FEC = 8 (formally d^8 compounds), have long Pt···Pt distances (only one compound is shown in Table 3 as an example).

TABLE 3. Some examples of rhombic binuclear complexes with two metal atoms in a square-planar geometry.

Compound	FEC	M--M (Å)	Ref
$[Pd_2Cl_4(\mu\text{-}CO)_2]^{2-}$	6	2.697	[33]
$[Pt_2(C_6F_5)_6]^{2-}$	4	2.714	[34]
$[Pt_2(C_6F_5)_4(\mu\text{-}PPh_2)_2]$	6	2.772	[36]
$[Pt_2(dppe)_2(\mu\text{-}PPh_2)_2]^{2+}$	8	3.699	[37]
$[Rh_2(P\{O^iPr\}_3)_4(\mu\text{-}H)]_2$	4	2.647	[35]

4.3. DIAMONDS WITH ML_4 OCTAHEDRAL FRAGMENTS

When a d^n-ML_4 fragment occupy a vertex in a diamond, the two neighboring vertices complete a octahedral coordination around the metal atom. The relevant fragment orbitals [38] of ML_4 are the t_{2g}-like, non-bonding set, and the symmetry-adapted combinations of the two hybrids pointing toward the cis vacant sites (4a and 4b). In principle, these fragment orbitals are well suited for framework bonding, but several new problems arise: (a) One of the t_{2g}-like orbitals of each metal atom (4c) has the same symmetry of, and can mix with, the sp hybrid usually involved in framework bonding (4a). (b) For several electron counts one must decide whether the FBO's have lower energy than the t_{2g} sets or *viceversa* in order to be able to apply the *aufbau* principle. (c) The combinations of the t_{2g} orbitals of the antipodal metal atoms interact through space, with metal-metal σ, π or δ bonding or antibonding character, and this bonding effects may be superimposed on the ring geometry preference imposed by the filling of the framework orbitals. (d) Partial filling of the t_{2g} block may also result in different spin states with only small energy differences, hence antiferro- or ferromagnetic behavior.

We would like to assign framework bonding character to the sp hybrids **4a** and non-bonding metal character to the d_z2 orbital **4c**. Unfortunately, this is not always possible, since the degree of symmetry-allowed mixing of both orbitals varies with the nature of the metal and ligands in the framework. However, the highest valence molecular orbital is always one with some framework bonding character and clearly M-M σ-antibonding. Therefore, for electron counting purposes, we can consider the lowest 3 orbitals as FBO's, at higher energy the two sets of t_{2g} orbitals, and the highest orbital as topologically equivalent to b_{1u} in Figure 4. In other words, the first 6 electrons must be assigned to the FEC; the next electrons, up to 12, are formally t_{2g} electrons, and the last two electrons are assigned to the FEC. The calculated geometries for model Cr complexes, as well as experimental data for related compounds are presented in Table 4. It is clear that, with this counting scheme, the FEC rules also account for the existence or not of cross-ring bond in this type of diamonds.

Substitution of the XR_n vertices by ML_4 fragments gives rise to tetranuclear M_4L_{16} clusters, such as $[Os_4(CO)_{16}]$ or $[Re_4(CO)_{16}]^{2-}$. The electron counting rules for such clusters [44], are equivalent to the FEC rule for such cases, disregarding the electrons donated by terminal ligands.

TABLE 4. Geometries of $[(CrL_4)_2(\mu-XR_m)_2]$ diamonds with different electron counts, calculated (upper part) and experimental (lower part).

Compound	t_{2g}	FEC	M--M (Å)	Ref
$[Cr(CO)_4(\mu-NH_2)]_2^{4+}$	8	6	2.62	
$[Cr(CO)_4(\mu-PH_2)]_2$	12	6	3.07	
$[Cr(CO)_4(\mu-PH_2)]_2^{2-}$	12	8	3.51	
$[Cr(CO)_4(\mu-SEt_2)]_2$	12	8	3.79	[39]
$[Cr(O_2CNEt)_2(\mu-NEt_2)]_2$	8	6	2.948	[40]
trans-$[CrCp(NO)(\mu-NMe_2)]_2$	12	6	2.670	[41]
cis-$[CrCp(NO)(\mu-NMe_2)]_2$	12	6	2.719	[41]
$[Cr(CO)_4(\mu-AsMe_2)]_2$	12	6	2.995	[42]
$[Cr(CO)_4(\mu-PMe_2)]_2$	12	6	2.902	[43]
			2.908	

5. Acknowledgements

Financial support to this work was provided by *DGICYT*. (grant PB92-0655), and *Fundació Catalana per a la Recerca* through a grant for computing resources in the *Centre de Supercomputació de Catalunya* (CESCA). The authors are indebted to J. Forniés for facilitating unpublished data and to M. A. Ciriano for clarifying discussions. G.A. and A.A.P. thank CIRIT and MEC, respectively, for doctoral grants.

6. References

1. Mann, K. R. and Gray, H. B. (1979) *Adv. Chem. Ser.* **173**, 225.

2. Smith, D. C. and Gray, H. B. (1990) *Coord. Chem. Rev.* **100**, 169-181.

3. Novoa, J. J., Aullón, G., Alemany, P. and Alvarez, S. Submitted for publication.

4. Oro, L. A., Pinillos, M. T., Tejel, C., Apreda, M. C., Foces-Foces, C. and Cano, F. H. (1988) *J. Chem. Soc., Dalton Trans.* 1927.

5. Cini, R., Fanizzi, F. P., Intini, F. P., Maresca, L. and Natile, G. (1993) *J. Am. Chem. Soc.* **115**, 5123-5131.

6. Blake, A. J., Holder, A. J., Roberts, Y. V. and Schröder, M. (1993) *J. Chem. Soc., Chem. Commun.* 260-262.

7. Mann, K. R., Gordon, J. G. and Gray, H. B. (1975) *J. Am. Chem. Soc.* **97**, 3553.

8. Aullón, G., Alemany, P. and Alvarez, S. To be published.

9. Thewalt, U., Neugebauer, D. and Lippert, B. (1984) *Inorg. Chem.* **23**, 1713.

10. Vagg, R. S. and Walton, E. C. (1978) *Acta Crystallogr., Sect. B* **34**, 2715.

11. Stephens, F. S. and Vagg, R. S. (1977) *Acta Crystallogr., Sect. B* **33**, 3159.

12. Stephens, F. S. and Vagg, R. S. (1980) *Inorg. Chim. Acta* **42**, 139.

13. Allen, F. H., Kennard, O. and Taylor, R. (1983) *Acc. Chem. Res.* **16**, 146.

14. Mota, F., Novoa, J. J., Losada, J., Alvarez, S., Hoffmann, R. and Silvestre, J. (1993) *J. Am. Chem. Soc.* **115**, 6216-6229.

15. Losada, J., Alvarez, S., Novoa, J. J., Mota, F., Hoffmann, R. and Silvestre, J. (1990) *J. Am. Chem. Soc.* **112**, 8998-9000.

16. Cotton, F. A. and Walton, R. A. (1993) *Multiple Bonds between Metal Atoms;* 2nd. ed.; Clarendon Press, Oxford.

17. Alemany, P., Alvarez, S., Zubkov, V. G., Zhukov, V. P., Pereliaev, V. A., Kontsevaya, I. and Tyutyunnik, A. (1992) *Butll. Soc. Cat. Cièn.* **13**, 251.

18. Alemany, P., Zubkov, V. G., Alvarez, S., Zhukov, V. P., Pereliaev, V. A., Kontsevaya, I. and Tyutyunnik, A. (1993) *J. Solid State Chem.* **105**, 27-35.

19. Aullón, G. and Alvarez, S. (1993) *Inorg. Chem.* **32**, 3712-3719.

20. Aullón, G. and Alvarez, S. To be published.

21. Schleyer, P. v. R. and Bremer, M. (1989) *Angew. Chem., Int. Ed. Engl.* **28**, 1226.

22. Bellitto, C., Flamini, A., Piovesana, O. and Sanazzi, P. F. (1980) *Inorg. Chem.* **19**,

23. Bellitto, C., Dessy, G., Fares, V. and Flamini, A. (1981) *J. Chem. Soc., Chem. Commun.* 409.

24. Bellitto, C., Bonamico, M., Dessy, G., Fares, V. and Flamini, A. (1987) *J. Chem. Soc., Dalton Trans.* 35.

25. Burke, J. M. and Fackler Jr., J. P. (1972) *Inorg. Chem.* **11**, 3000.

26. Kawamura, T., Ogawa, T., Yamabe, T., Masuda, H. and Taga, T. (1987) *Inorg. Chem.* **26**, 3547.

27. Alemany, P. and Alvarez, S. (1992) *Inorg. Chem.* **31**, 4266-4275.

28. Aullón, G., Alemany, P. and Alvarez, S. (1994) *J. Organomet. Chem.* In press.

29. Palacios, A., Alemany, P. and Alvarez, S. To be published

30. Aullón, G., Alemany, P. and Alvarez, S. Unpublished results.

31. Jones, R. A., Wright, T. C., Atwood, J. L. and Hunter, W. E. (1983) *Organometallics* **2**, 470-472.

32. Jones, R. A. and Wright, T. C. (1983) *Organometallics* **2**, 1842-1845.

33. Goggin, P., Goodfellow, R. J., Herbert, I. R. and Orpen, A. G. (1981) *J. Chem. Soc., Chem. Commun.* 1077.

34. Usón, R., Forniés, J., Tomás, M., Casas, J. M., Cotton, F. A., Falvello, L. R. and Llusar, R. (1988) *Organometallics* **7**, 2279.

35. Teller, R. G., Williams, J. M., Koetzle, T. F., Burch, R. R., Gavin, R. M. and Muetterties, E. L. (1981) *Inorg. Chem.* **20**, 1806.

36. Forniés, J., Fortuño, C. and Tomás, M. Unpublished results.

37. Carty, A. J., Harnstock, F. and Taylor, N. I. (1982) *Inorg. Chem.* **21**, 1349.

38. Albright, T. A., Burdett, J. K. and Whangbo, M.-H. (1985) *Orbital Interactions in Chemistry*; J. Wiley, New York.

39. Bremer, G., Klufers, P. and Kruck, T. (1985) *Chem. Ber.* **118**, 4224.

40. Chisholm, M. H., Cotton, F. A., Extine, H. W. and Rideout, D. C. (1978) *Inorg. Chem.* **17**, 3536.

41. Bush, M. A. and Sim, G. A. (1970) *J. Chem. Soc. A* 611.

42. Vahrenkamp, H. and Keller, E. (1979) *Chem. Ber.* **112**, 1991.

43. Vahrenkamp, H. (1978) *Chem. Ber.* **111**, 3472.

44. Mealli, C. and Proserpio, D. (1990) *J. Am. Chem. Soc.* **112**, 5484-5496.

PERIODIC HARTREE-FOCK CALCULATIONS OF THE ADSORPTION OF SMALL MOLECULES ON TiO_2

C. Minot, A. Fahmi and J. Ahdjoudj
Laboratoire de Chimie Organique Théorique; U.R.A. 506 CNRS
Université P. et M. Curie. Boîte 53; Bât. F 642,
4 Place Jussieu 75252 Paris Cédex 05, FRANCE.

1. Introduction

We present *ab-initio* Periodic Hartree-Fock calculations of small molecules on TiO_2. These are effective core pseudopotential calculations using the code CRYSTAL from the group of Torino [1]. The dimension for the periodicity for surface studies is two; the surface is represented by one or several layers and the adsorbate is repeated. The thickness of the slab is a limitation similar to that for the size of a cluster used as surface model. The size of the calculation depends on the coverage and on the symmetry of the adsorbate. At high coverage ($\theta=1$) the unit cell is that of the bare surface. Considering for example two layers of MgO implies a unit cell with four atoms, each of which has a coordination number of five. A comparable calculation with a cluster model would involve eight atoms in a cube. Then each atom would only have coordination number of three. For rutile (anatase), a linear (zigzag) one-dimension polymer can be used in a first approach. In this simplified model the titanium atoms have a coordination number of four with an oxidation number +IV. The main crystallographic faces of rutile such as (110) and (100) can be obtained by different coupling of the linear polymers.

The calculations are performed at the SCF level with a restricted Hartree-Fock Hamiltonian. The pseudopotentials are those from Durand-Barthelat [2]. The basis sets for the atoms of the second or third period are the PS-31G basis sets. For titanium atoms [3], the basis set consists of the d functions contracted to (4/1) basis set and a single 4sp shell with an exponent of 0.484 for polarization purposes.

The adsorption energies are defined to be positive when the adsorption is exothermic: $E_{ads} = E_{adsorbate} + E_{TiO_2} - E_{(adsorbate+TiO_2)}$. They are expressed in kilocalories per mol.

L. J. Farrugia (ed.), The Synergy Between Dynamics and Reactivity at Clusters and Surfaces, 257–270.
© 1995 Kluwer Academic Publishers.

2. Molecular adsorptions

The oxide TiO_2 is classified as an amphoteric species with acidic centers (Ti^{4+}) and basic centers (O^{2-}). The acidity and basicity of the oxides have been determined in particular by the measurements of the differential heats of adsorption of basic (NH_3) and acidic (CO_2) gas probe molecules [4]. In this paper, we will investigate the adsorption of a series of compounds that range from carbon dioxide to ammonia according to their adsorption energies calculated for the molecular adsorption.

TABLE 1. The proton affinities and the molecular adsorption energies of the various compounds in kcal/mol. The coverage is indicated between parentheses. The best adsorption mode is indicated in the third columns (M for molecular, D for dissociative; - when the adsorption may take place on a different atom, the adsorption site is underlined). Experimental values from ref [5] or from ref [6] when labelled*) are between parenthesis

	Proton affinities		Molecular Adsorption	
$\underline{C}O_2$	0.95		6 ($\theta=1/2$)	-
$C\underline{O}$	117.5		16.5 ($\theta=1/2$)	-
$C\underline{O}_2$	130.25	(126.8)	19.0 ($\theta=1/2$)	M
$\underline{C}O$	138.9	(139)	19.3 ($\theta=1/2$)	M
H_2S	157.3	(176.6)	21 ($\theta=1/2$)	D
$HCO\underline{O}H$	166	(178*)	26.2 ($\theta=1$)	-
$HC\underline{O}OH$	184	(182.8)	32 ($\theta=1$)	D
H_2O	182	(173)	36.5 ($\theta=1$)	D
MeOH	199	(184.9)	39.6 ($\theta=1$)	D
NH_3	220.6	(205)	48 ($\theta=1$)	M

As shown in Table 1, there is a correlation between the molecular adsorption energies and the proton affinities. The ammonia strongly adsorbs on the acidic titanium center of the surface while the carbon dioxide weakly adsorbs on the same place. The molecular adsorption therefore reveals the basic property of the adsorbate. The basicity provides an indication of the adsorption site when two sites can be involved. The atom that bears the most basic electron pair is adsorbed : the oxygen atom from the carbonyl group for the formic acid vs that from the hydroxyl group; the carbon atom for the carbon monoxide vs the oxygen atom; the terminal oxygen atom for the carbon dioxide vs the central carbon atom.

3. Dissociative adsorptions

If the adsorption is strong enough, the adsorbate may be distorted or, in the extreme case, may be dissociated into fragments. In contrast to molecular adsorption, dissociative adsorption involves simultaneously the acidic *and* the basic properties of both the adsorbate and the surface [7]. One bond of the adsorbate is broken while two bonds are built between the fragments and the surface atoms. In table 2 and 3 are reported the gas phase cleavage of the adsorbates. Clearly, H_2S is the strongest Brönsted acid and is consequently easily dissociated on the surface [8]; NH_3 is the poorest Brönsted acid and remains uncleaved on the surface [9]. Water is intermediate and dissociates on the surface.

TABLE 2. The energies for the acidic cleavages in gas phase (first column, the experimental result is given between parentheses from ref. [5]) and on the surfaces (second column) for the various compounds in kcal/mol. The adsorption energies for the dissociative adsorptions are given in the last column.

	Gas phase cleavage	Surface dissociation	Dissociative adsorption
$H_2S \rightarrow H^+ + SH^-$	354.9 (353.4)	-43.3	64.3
$MeSH \rightarrow H^+ + MeS^-$	360.6 (359-361.1)	-38	57
$H_2O'' \rightarrow H^+ + OH^-$	423.84 (390.8)	-26.9	52-63.4
$HCOOH \rightarrow H^+ + HCOO^-$	356.7 (345.2)	-12.7	44.7
$MeOH \rightarrow H^+ + MeO^-$	406.4 (378.2-379.2)	-11.8	51.4
$NH_3 \rightarrow H^+ + NH_2^-$	447 (403.6)	+6	42

TABLE 3. The energies for the basic cleavages in gas phase (first column; the experimental result is given between parenthesis a :[5] ;, b : [10], c :[11]) and on the surfaces (second column) for the various compounds in kcal/mol. The adsorption energies for the dissociative adsorptions are given in the last column.

	Gas phase cleavage	Surface dissociation	Dissociative adsorption
$MeSH \rightarrow Me^+ + SH^-$	455.0	-36	55
$H_2O \rightarrow H^+ + OH^-$	423.84 (390.8[a])	-26.9	52-63.4
$MeOH \rightarrow Me^+ + OH^-$	289.2 (271.4[b]-277.4[c])	-22.6	62.2
$HCOOH \rightarrow HCO^+ + OH^-$	290.8 (259[b])	-19.9	52

An understanding of the adsorption (molecular or dissociative) can be obtained from the scheme below where a proton either goes on a bridging oxygen atom (Ti-O-Ti) leading to the dissociation of the water molecule, or goes on an adsorbed hydroxyl group (Ti-O-H) leading to molecular adsorption. The larger basicity of Ti-O-Ti explains the dissociation. When the adsorbate is a weaker Brönsted acid such as ammonia, a stronger basic group (NH_2) replaces the OH group on the surface and molecular adsorption is obtained. If the basicity of the oxygen atom from the surface is increased, the adsorption is shifted toward more dissociation. This is the case for the zigzag chain that represents the anatase and involves a singly-coordinated oxygen atom which is very basic. The dissociation energy on that polymer is much larger (51.2 kcal/mol) than obtained on the linear polymer (26.5 kcal/mol).

In several cases (methanethiol, methanol, formic acid) [12] two dissociations can be considered in gas phase. For the sulfur compound (MeSH), the acidic cleavage (S-H cleavage) is easier than the basic cleavage (S-C cleavage) while, for the oxygen compounds, the situation is opposite; the O-C bond cleaves more easily than the O-H bond. This also corresponds to the energies calculated on the linear polymer (see Tables 2 and 3). As the polymer model for anatase is more basic, the acidic cleavage is slightly favored in the cases of the oxygen compounds.

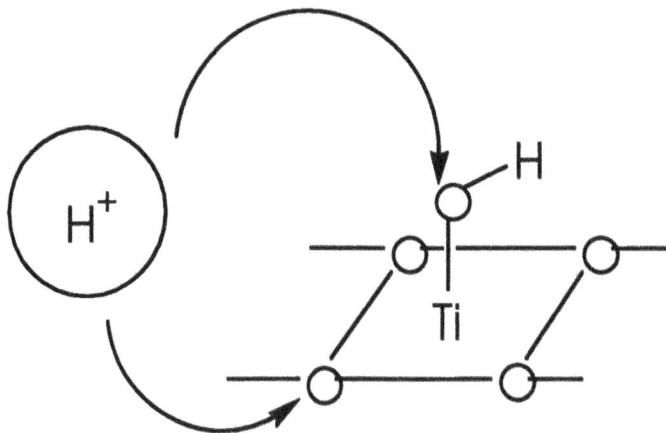

The factors controlling the comparative adsorptions of methanol and formic acid with water are not the molecular cleavage energies but the ease of adsorption of the anion. Methanol and formic acid do not dissociate on the surface as easily as water molecule since $HCOO^-$ and CH_3O^- are less basic than HO^-. The conjugation that stabilizes the ions for the gas phase reaction, and hence facilitates the cleavage, prevents them being reactive. It is not possible to calculate the adsorption of these ions on neutral surfaces without a corresponding adsorption of a counterion, since with the periodicity, this would make the number of charges infinite. The trends in the basicity of the ions may be estimated by considering the solvation energies of the ions by a single molecule of water. These energies are 20.1, 30.4 and 39.8 kcal/mol for $HCOO^-$ CH_3O^- and HO^- respectively. The dissociation on the surface of these three compounds correlates with this order.

4. Criteria for the reactivity

Usual indices for the acidity or basicity are the charges, the Mulliken Overlap population and the Frontier Orbitals. We will use two examples to emphasize that, with *ab-initio* calculations, the latter index better reveals the reactivity.

Let us first consider as an example the basic property of the surface in the dissociative water adsorption on the model anatase. The OH group adsorbs on the titanium atom while the protons may choose between the singly-coordinated oxygen atom O_1 or the triply-coordinated one, O_3. It obviously chooses O_3. Considering the charges, O_3 has a large charge, -1.52, three times larger than that of O_1, -0.62. This large charge is due to the presence of three adjacent counterions which stabilize the excess electron density. It therefore appears to be an index of stability rather than an index of reactivity. The overlap populations express the covalent nature of the bond, and, as the single Ti-O_1 bond is covalent, the corresponding overlap population exceeds the sum of those for the three ionic Ti-O_3 bonds. The basic property of the two atoms appears in the projected density of states. Figure 1 shows that O_1 peaks higher than O_3 and is closer to the Fermi level. The O_3 states are pushed down in energy by the conjugation with the three cations. O_1 is therefore more reactive (i.e. having a higher HOMO) than O_3.

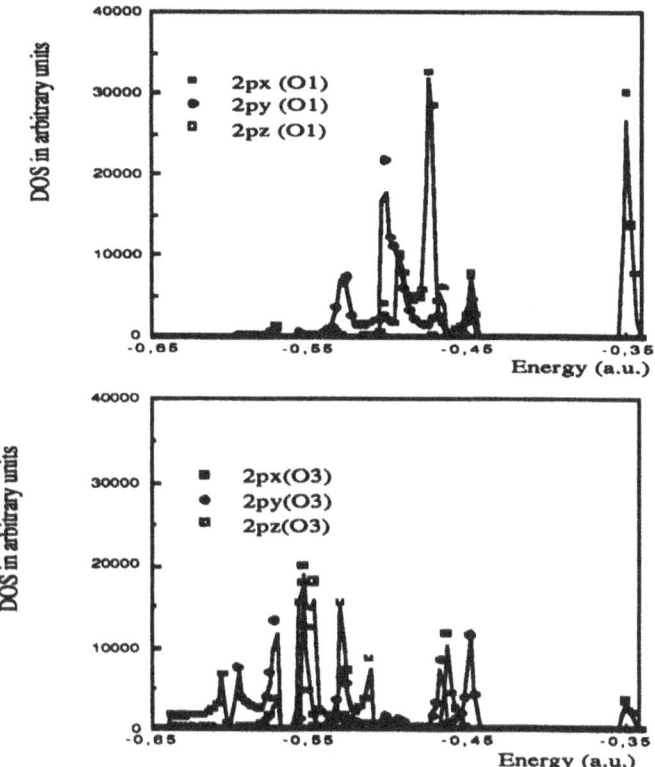

Figure 1 : The projected DOS for the p bands of the singly-coordinated oxygen atom and of the triply-coordinated one. Close to the Fermi level (at -0.35 a.u. at the right hand side of the figure) the highest peak mainly represents the 2p electron pair on O_1.

Similarly the frontier orbitals reveal the acidic properties of the oxide. Let us compare a four-coordinated titanium atom (Ti_4) to a five-coordinated one (Ti_5). If one considers a double polymer (figure 2), the surface atoms are pentacoordinated and the adsorption energies decrease by one third. A comparison of the positive charges on the titanium atoms, which are larger for Ti_5 (cf 2.38 vs 2.32) does not show this decrease. The charge increase corresponds to an increase of the ionicity with the coordination and is not to be associated with an increase of the reactivity. The DOS shows that the Ti5 atoms have higher unoccupied states than the Ti4 atoms, and this explains the decrease in reactivity (see Figure 3).

Two experimental results [13] are explained by the presence of two sites of adsorption : Ti4 or Ti5. In the water adsorption, OH groups compete with water molecules to adsorb on the most acidic centers. The more basic OH⁻ ion goes to the more acidic center, i.e. on Ti_4 , while the water molecules go to Ti_5

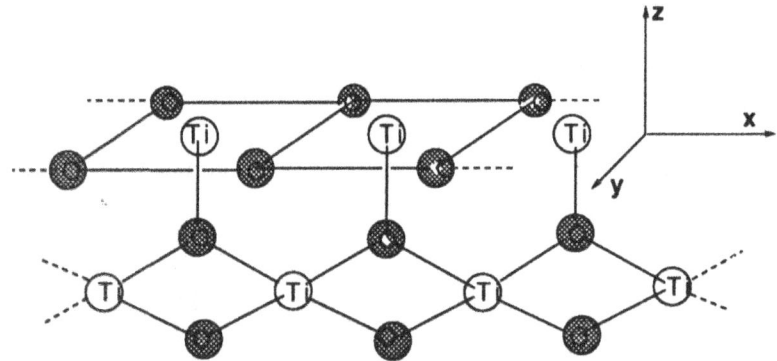

Figure 2 : The double polymer with surface pentacoordinated surface atoms.

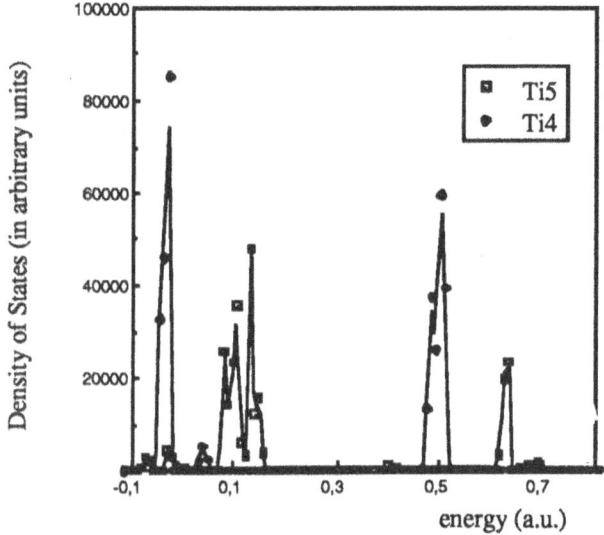

Figure 3 : Unoccupied states for Ti4 and Ti5. Ti4 has lowest unoccupied levels; the lowest unoccupied peak at the left hand side of the figure corresponds to Ti4 .

In the CO adsorption, the adsorption on Ti_4 induces a shift of the vibrational frequency to higher values with respect to the adsorption on Ti_5. This is also attributed to an increased acidity of Ti_4 (see section 7 below).

5. Lateral effects.

At high coverages, the adsorbate molecules are close to each other. Therefore they interact through van der Waals interactions or through H-bonds. When their relative positions are determined to optimize the adsorbate-substrate interactions, they often repel each other. This can be easily verified by doing a calculation of a layer of adsorbate with no substrate. If the adsorption energy is weak (CO and CO_2) a better adsorption mode can be determined by optimizing the adsorbate-adsorbate interaction. On the contrary, for the strong adsorptions resulting from acid-base interactions, the adsorption can be enhanced by lateral effect allowing the formation of H-bonds. When there is only one kind of adsorbate, high coverages are more favorable than low coverages since adsorbates are close to each other. If the adsorbates are different, H-bonds imply coadsorption effects. Finally, H-bonds can take place between the adsorbate and the substrate, inducing face-specificity.

6. CO_2 adsorption.

According to Morterra [14], linear CO_2 chemisorbs at those sites responsible for CO chemisorption and is sensitive to lateral effects. According to our calculations, CO_2 indeed chemisorbs vertically above the titanium ions. Besides the adsorption energy, the rather long O-C distance, 2.24Å, and the fact that the geometry of the adsorbed CO_2 is nearly unperturbed revealing the weakness of the adsorption.

At $\theta=1/2$, a symmetrical adsorption mode is competitive with the perpendicular one (16.35 kcal/mol vs 19. kcal/mol). The molecule is parallel to the surface and the two oxygen atoms are bound to two titanium ions from the surface. This mode also involves the basic property of the adsorbate. CO_2 insertion through an electron donor-acceptor interaction between the oxygen from the surface and the CO_2 carbon, as might be invoked considering the acidic property of CO_2 [15], is not favorable. The adsorption takes place only when the C-O distances are large, 2.4 Å, corresponding to van der Waals interactions and the adsorption energies remain weak, a third of the value for the perpendicular adsorption.

As the adsorption is weak, it is sensitive to adsorbate-adsorbate interactions. The saturation of the polymer is reached despite a strong adsorbate-adsorbate repulsion of 3.7 kcal/mol.

$CO_2(\theta =1/2) \rightarrow CO_2 (\theta =1)$ $\Delta H = -2.$ kcal/mol

Lateral effects can be inferred from consideration of the dimer structure. The rotation by 16° relative to the direction normal to the C..C axis leads to a stabilization of 2.9 kcal/mol with respect to two molecules. This suggests that if a CO_2 molecule stands is adsorbed with a larger Ti-O distance than its two neighbors, reproducing locally the geometry of the dimer, it would benefit from an attractive lateral interaction.

In figure 5 the energy profile for the desorption of one half of the CO_2 molecules is represented. After passing an activation barrier, the system is again stabilized when half of the molecules are adsorbed on the titanium atoms and the other half are bound to them through van de Waals interactions.

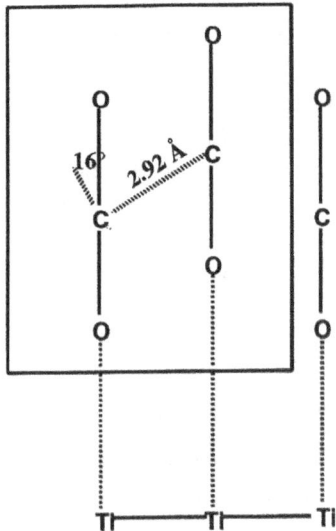

Figure 4 : The geometry of the dimer is shown in the frame; on the surface, such geometry corresponds to parallel CO_2 orientations with successive short and long TiO distances.. In this case, the distance between the CO_2 molecules is imposed by the cell vector.

7. CO adsorption on TiO2 and MgO.

CO is a soft base used as probe for acid sites on oxides [16]. The main features of the adsorption are the following [17] : i) CO is vertically adsorbed over a titanium ion of the surface. ii) the difference in energy between the Ti-CO orientation (the most favorable one) and the Ti-OC orientation is small. iii) the bending of the Ti-C-O angle for the first orientation is much more expensive in energy than the bending of the Ti-O-C angle for the second one. Despite very weak charge transfers, these results are well explained by the Dewar-Chatt model [18] which involves two frontier orbitals : a σ-donation and a π-back-donation. The σ-donation favors the Ti-CO orientation since the highest highest σ-orbital is located on the carbon atom. When this level is depopulated, the $2s_C$-$2p_O$ antibonding character decreases and this induces a shift of the vibrational frequency to high values. The π-back-donation also favors the Ti-CO orientation since the lowest π*-orbitals are located on the carbon atom.

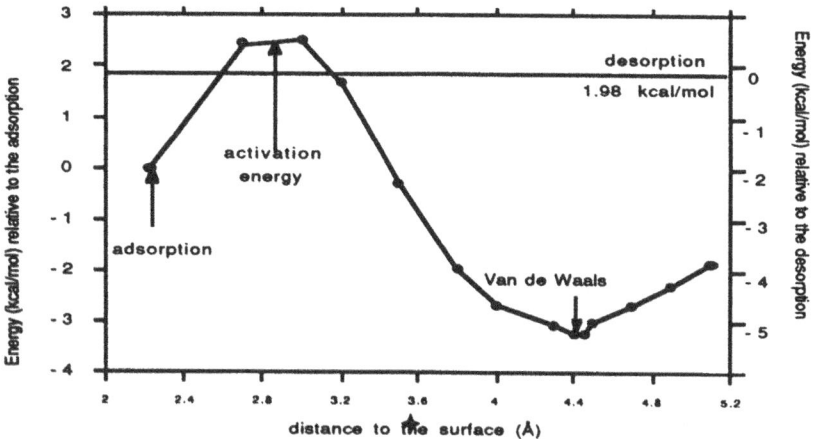

Figure 5: The energy profile for the reaction $CO_2(\theta = 1/2) \rightarrow CO_2 (\theta = 1)$.

When these levels are populated, the CO antibonding character increases and the CO vibration is shifted to small frequencies. With respect to neutral metal atoms, the π-back-donation is reduced. If the π-states of the metal were unoccupied, they only could be involved in a π-donation from the CO and this would favor the Ti-OC orientation.

For the adsorption on a single ion, Li^+, or on very ionic species such as LiF[19], Li-OC is the best orientation. Then, a bending of the TiOC angle mixes the two donations to the metal; this is slightly unfavorable since the overlap decreases but this is not very expensive in energy. For the adsorption on the titanium dioxide, the Ti-CO orientation remains more favorable. A bending of the TiCO angle by mixing the MOs responsible for the donation and the back-donation introduces some 4-electron repulsion which is much more expensive in energy than in the previous case.

A purely ionic description of the adsorption is not very predictive. The main coulombic term for the CO molecule which has a small dipole moment is the quadrupole moment. As previously shown (see section 4) the charge of the titanium atom is not a good criterion for reactivity.

The adsorption on MgO is similar but weaker than on TiO_2, so it is more sensitive to lateral effects. At low coverages, CO is vertical above the surface [20]. This implies parallel adsorbed CO molecules. At higher coverage, the repulsion between the adsorbates is large and the best adsorption mode changes. At $\theta = 1/2$, the best adsorption

266

mode is obtained when the CO molecules are bridging a Mg-Mg bond. In figure 6 the adsorption pattern is represented with and without the substrate. When the CO molecules are aligned along the Mg-Mg directions, they stabilize each other and the adsorption energy is the largest; when they are oriented along the MgO directions, they repel and the adsorption is not favorable.

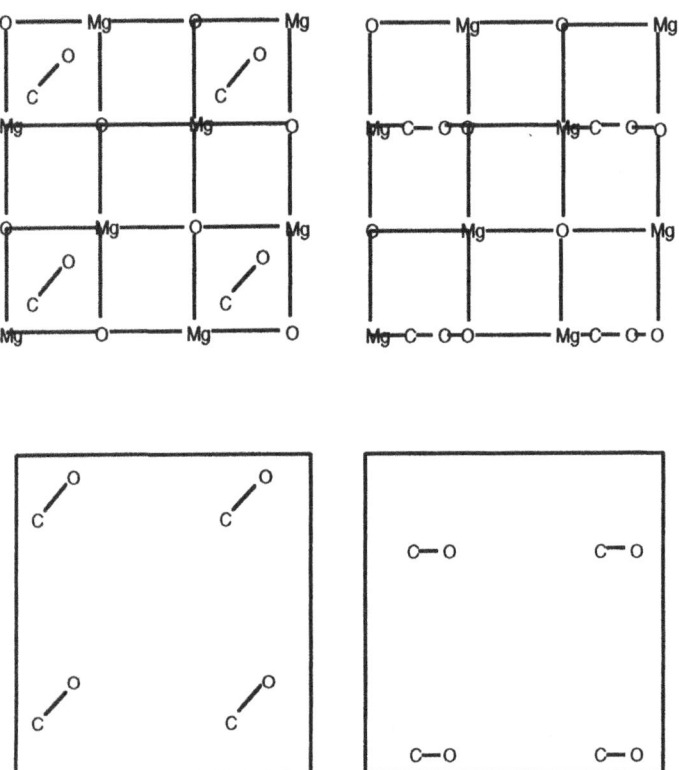

Figure 6: CO/MgO at θ=1/2. The adsorbate-adsorbate interactions are favorable when the molecules are oriented along the Mg-Mg directions.

Calculations [21] also verify the experimental results [22] obtained for θ=3/4 which are sketched in figure 7. The three adsorbed molecules are different; one is bridging a Mg-Mg bond, another is perpendicular and the last one is bent. This is much better than having all CO molecules perpendicular. All the motions of the CO adsorbates leading to this optimal geometry can be followed by similar motions in the absence of substrate. They correspond to an improvement in the interaction between the CO molecules.

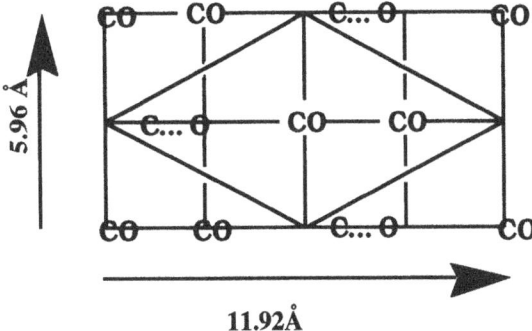

Figure 7: CO/MgO at θ=3/4. The space group is C1m1.

8. H_2O and H_2S adsorption.

In the case of molecular adsorption, water is vertically adsorbed over a titanium center [7,23]. This optimizes the σ-donation to the titanium ion and allows a π-donation from the b_1 electron pair of the water. No H-bond is possible for this geometry except for the 100 face of rutile where there is an empty space between the polymers that allow the adsorption on one of them and the formation of an H-bond with the other. This face corresponds to the optimal situation for the molecular adsorption as is experimentally found [24]. Since H_2S is weakly adsorbed, the molecule can lie over the surface to benefit from an H-bond with the O^{2-} ions from the surface.

For the dissociative adsorption of water on the linear polymer, the best situation corresponds to a sequence of H-bonds between the successive adsorbed OH groups. Thus, the dissociation up to a coverage of θ=1 is best. Going from θ=1/2 to θ=1 is better than going from θ=0 to θ=1/2.

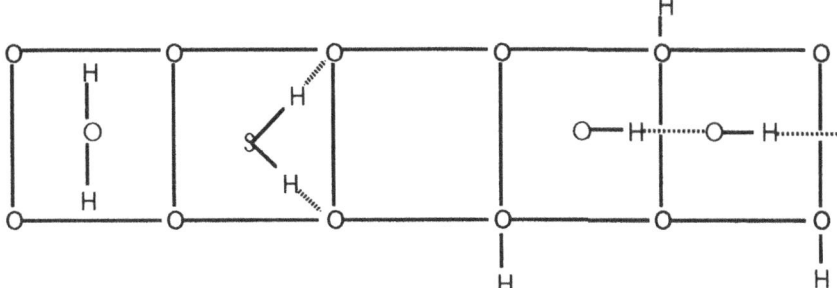

Figure 8: Represented here is the geometry of adsorption of the linear polymer of H_2O (molecular) H_2S (molecular) and H_2O (dissociative with the sequence of H-bonds).

268

Adsorption of a second layer of water through H-bonds to adsorbed molecules is weak [7]. The largest interaction (an adsorption energy of 11.3 kcal/mol) is that involving the protons from adsorbed water molecules and the oxygen atoms from the water molecules of the second layer.

9. NH₃ adsorption.

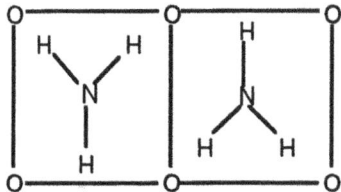

Ammonia is a strong base and a poor Brönsted acid so the adsorption is molecular [9]. Going from $\theta=1/2$ to $\theta=1$ is exothermic by 26 kcal/mol. Ammonia can co-adsorb with water. In the best mode, water is dissociated and ammonia remains as a molecule. The co-adsorption allows the formation of H-bonds. In the diagram below, it can be seen that the second adsorption clearly benefits from the presence of the first adsorbate.

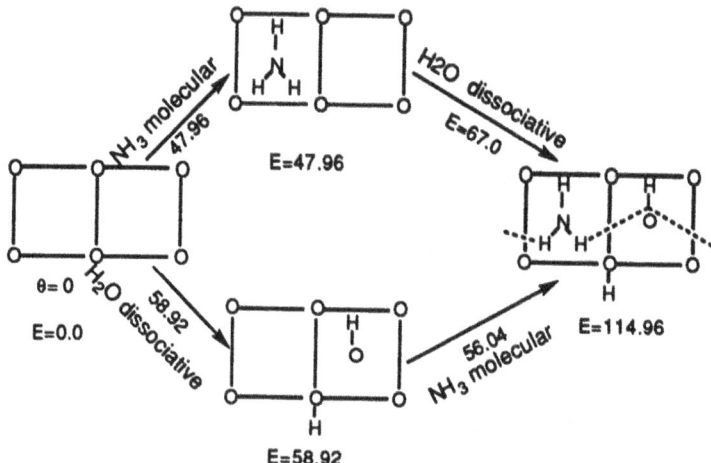

Figure 9 : Co-adsorption of water and ammonia. The adsorption energy (in kcal/mol) is larger in presence of the co-adsorbate.

10. Conclusions

The various examples show the importance of the acidic and basic properties of the adsorbates and of the substrate to control the adsorptions. The Frontier orbitals successfully explain these properties. Periodic calculations are adequate for the calculation on surfaces with no defects when the coverage is high. In this case, it is important to consider the adsorbate-adsorbate interactions.

References

1) (a) Dovesi R., Pisani C., Roetti C., Causà M., Saunders V. R., (1989) Crystal 88, *QCPE program* N°577, *Bloomington, Indiana.* (b) Pisani C., Dovesi R., Roetti C., (1988) *Lecture Notes in Chemisty,* **48**. Springer, Heidelberg.

2) Durand P., Barthelat J. C.,(1975) *Theor. Chim. Acta (Berl.)* **38**, 283-302.

3) (a) Silvi B. , Fourati N., Nada R., Catlow C. R. A., (1991) *J. Phys. Chem. Solids* **52** 1005-1009. (b) Fahmi A. , Minot C., Silvi B., Causà M.,(1993) *Phys. Rev. B* **47**, 11717-11724

4) (a) Cervasini A. and Auroux A. (1991) *J. Therm. Analysis* **37**, 1737-1744.(b) Auroux A. and Cervasini A. (1990) *J. Phys. Chem.* **94**, 6371-6379

5) Janousek B. K. and Brauman J. I. , (1979) Gas Phase Ion Chemistry *Vol* **2** *Browers M.T. Acad. Press. Inc. N.Y.*

6) Yamdagni R. and Kebarle P. (1976) *J. Amer. Chem. Soc.* **98** 1320

7) Fahmi A. , Minot. C. (1994) *Surf. Sci.* **304** 343-359

8) Barbier J., Marecot P. , Tifouti L., Guenin M. Frety R. (1985) *Applied Catalysis* **19** 375-385

9) (a) Pittman R.M. and Bell A.T. (1994) *Catalysis Letters* 24 1-13 .(b) Boddenberg B. and Eltzner K. (1991) *Langmuir* **7** 1498-1505; Parfitt G.D.(c) Ramsbotham J. and Rochester C.H. (1971) *Trans Farad. Soc.* **67** 841-847

10) (a) Hiraoka K. and Kebarle P. (1977) *J. Amer. Chem. Soc.* **99** 360-366. (b) Hiraoka K. and Kebarle P. (1977). *J. Amer. Chem. Soc.* **99** 366-370

11) (a) Bartmess, J. E., McIver R.T. Jr. (1979) in Gas Phase ion Chemistry ; Vol 2 Bowers, M.T.; Acad Press. Inc. NY 87-121. (b) (1977) J. Phys. Chem. Ref. Data Vol. 6, Sup n°1

12) (a) von Boehm, H. P.; Herrmann M., (1967) *Z. Anorg. Allg. Chem.* , **352**, 156. (b) Boehm H. P. , (1971) *Discuss. Faraday. Soc.* **52** 264-275 (c) Rossi, P. F.; Busca, G.; Lorenzelli, V.; Saur, O.; Lavalley, J. C. (1987) *Langmuir* **3**, 52-58. (d) Kim, K. S.; Barteau, M. A.; Farneth W. E. (1988) *Langmuir* **4**, 533-543. (e) Rossi, P. F.; Busca, G. (1985)*Colloids Surf.* **16**, 95-102; Munuera, G.; Stone, F. S. (1971) *Discuss. Faraday Soc.* **52**, 205. (f) Carrizosa, I.; Munuera, G. (1977) *J. Catal.* **49**, 174-188; *J. Catal.* **49**, 189-200. (g) Carrizosa, I.; Munuera, G. and Castañar S. (1977 *J. Catal.* **49**, 265-277. (h) Onishi, H.; Aruga, T.; Egawa, C.; Iwasawa, Y. (1988). *Surf. Sci.* **193**, 33-46. (i) Primet M., Pichat P., and Mathieu M.V. (1971) *J. Chem. Phys.* **75**, 1216-1221; *J. Chem. Phys.* **75**, 1221-1226. (j) Kim, K. S.; Barteau, M. A. (1990) *Langmuir* **6**, 1485-1488. (k) Kim, K. S.; Barteau, M. A. (1988). *Langmuir* **4**, 945-953. (l) Kiselev A.V. and Uvarov A.V. (1967) *Surface Sci.* **6** 399-421.

13) Garrone E., Bolis V., Fubini B., Morterra C. (1989) *Langmuir* **5** 892-899.

14) Morterra C., Ghiotti G., Garrone E. and Fisicaro E. (1980) *J.C.S. Faraday I,* **76** 2102-2113 also see Ramis G., Busca G. and Lorenzelli V. (1991) *Material Chemistry and Physics,* **29** 425-435. (b) Raupp G.B. and Dumesic J.A. (1985) *J. Phys. Chem.* **89**, 5240-5246. (c) Göpel W., Rocker G. and Feierabend R. (1983) *Phys. Rev. B* **28** 3427-3438.

270

15) Hoggan P.E., Bensitel M., Lavalley J.C. (1994). J. Mol. Struct.*(Theochem)* **320** 49-56

16) Knözinger H. Z.,(1989) Probing Acid sites by carbon monoxide: *Acid-Base Catalysis* , K. Tanabe et al. Eds., 147-167

17) Fahmi A., Minot. C. (1994) *J. Organometallic Chem.* in press

18) Dewar M.J.S. , (1950) *Bull. Soc. Chim. Fr.* **C71** 18 . (b) Chatt J., (1950) *Nature* **165**, 859. (c) Chatt J., (1951) *Research* **4**,180. (d) Chatt J. , Duncanson L. A., (1953). *J. Chem. Soc.* 2939-2947. (e) Blyholder G. , (1962) *J. Chem. Phys.* **36** 2036; (1966) *J. Chem. Phys.* **44** 3134-3136. (f) Veillard H. (1977. *Nouv. J. Chim.* **2** 217-227. (g) Daoudi A., Suard M., Berthier G., (1990) *J. Mol. Struct. (Theochem)* **210** 139-145.

19) Causà M., Dovesi R., and Ricca F. , (1993) *Surf. Sci.* **280** 1-13.

20) (a) Dovesi R., Orlando R., Ricca F. , and Roetti C.,(1987) *Surf. Sci.* **186** 267-278. (b) Causà M., Kotimin E., Pisani C. and Roetti C.,(1987) *Solid State Phys.* **20** 4991

21) (a) Minot C. and van Hove M.A. to be published. (b) Girardet C. and Hoang to be published.

22) (a) Audibert P., Sidoumou M. and Suzanne J. (1992) *Surf. Sci. Lettr.* **273** L467. (b) Heidberg J., Kandel M. Meine D. and Wildt U. (1994) Post-Conference satellite Meeting of the 8th ICQC : Quantum Chemical Aspectds of heterogeneous catalysis.

23) (a) Burdett J. K. ,(1985) *Inorg. Chem.* **24** 2244. (b) Smith P. B. , Bernasek S. L. (1987) *Surface Sci.* **188** 241. (c) Henrich V. E. , (1985) *Rep. Prog. Phys.* **48** 1481. (d) Henrich V. E. (1987), *Phys. Chem. Minerals* **14** 396. (e) Munuera G., Stone G. S., (1971) *Discuss. Faraday Soc.* **52** 205.

24) (a) Lo J. W., Chung Y. W. , Somorjai G. A., (1978) *Surface Sci.* **71** 199-219. (b) Suda Y., Morimoto T., (1987) *Langmuir* **3** 786-788.

AN OVERVIEW OF THE MO ARCHITECTURES OF METAL CLUSTERS USING GRAPHIC TOOLS

CARLO MEALLI

Istituto per lo Studio della Stereochimica ed
Energetica dei Composti di Coordinazione del CNR
(ISSECC) Via J.Nardi 39,
50132 Firenze (Italy)

ABSTRACT

The presentation illustrates an approach to the understanding of the MO structure of transition metal clusters through the graphical interpretation of EHMO calculations. The interaction and the correlation diagrams, as well as the 3D MO drawings, depicted by the program CACAO on the screen of a PC, help to gain an overview of the whole MO architecture. Using chemical intuition, new insights about the electronic structure can be obtained by directly evaluating the symmetry properties and the topology of the MOs in terms of the basic principles of Perturbation theory. Thus, it becomes relatively easy to extrapolate the role of each single MO in the cluster. At variance with the non-transition elements, the metals may have *excess* orbitals which behave as *lone pairs*. Thus, repulsions between two metals are triggered even if other orbitals are already engaged in the proper bonding/antibonding interactions which fix the formal M-M bond order equal to one. In some cases, the M-M bond is broken but alternative examples exist in which the bond is only elongated. Highly symmetrical clusters such as octahedra can accomodate a very different number of electrons. When all of the M-M bonding orbitals are filled and the corresponding antibonding MOs are empty (84e species) the number of bonds (twelve) matches that of the intermetal connectivities. In such a scheme, the atomic d orbitals with δ symmetry are seen to play a fundamental role which has been probably overlooked up to now. Additional electron pairs can enter the general MO scheme without destroying the primary octahedral geometry of the cluster. In 98e species such as $[Co_6(\mu_3\text{-}S)_8L_6]$, the coexistence of filled bonding and filled antibonding MOs amounts to an equivalent number of *lone pairs* (14) which are strongly repulsive towards each other. Hence the expansion of the cluster skeleton occurs, but not its rupture because of the five surviving Co-Co bonds.

L. J. Farrugia (ed.), The Synergy Between Dynamics and Reactivity at Clusters and Surfaces, 271–284.
© 1995 *Kluwer Academic Publishers.*

INTRODUCTION

The analysis of MO calculations for metal clusters of medium or high nuclearity becomes a complicated task if one wishes to describe the contribution of each single level to the overall electronic structure. Whereas for mononuclear (or binuclear) metal complexes a few frontier MOs depict the origin of the observed molecular geometry quite well, and provide useful hints for deformational trends and reactivity, important skeletal MOs of polynuclear clusters are buried in thick energy bands. Moreover, the HOMO and the LUMO alone (their identification is sometimes dubious because of the close-packing of the frontier MOs) are generally insufficient to describe the inter-metal bonding network.

During the years many useful concepts have been developed to simplify the interpretation of the chemical bonding in clusters, [1] most of them being based on the results of MO calculations (performed at all levels of sophistication). Thus, in addition to the principles of Perturbation Theory, [3] the constraints of the symmetry, the rules of the electron counting [2] the application of isolobal analogies with regard to the local geometry at each metal center, [4] the empirical correlations between cluster topology and the electron count [5] have all proved to be valuable tools. However for clusters of medium and higher nuclearity (*e.g.* the classical octahedra), there is still a large gap between the simplified schemes which provide an insight at the electronic structure and the more complex information derivable from MO calculations. Here, a few guidelines are indicated on how to extract a good deal of chemical infomation from the direct evaluation of symmetry, composition and topology of all the MOs through the usage of user-friendly visual aids.

The computer package CACAO, [6] allows the graphing of the relevant numerical results from EHMO calculations.[7] Walsh and interaction diagrams as well as 3D drawings of the MOs and their components can be viewed on the screen of a PC. The ease by which the calculations can be repeated and cross-checked helps to interpret the the chemical bonding in a rather *empirical* way. Essentially, the global MO architecture of clusters can be described in terms of the prevailing characters of each level, *i.e.* metal-ligand bonding/antibonding (*b/a*), intra-ligand *b/a*, metal-metal *b/a*, metal and/or ligand *lone pairs* (*lp*). These pieces of chemical information are not necessarily implicit in the common electron counting rules. On the other hand, the rules themselves have a MO basis, so it is ultimately interesting to make a correlation between them and the newly gained vision of the MO architecture.

Useful Guidelines to Analyze the MO Pictures

Each of the nine atomic metal orbitals is an important brick for building the MO architecture of any cluster and a good interpretation requires the understanding of the distribution and the role of all of the atomic orbitals. The possibility offered by CACAO of overviewing the MOs is particularly useful to describe the electronic structure of clusters in terms of the localized properties of the functional groups. This approach, which is most familiar to organometallic chemists, appears of dubious application in clusters. In any case, the extension of the *isolobal analogy* [4] to clusters requires some caution. Consider for example the isolobal methylenic groups and d^8 metal fragments with C_{2v} symmetry (see **I**). The existence of three-membered or four membered organic (cycloalkanes) and inorganic [*e.g.* $M_3(CO)_{12}$ or $M_4(CO)_{16}$] rings is easily explained in view of the parallel contributions of the σ and π FMOs.

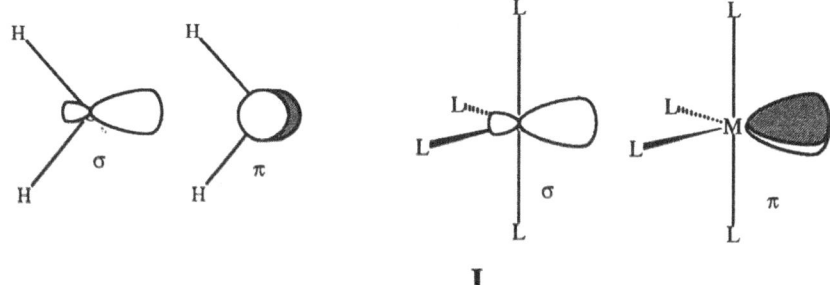

I

However, an important difference is often underestimated. While each carbon atom utilises all of its atomic orbitals in the bonding interactions (beside the σ and π orbitals in **I**, the remaining two atomic orbitals are devoted to C-H bonding), each metal fragment carries three *excess* orbitals (beside the two at the right side in **I** and the four already engaged in M-L bonding). In view of the local *pseudo*-octahedral symmetry, the three *excess* orbitals can be regarded as those of a t_{2g} set which, if populated in the cluster, may behave as metal *lone pairs* repelling each other. Although mitigated by the backdonation toward the terminal ligands (especially the π-acceptors ligands such as carbonyls), the repulsion may be especially effective between the in-plane *radial* orbitals (x^2-y^2) and it is a definite cause for the stretching of the skeletal bonding in the cluster. This effect does not feature in the clusters of the non transition elements.

Based on the above considerations, it becomes important to track down the presence and the stereo-distribution of the metal *lone pairs*. A prediction of their number and constitution can also be attempted by referring to the elementary electron counting rules. Thus, an extension of the Effective Atomic Number rule (EAN) stems from the considerations summarized in Scheme **II**. If L metal orbitals, out of a total of $9 \times V$, are involved in M-L bonding (terminal or bridging), it is reasonable to assume that the remaining orbitals are available in either M-M bonding interactions (**m** bonding and **m** antibonding combinations) or are localized at the metals as *lone pairs* (= **n**).

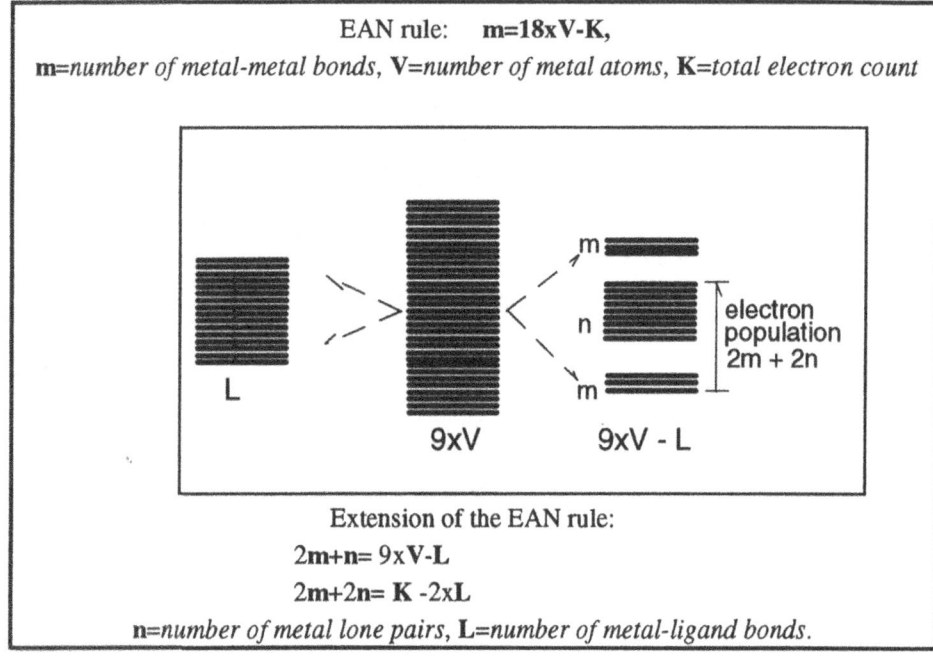

EAN rule: **m=18xV-K,**

m=*number of metal-metal bonds,* **V=***number of metal atoms,* **K=***total electron count*

$9xV$

$9xV - L$

L

m

n

m

electron population $2m + 2n$

Extension of the EAN rule:

$2m+n= 9xV-L$

$2m+2n= K -2xL$

n=*number of metal lone pairs,* L=*number of metal-ligand bonds.*

SCHEME II

Two equations can be written by assuming that each M-L and M-M bond is of the canonical type (*i.e.*, two-electron/two centers) and that of all of the available **K** electrons, 2xL are involved in metal-ligand bonding, whereas the remaining ones occupy only the metal-metal bonding levels as well as the non-bonding metal orbitals (*lone pairs*). The solution of the two equations does not only reproduces the **m** value easily obtainable from the EAN rule but it allows an estimation of the number of non-bonding electrons at the metals (=2xn). This knowledge is a useful guideline for better analyzing the MO results.

As an example, consider the classical $M_3(CO)_{12}$ carbonyl clusters of the triad Fe, Os, Ru. It is well known that while Os and Ru have no bridging carbonyls, the structure of the iron derivative has two bridging carbonyls.[8] The solution of the system of equations in Scheme **II** (see Scheme **III**) suggests a lower number of lone pairs (7 *vs.* the 9 of Ru and Os), hence a stronger M-M bonding in the iron derivative. It is likely that the unequal distribution of *lone pairs* may determines the asymmetry of the bridged Fe-Fe bond (shorter) *vs.* the bridged ones (longer).

Another interesting example of the importance of the metal *lone pairs* comes from another trinuclear cluster, namely $[Pt_3(\mu-PPh_2)_3Ph(PPh_3)_2]$ with 44 electrons. X-ray studies have shown the existence of two geometric isomers.[9] Depending only on the conditions of crystallization, one sample differs from the other in having two short (ca. 2.75 Å) and one long (3.59 Å), vs. three approximately equal Pt-Pt vectors (ca. 3.0 Å). By considering that the skeleton of the cluster (see **IV**) is planar and that one metal

275

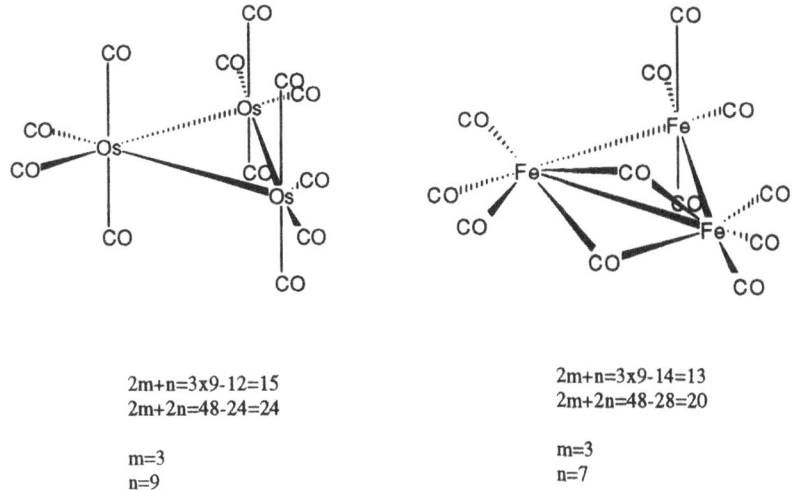

$$2m+n=3\times9-12=15$$
$$2m+2n=48-24=24$$

$$m=3$$
$$n=9$$

$$2m+n=3\times9-14=13$$
$$2m+2n=48-28=20$$

$$m=3$$
$$n=7$$

SCHEME II

orbital per atom (p_z) is extraneous to bonding, the two equations **II** take the form:

$$2m + n = 8\times3 - 9 = 15$$
$$2m + 2n = 44 - 18 = 26$$

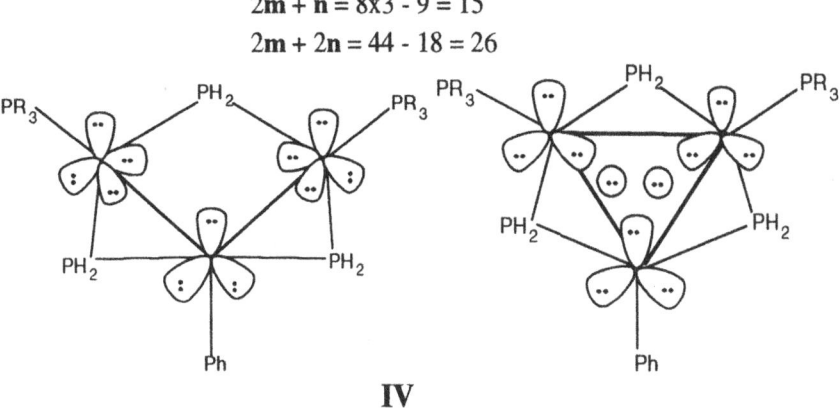

IV

and the solutions for **m** and **n** are 2 and 11, respectively. The MO calculations [10] indicate that the energetics of the two forms are close, but the local geometry at the metals induces a different distribution and stereochemistry of the *lone pairs*. Several well known 42 electron species,[10] all having equilateral trianglar geometry, show a MO pattern of three filled M-M bonding levels below three empty antibonding partners. The components are radial (σ) and tangential (π) orbitals at each center which intermix strongly (see SCHEME V).

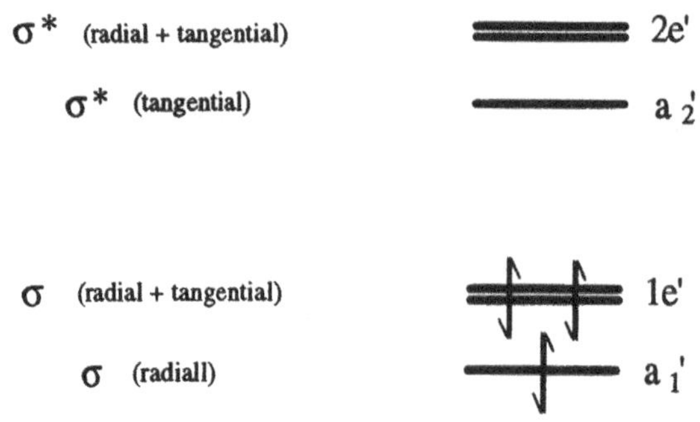

σ^* (radial + tangential) ══════════ $2e'$

σ^* (tangential) ─────────── a_2'

σ (radial + tangential) ⇅⇅ $1e'$

σ (radiall) ↿⇂ a_1'

SCHEME V

A similar situation in cyclopropane has been described by Dewar as a case of σ-aromaticity.[11] In the 44e metal cluster, the addition of two electrons in the a_2' orbital increases the number of *lone pairs* by two units. If these localize at only two of the metals, the repulsive effects between the latter become intolerably high and the cleavage of one M-M bond is induced (see the leftmost sketch in **IV**). As another option, the two additional electrons populate the overall M_3 antibonding level a_2' with expansion but no distorsion of the equilateral triangle. The orbital delocalization is such that the bonding and antibonding MOs maintain the topological features of the 42e species. The electrons in a_2' cancel the bonding effects of one of the lower electron pairs (bonding). The average M-M bond order becomes 2/3 for each M-M connection.

The latter situation is not unusual for clusters of higher nuclearity especially the highly symmetric ones, such as the octahedra. It is reasonable that the MO architecture of the latter, with highly delocalized MOs, allows a large variety of electron counts. For each electron pair populating antibonding MOs, the number of *lone pairs* increases by two units: These may not necessarily localize at adjacent atoms so as to induce bond cleavage. Ultimately, the corresponding M-M connectivity is maintained although a cluster expansion is observed.

Octahedral Clusters

Remarkably, a variety of octahedral clusters with formula $M_6(\mu_3-X)_8L_6$ are known to be stable, but with very different electron counts. The basic structure, with six terminal σ–donor ligands and all of the octahedral faces capped by single atoms or ligands (X= π donor or acceptor) is shown in **VI**. Table I summarizes some of the most classical species which have been characterized by X-ray analyses.

Notice that in the sketch as many as nine connectivities depart from each metal atom. Can all these be described as two-center/two electron bonds? In particular, can the total number of M-M bonds ever be equal to 12 (the number of connectivities)?

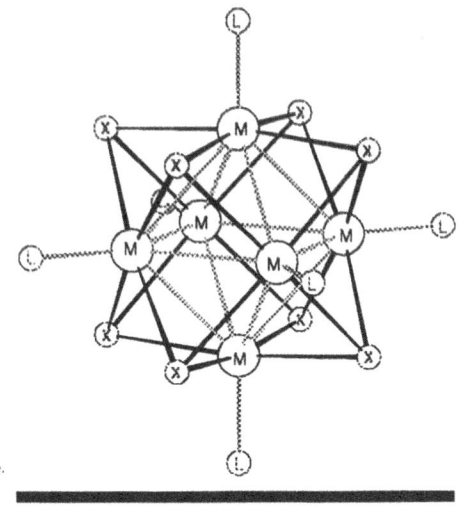

VI

It is worth mentioning that according to the classic Wade-Mingos rules [3] the number of cementing electron pairs in an octahedral cluster is 7 while 12 bonds would imply 24 bonding electrons.

TABLE I.
<u>SELECTED EXAMPLES OF OCTAHEDRAL CLUSTERS WITH FORMULA</u>
$M_6(\mu_3\text{-}X)_8L_6$

a) 84e species, X=π-donor (weak)		
$Mo_6(\mu_3\text{-}X)_8L_6$, X=L=Cl; X=L=OR, *dianionic*[12]		X=Cl, L_4=Cl,
L_2=PR$_3$[13]		
b) 80, 81e species, X=π-donor (strong)		
$[Mo_6(\mu_3\text{-}X)_8L_6]^{0,-1}$ X=S, Se L=PR$_3$[14]		
c) 97, 98e species, X=π-donor (strong)		
$[Co_6(\mu_3\text{-}X)_8L_6]^{+1,0}$ X=S,Se L=PR$_3$, CO [15]		
d) 86e species, X=π-acceptor		
$[Co_6(\mu_3\text{-}X)_8L_6]^{-4}$ X=CO, L=CO [16]		
e) 86, 92e species, X=O, Cp terminal ligands		
$[M_6(\mu_3\text{-}X)_8L_6]$ X=O, L=Cp, M=Ti, [17]		M=V[18]

Compounds containing weak π-donor capping ligands such as $[Mo_6(\mu_3\text{-}X)_8Y_6]^{2-}$, (of the type **a** in Table I) conform best with the existence of twelve two center/two electrons M-M bonds which may be deduced from a generalized MO picture of 84e octahedra. In fact, the equations in **II** hasve solutions **n=0, m=12**. In terms of symmetry, the nine atomic orbitals of each metal are classified as three *radial* (s, p_z and z^2), four *tangential* (p_x, p_y and xz, yz) and two δ (xy and x^2-y^2) functions.

A set of metal *radial* orbitals (p_z, in a first approximation although some s-p mixing is allowed) is devoted to the formation of six M-L$_{term.}$ bonds (a_{1g}+e_g+t_{1u}). In order to learn about the bonding network between metals and capping ligands, the symmetries spanned by the $X_{capp.}$ atoms [with one s- (*radial*) and two p- (*tangential*), each] are needed. They are conveniently summarized in **VII**. The overlap populations between FMOs confirm sufficiently good matches between the 24 $X_{capp.}$ functions and a corresponding number of isosymmetric metal combinations. For the sake of brevity, we do not show the drawings of the latter metal FMOs which are essentially formed by *radial* (sp hybrids), *tangential* (linear combinations of p_x and p_y with lobes at 45° from the main axes) and δ (xy). In particular, xy seems the only metal d orbital used for the M-L bonding network which otherwise involves the diffuse orbitals s and p.

Symmetry Combinations of Capping Ligands
Radial σ *Tangential* π

| a_{1g} + a_{2u} + t_{1u} + t_{2g} | e_g + e_u + t_{1u} + t_{2u} + t_{1g} + t_{2g} |

VII

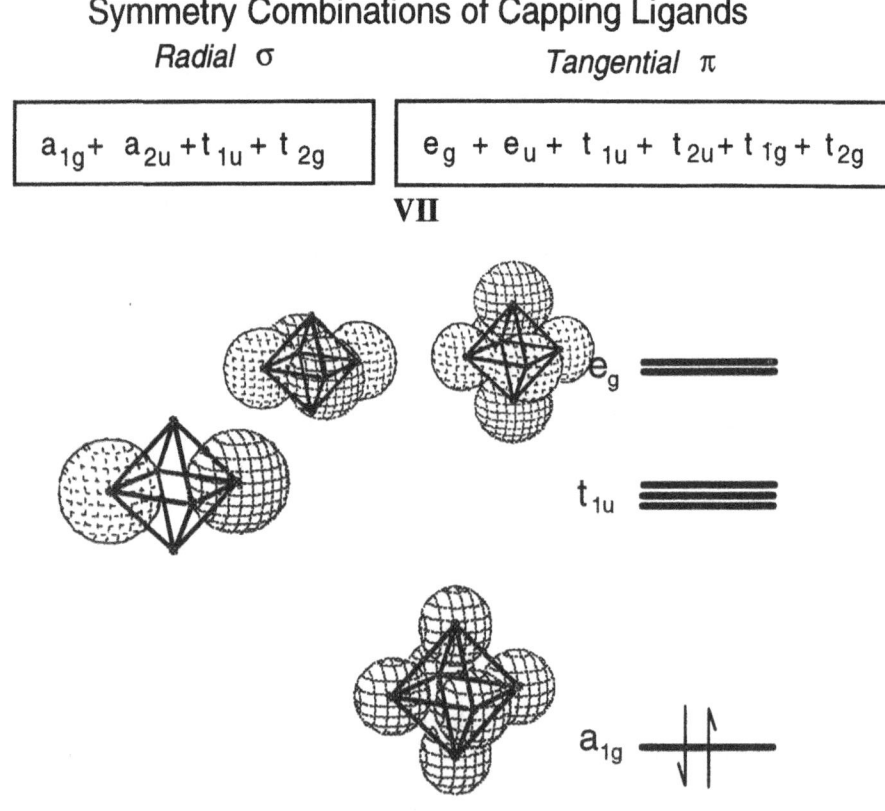

SCHEME VIII

The remaining 24 orbitals (so called *excess* orbitals: one *radial*, two *tangential* and one δ per metal) are used in the M_6 bonding network. It is important to see how the different s, p and δ components behave separately. Thus, by referring to a hypothetical

H_6 octahedron, the s orbitals (*radial*) give rise to the symmetry combinations a_{1g}, e_g and t_{1u}. As shown in Scheme **VIII**, only the first set is evidently bonding so there should be only two cementing electrons with a total charge +4 for the cluster.

The previous scheme can then be integrated with *tangential* orbitals (the two possible basis pairs are p_x, p_y and xz, yz), which in O_h symmetry separate into the four triply degenerate classes t_{1u}, t_{2g}, t_{2u} and t_{1g}. The trends are bonding for the two former combinations and antibonding for the latter. The overall MO pattern is that typical of the cluster $B_6H_6^{-2}$, characterized by the well known 7 bonding and 11 antibonding MOs (see Scheme **IX**). [1]

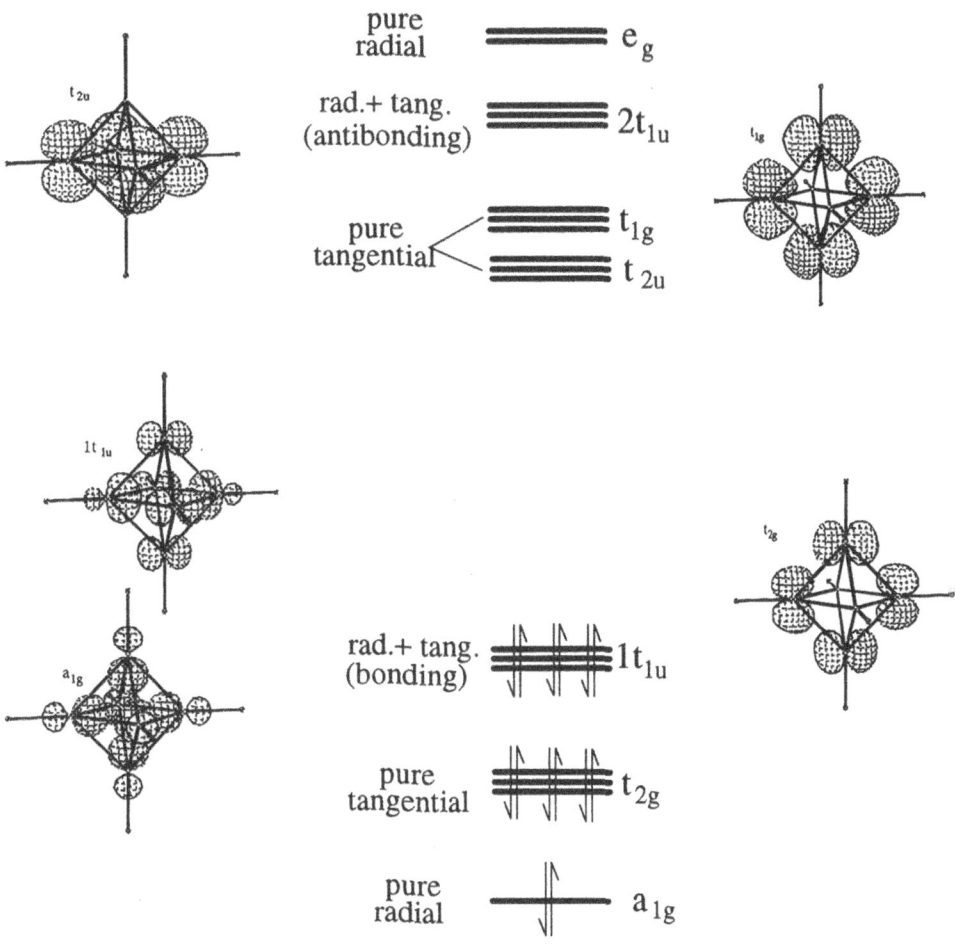

SCHEME IX

Importantly, since both the *radial* and the *tangential* atomic orbitals originate as irreducible representations of the type t_{1u}, they can mix together. The implicit bonding character of the t_{1u}, *tangential* combinations, is enhanced as it is the antibonding character of the t_{1u} originally *radial* combinations (see Scheme **VIII**). It is evident that

280

only seven bonding electron pairs can be accommodated into the MO scheme **IX** whereas the B-B linkages (or connectivities) are as many as twelve.

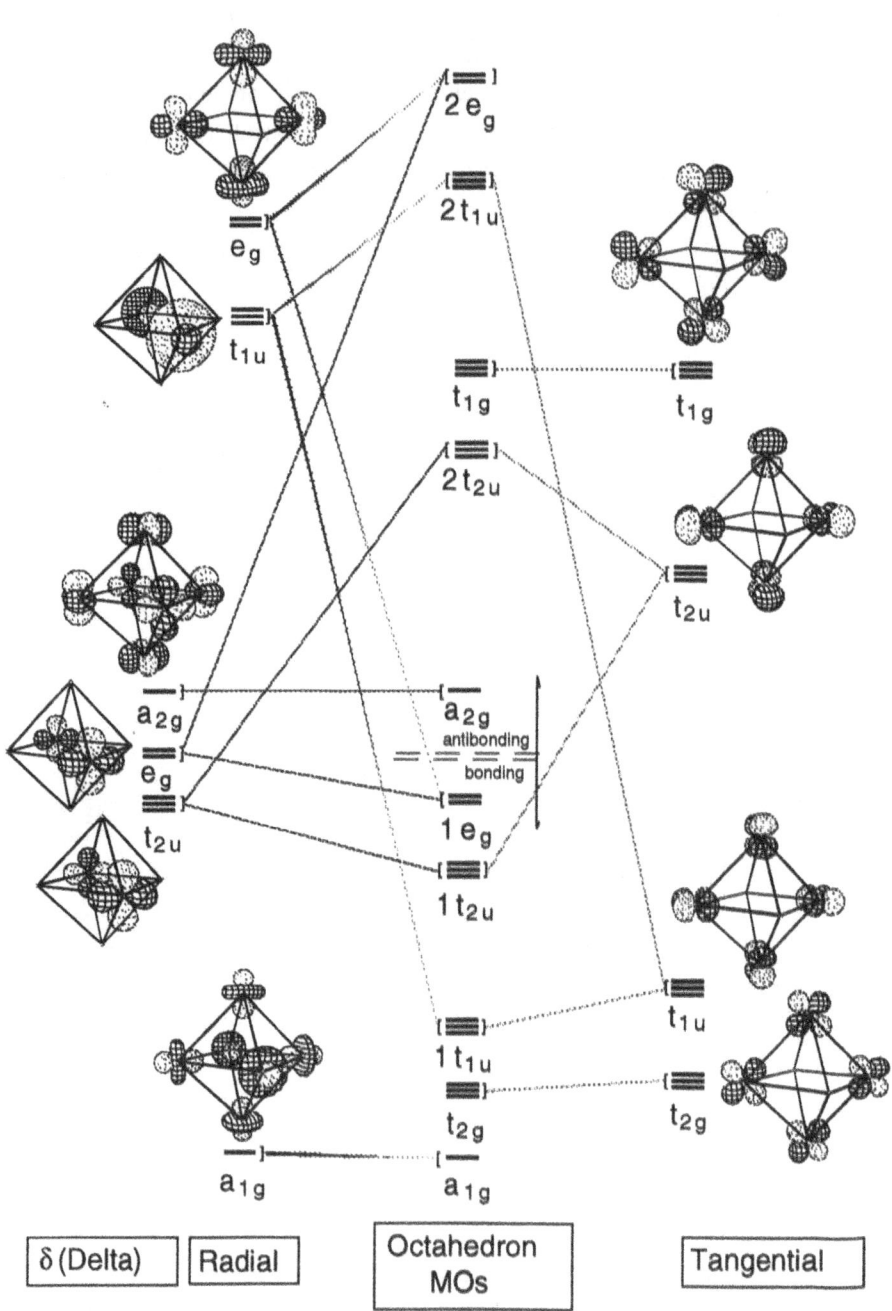

δ (Delta) | Radial | Octahedron MOs | Tangential

SCHEME X

The concept of *hypervalency* can be invoked also because each B atom has only four atomic orbitals whereas as many as five connectivities are shown to depart from it. The idea that the number of skeletal bonding electron pairs are also seven in octahedral metal clusters has inspired theories such as PSEPT.[19] A scheme similar to **IX** for the intermixing of *radial* and *tangential* orbitals can be constructed, although the s and p character of the atomic orbitals is now largely substituted by z^2 (*radial*) and xz, yz (*tangential*) functions. Although not always true (see $[Co_6(CO)_{14})]^{-4}$), the s and p metal orbitals are mainly involved in ligand-metal donor-acceptor interactions.

At this point it is important to highlight the role of the specific set of δ orbitals which cannot be involved in σ-bonding to any ligand (those of type x^2-y^2 have lobes eclipsing the octahedal edges). Five out of the six δ combinations are allowed by the symmetry to mix with the *radial* or *tangential* combinations, of the type shown in **IX**. The MO drawings in **IX** show clearly *radial*-δ (e_g) / *tangential*-δ (t_{2u}) mixings. Only the δ combination a_{2g} is unique and is clearly M_6 antibonding. Even if the extent of the mixing is not large, the MOs $1t_{2u}$ and $1e_g$, centered on the metal δ combinations, acquire a formal M_6 bonding role. For a proper electron count, not only do the δ orbitals lose the character of *lone pairs* , but they raise the number of bonding interactions to match that of the M-M connectivities (12). This is clearly unattainable by the non-transition elements.

The viewpoint is highlighted by the general MO scheme **X**. In particular, if all of the MOs up to $1e_g$ (bonding) are filled and the upper 12 MOs (antibonding) are empty, the number of bonds sums up to twelve. The MO a_{2g} is most critical, because, in spite of its overall M_6 antibonding character (δ^*, see **X**), it lies at relatively low energy (the overlap is poor because its pure d components are generally contracted). Thus an electron vacancy in the a_{2g} is unlikely, which is the first evident reason why octahedral carbonyl clusters stabilize with the 86 electron count. In reality, the electronic structure of the tetraanion $[Co_6(\mu_3\text{-}CO)_8(CO)_6]^{-4}$ [16] with 86 electrons (type **d** in Table I), is more complicated. Essentially, there is a reverse role between d and sp metal orbitals. A good part of the M-CO bonding interactions involves d orbitals (*via* backdonation), whereas the high lying sp metal orbital combinations are more M_6 antibonding than metal-ligand antibonding in character. Nonetheless, a scheme similar to **X** does still apply with 12 identifiable M_6 bonding orbitals (all populated), and 12 antibonding ones, one of which is also filled (a_{2g}). Accordingly, one of the bonding interactions vanish while the coexistence of a pair of filled bonding/antibonding MOs can be also viewed as raising the number of metal *lone pairs* by two units. The interested reader may consult reference 20, where a new MO analysis of $[Co_6(\mu_3\text{-}CO)_8(CO)_6]^{-4}$ is presented in detail (to be compared with an earlier one [21]).

The idea, that the presence of electrons in some antibonding MOs of scheme **X** does not necessarily deform the primary geometry **VI** but forces its expansion, seems nicely confirmed by the existence of the 98e clusters of type **c** (Table I). The species [$Co_6(\mu_3\text{-}S)_8L_6$] conform to a picture where all of the levels up to t_{1g} are all occupied (because of the long Co-Co separations of ca. 2.8 Å, the order of the filled levels may be different with respect to **X** but the substance does not change). In the global MO architecture of octahedral clusters, such a situation (12 bonding MOs filled as well as 7 of the antibonding MOs) suggests the formal coexistence of 5 M-M bonds and 14 lone pairs. The repulsion between the latter can be also taken as the cause of the cluster expansion.

There are also cases where not all of the bonding MOs are filled (*e.g.* the 80 electron compounds such as $Mo_6(\mu_3\text{-}S)_8(PR_3)_6$ or its 81e anion (type **b** in Table I). According to the MO scheme **X**, the e_g levels are now empty or singly occupied. This is a very peculiar case where the metal-sulfur and metal-metal bonding/antibonding framework cannot be effectively separated. Although the MO architecture remains that described up to now, more subtle considerations need to be made and once again the reader is invited to consult reference 21 for a more detailed analysis.

Conclusions

In the report, presented at the NATO Advanced Research Workshop (Drymen, Scotland 1994) a powerful graphic approach to the understanding of the MO structure of classic metal clusters, such as the face-capped octahedra, has been illustrated. A major point is that the metal basis set, which includes d orbitals, offers more bonding capabilities than the basis set of main group elements. Thus, in the octahedral clusters of the latter elements, some atomic components have to be shared in the interactions with several neighbour atoms (*hypervalence*) whereas this does not necessarily apply to metals. In fact, one *radial*, two *tangential* and one δ orbital per atom are what is needed to make as many different interactions as the number of edges departing from the atom itself. Obviously, in order to transform all the interactions into bonds, an appropriate number of electrons must populate only the bonding MOs. *Excess* electron pairs which can populate some M_6 antibonding MOs, not only nullify an equivalent number of bonds but trigger destabilizing four-electron repulsions. Under these circumstances, the clusters does usually expand while maintaining the primary, highly symmetric, structure.

If the electron distribution in the clusters is qualitatively known through *ad-hoc* MO studies it becomes also possible to assign a formal bond order to the M-M linkages. In any case, the bond order of $^7/_{12}$ (as in main group octahedral clusters) does not seem to be a general rule of thumb for metal octahedra.

Aknowledgements

Many of the concepts presented here were developed in discussions and collaborative work with Dr. J.A. Lopez of the University of Zaragoza (Spain), Dr. Y. Sun of the

University of Waterloo (Canada) and with Professor M.J.Calhorda of the Istituto Superior Técnico de Lisboa (Portugal).

References

(1)   Mingos D. M. P. and Wales D. J. (1990) *Introduction to Cluster Chemistry*, Prentice-Hall, Englewood Cliffs, N.J. (b) Albright T. A., Burdett J. K. and Whangbo, M.-H. (1985) *Orbital Interactions in Chemistry*, Wiley, New York.

(2)   Hoffmann R. (1971) *Acct.Chem.Res.* **4**, 1 and references therein.

(3)   (a) Wade K. (1971) *J.Chem.Soc., Chem.Commun* 792. (b) Wade K (1976) *Adv.Inorg.Chem.Radiochem.*, **18**, 1. (c) Mingos D. M. P. (1972) *Nature (London) Phys.Sci.* **236**, 99. (d) Mingos D. M. P. (1984) *Accts. Chem.Res.* **17**, 311. (c) For a review see: Owen S. M. (1988) *Polyhedron* **7**, 253.

(4)   (a) Hoffmann R. (1982) *Angew.Chem, Int.Ed.Engl.* **21**, 711. (b) Evans D. G. (1983) *J.Chem.Soc., Chem.Commun.* 675.

(5)   Teo B. K. (1984) *Inorg.Chem.* **23**, 1251.

(6)   Mealli C. and Proserpio D. M. (1990) *J. Chem. Ed.*, **67**, 399.

(7)   (a) Hoffmann R. (1963) *J. Chem. Phys.* **39**, 1397. (b) Hoffmann R. and Lipscomb W. N. (1962) *J. Chem. Phys.* **36**, 2179. (c) Hoffmann R. and Lipscomb W. N. (1962) *J. Chem. Phys.* **36**, 2872.

(8)   (a) Mason R. and Rae A. (1968) *J.Chem.Soc. A.* 778. (b) Dahl L.F. and Corey E.R. (1962) *Inorg.Chem.* **1**, 521. (c) Wei C. H. and Dahl L.F. (1969) *J.Am.Chem.Soc.* **91**, 1351.

(9)   (a) Taylor N. J., Cieh P. and Carty A. J. (1975) *J.Chem.Soc. Chem.Commun.* 448. (b) Bender R., Braunstein P., Tiripicchio A. and Camellini M. (1985) *Angew. Chem. Int. Ed. Engl.* **24**, 861.

(10)   (a) Mealli C (1985) *J. Am. Chem. Soc. 107*, 2245. (b) Mealli C. and Proserpio D. M. (1990) *J. Clusters Science* **1**, 93.

(11)   Dewar M. J. S. (1984) *J. Am. Chem. Soc. 106*, 669

(12)   (a) Ouahab L., Batail P., Perrin C. and Garrigou-Lagrange C. (1986) *Mater. Res. Bull.*, **21** 1223. (b) Chisholm M. H., Heppert J. A. and J. C. Huffman (1984) *Polyhedron*, **3**, 475.

(13)   Saito T., Nishida M., Yamagata T., Yamagata Y. and Yamaguchi Y. (1986) *Inorg.Chem.* **25**, 1111.

(14)   (a) Saito T., Yamamoto N., Yamagata T. and Imoto H. (1988) *J. Am. Chem. Soc.* **110**, 1646. (b) Saito T., Yamamoto N., Nagase T., Tsuboi T., Kobayashi K., Yamagata T., Imoto H. and Unoura K. (1990) *Inorg. Chem.* **29**, 764.

(15)   (a) Cecconi F., Ghilardi C. A., Midollini S., Orlandini A. and Zanello P. (1986) *Polyhedron*, **5**, 2021. (b) Diana E., Gervasio G., Rossetti R., Valemarin F. Bor G. and Stanghellini P.L. (1991) *Inorg. Chem.* **30**, 294. (c) Fenske D., Ohmer J. and Hachgenei J. (1985) *Angew. Chem. Int. Ed. Engl.*

24, 993. (d) Fenske D., Ohmer J. and Merzweiler K. (1987) *Z. Naturforsc.*, *Tail B*, **42**, 803. (e) Mao-Chun H. , Zhi-Ying H., Xin-Jian L., Guo-Wei W., Bei-Sheng K., Han-Qin K. and J. Huaxnef (1989) *Inorg. Chim. Acta* **159**, 1.

(16)   Albano V., Bellon L.. Chini P. and Scatturin V. (1969) *J. Organomet. Chem.*, **16**, 461.

(17)   Huffman J. C:, Stone J. G., Krusell W. C. and Caulton K. G. (1977) *J. Am. Chem. Soc.* **99**, 5829.

(18)   (a) Bottomley F., Paez D. E. and White P. S. (1985) *J. Am. Chem. Soc.* **107**, 7226. (b) Bottomley F., Drummond D. F., Paez D. E., and White P. S. (1986) *J. Chem. Soc., Chem. Commun.* 1752.

(19)   Mingos D. M. P. (1974) *J. Chem.Soc., Dalton Trans.* 133.

(20)   Mealli C., Lòpez J.A, Sun Y. and Calhorda M. J. (1994) *Inorg.Chim.Acta*, **213**, 199.

(21)   Mingos D. M. P. and Forsyth M. I. (1977) *J. Chem.Soc., Dalton Trans.* 610.

MOLECULAR ORBITAL APPROACH OF SKELETAL ISOMERISM AND POLYHEDRAL REARRANGEMENTS IN SOME ORGANOMETALLIC CLUSTERS.

J.-Y. SAILLARD, M. T. GARLAND, S. KAHLAL AND J.-F. HALET
Laboratoire de Chimie du Solide et Inorganique Moléculaire
URA-CNRS 1495,
Université de Rennes 1,
35042 Rennes-Cedex, France.

1. Introduction

The bonding and structure in transition-metal clusters is generally well understood with the help of the Polyhedral Skeletal Electron Pair (PSEP) theory [1]. This theory provides simple rules, known as the Wade-Mingos rules within the organometallic community, which describe the relationship between the geometry of the cluster and its number of valence electrons. The theoretical basis of these rules have been developped later by Stone within the framework of the Tensor Surface Harmonic (TSH) theory [2]. Historically they have been established first for simple main-group clusters, such as boranes and carboranes, and subsequently generalized to transition-metal clusters with the help of the isolobal analogy [3].

A simple example of these electron counting rules concerns the clusters having a skeleton core based on an octahedron. According to the PSEP theory, they all are expected to be stable if they possess seven Polyhedral Skeletal Pairs (SEP), that is, fourteen electrons mainly delocalized on the cluster core and responsible for its stability and bonding. For instance, this is the case of the 6-vertex compounds $[B_6H_6]^{2-}$ and $[Os_6(CO)_{18}]^{2-}$ [4]. This is also the case of 5-vertex square pyramidal clusters, such as $[C_5H_5]^+$, which can be described as having an octahedral core with five occupied and one vacant vertices. According to the capping principle established by Mingos [5], the capping of a triangular face of a cluster by a conical fragment does not change the favored electron count. Therefore, the capped octahedron and the capped square pyramid are also characterized by a 7-SEP count. For example, the 6-vertex compound $H_2Os_6(CO)_{18}$ adopts the geometry of a capped square pyramid [4].

It is noteworthy that, although being isoelectronic, the dianion $[Os_6(CO)_{18}]^{2-}$ and its protonated derivative $H_2Os_6(CO)_{18}$ adopt different skeletal arrangements in the solid state. These strongly chemically related compounds can be considered as being almost

L. J. Farrugia (ed.), The Synergy Between Dynamics and Reactivity at Clusters and Surfaces, 285–295.
© 1995 *Kluwer Academic Publishers.*

skeletal isomers, and address the possibility of skeletal rearrangements in solution for species which can *a priori* exist with different core geometries.

The comparative stability of skeletal isomers, and the possibility of their interconversion lie beyond the capability of the PSEP theory. MO calculations are necessary to tackle these problems, as well as to study clusters which do not obey the Wade-Mingos rules because of the limitations of the isolobal analogy. In this paper we describe some extended Hückel studies on problems related to skeletal isomerism, focussing the discussion on the dynamics of polyhedral rearrangements.

2. Intramolecular Conversion of an Azoalkane Ligand to Two Nitrene Ligands on a Tri-iron Cluster

The intramolecular reaction indicated below has been shown to occur upon heating by Vahrenkamp and coworkers [6]. It has been described as the conversion of an azoalkane ligand to two nitrene ligands on a tri-iron cluster. Alternatively, it can be viewed as an interconversion beween two 7-SEP distorted square pyramids. In 3, the distorted basal Fe_2N_2 square presents one N-N bond and three Fe-N bonds, while in 4 it is made of four Fe-N bonds. Kinetic experiments suggest that the reaction proceeds first by elimination, followed by the capture of a CO ligand [6].

We have explored the potential energy surface of this reaction. One of the simplest mechanisms for the interconversion between two square pyramids involves a trigonal bipyramid as an intermediate structure. This pathway is schematized overleaf, in a three-dimensional view and in projection of the metallic triangle. Assuming first no CO exchange, this mechanism is forbidden due to the high energy of the trigonal bipyramidal $Fe_3(CO)_{12}(N_2R_2)$ species (6) for the count of 7 SEP's. Indeed, a trigonal bipyramidal core is stable for the count of 6 SEP's. Therefore, the extra electron pair has to be housed in a high-lying antibonding skeletal MO. Moreover, a HOMO/LUMO level crossing occurs during the second part of the pathway, rendering the system symmetry-forbidden. There are two ways of stabilizing 6. One is to get rid of the extra electron pair, by losing a 2-electron CO ligand. The other one is to stabilize the antibonding occupied level by breaking a bond. These two ways lead to the two possible pathways described overleaf.

The first low-energy pathway involves the loss of a CO ligand at the first step of the reaction, leading to the formation of the 6-SEP trigonal bipyramidal $Fe_3(CO)_{11}(\mu_3-\eta^2-N_2R_2)$ intermediate (8). Calculations indicate that 8 can easily rearrange to give the more stable isomer $Fe_3(CO)_{11}(\mu_3-NR)_2$ (9), through a swinging motion of one NR group. 9 has also a trigonal bipyramidal geometry, in agreement

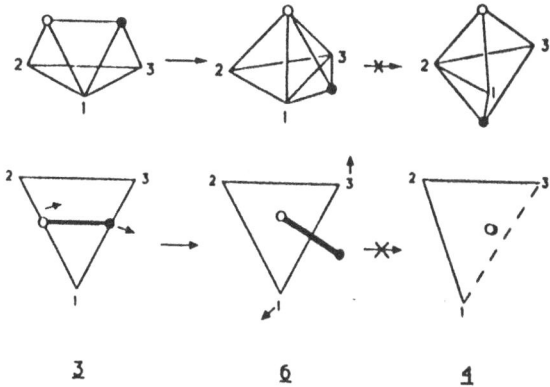

with its electron count. Finally, **9** can add one CO ligand to give the final product by opening one of its Fe-Fe bonds. This pathway is schematized below, in projection over the metallic triangle. From our simple MO calculations, it is not possible to predict if the last two steps of this scheme are really occuring subsequently or in a concerted way. This mechanism is the more likely to occur, since it fits the best with experiments [6].

The other low-energy pathway does not involve CO exchange. It can be related to the mechanism of skeletal isomerization of $[C_5H_5]^+$ suggested a long time ago by Stohrer and Hoffmann [7] and shown below.

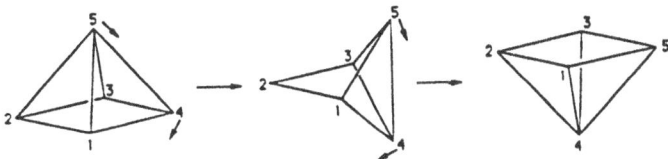

They showed that, since the trigonal bipyramid is highly unstable for this electron count, it is avoided by the process which involves a transition state of C_s symmetry. This polytopal rearrangement is identical to the mechanism of vertex migration on the fundamental octahedron (of which one vertex is unoccupied) of the cluster, which was proposed by McGlinchey et al [8] to account for the fluxional behavior of some 7-SEP square pyramidal clusters. This vertex migration mechanism is shown below.

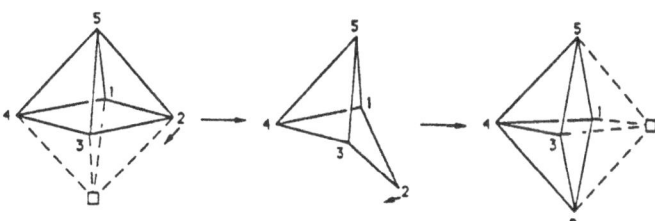

288

One has to apply twice this motion to our system. It involves the formation of an intermediate square pyramid (5), in which one nitrogen atom occupies the apical position. This pathway is shown below.

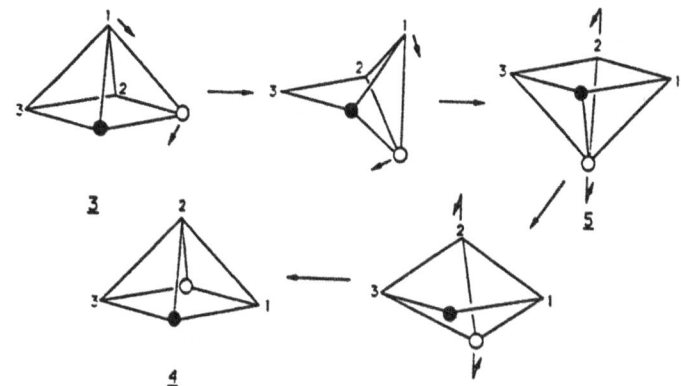

In fact, this process can also be described as a slipping of the N-N vector associated with the opening of a Fe-Fe bond, subsequently followed by a rocking motion of the black nitrogen atom associated with the breaking of the N-N bond, as shown below.

The difference between 5 and 6 is that there is no Fe$_1$-Fe$_3$ bond in 5, rendering this structure more stable by stabilization of its HOMO. The calculated energy profile indicates that the pathway is symmetry-allowed and that the distorted square pyramid 5 is better described as a transition state than a high-energy intermediate.

It is interesting to note that a similar mechanism of vertex migration on the fundamental octahedron can explain the fluxional behavior of isoelectronic M$_3$L$_{12}$(C$_2$R$_2$) clusters [9] which has been shown to exist in solution by Deeming [10]. This process is topologically equivalent to the windshield wiper motion proposed by Schilling and Hoffmann [11], as shown below.

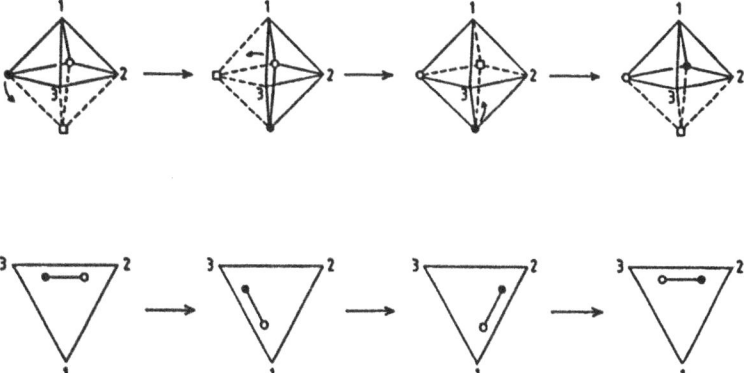

3. Polyhedral Rearrangements Involving Different Coordination Modes of a Main-Group E_2 (E = N, NR, CR) Ligand in Tetrametallic Butterfly Clusters

Six-vertex clusters, the core of which is made of four metal (M) and two main-group atoms (E), can adopt different geometries depending on their electron count and on the nature of their main group constituents. The four different structure-types, shown on the next page, have been reported so far for the cases of E = N, NR or CR [12, 13]. The octahedral cluster [12] exemplifying the structure-type I is the member of a large family of compounds in which E_2 is an alkyne ligand [14]. Bearing 7 SEP's, they conform with the Wade-Mingos rules. This is also the case of $Fe_4(CO)_{12}(NCPh)$ (Type II), which possesses 7 SEP's, and has the connectivity of a capped square pyramid [15]. The structure of the 8-SEP $Ru_4(CO)_{12}(N_2Et_2)$ compound (Type III) is rather unexpected from the point of view of the PSEP theory [12]. However, it can be considered as being electron-precise. On the other hand, the isoelectronic $Ru_4(CO)_{10}(C_2Ph_2)(NPh)_2$ [13] obeys the Wade-Mingos rules, since it can be derived from a pentagonal bipyramid having one vacant vertex. Structure-types III and IV can be derived from the octahedron by breaking two and one Ru-N bond, respectively. Structure-types I, II and III can be generated one from another by a simple displacement of the E-E vector over the M_4 butterfly, associated with a variation of the butterfly angle [12].

Structure-types I and II can be considered as being alternative skeletal isomeric geometries for 7-SEP M_4E_2 species. The same situation occurs for the count of 8-SEP species with structure-types III and IV. We have studied the electronic structure of these isomers, and in particular the various factors favoring their respective stability [14]. In this paper we discuss the possibility of interconversion between these skeletal isomers. The models used in the calculations are $Fe_4(CO)_{12}N_2$ (7 SEP's) and $[Fe_4(CO)_{12}N_2]^{2-}$ (8 SEP's). Calculations were also made on models in which one or two nitrogen atoms were replaced by isolobal NH^+ or CH units. They basically lead to the same qualitative conclusions.

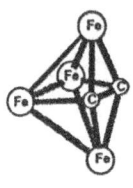 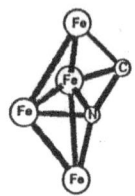 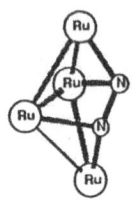 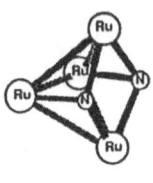

$Fe_4(CO)_{12}(C_2Me_2)$	$Fe_4(CO)_{12}(NCPh)$	$Ru_4(CO)_{12}(N_2Et_2)$	$Ru_4(CO)_{10}(C_2Ph_2)(NPh)_2$
7 SEP's	7 SEP's	8 SEP's	8 SEP's
Type I	Type II	Type III	Type IV

Our calculations indicate that the easiest pathway for the interconversion between the 7-SEP isomers, which have comparable energies, is the least-motion process schematized below, with the transition state in the middle. The octahedron (Type I) is transformed into the capped square pyramid (Type II) by a simple displacement of the E_2 ligand (roughly a combined rotation + translation motion) in the groove of the Fe_4 butterfly. The butterfly angle varies from ~100° to ~135°.

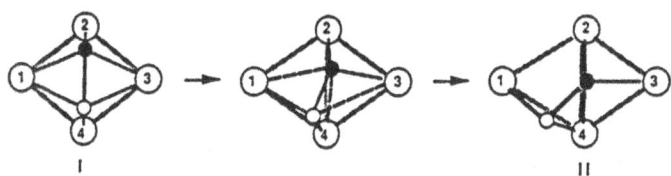

During the process the symmetry is C_1 and therefore the reaction is formally symmetry-allowed. During the transit, two 'bonds' are broken, while one 'bond' is created. As a consequence, a HOMO/LUMO avoided crossing occurs, which is responsible for a computed energy barrier of ~0.7 eV. At the level of our calculations, this qualitative value suggests strongly that structure-types I and II can interconvert in solution. In particular, our calculations indicate that compound $Fe_4(CO)_{12}(NCPh)$ [15] should undergo a fluxional behavior at a moderate temperature [14]. These results also suggest that if the octahedron <-> capped square pyramid polyhedral rearrangement can occur in the case of some homonuclear six-vertex 7-SEP clusters, it should involve the same pathway as for the M_4E_2 species.

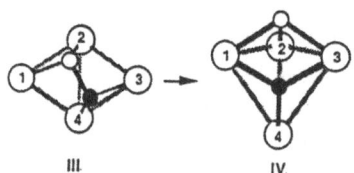

The least motion pathway for the interconversion between the two 8-SEP skeletal isomers (Type III and Type IV) is shown above. It can be described as a rotation of the E_2 vector associated with the E-E bond breaking. Calculations indicate that it is a symmetry-forbidden process [14]. Indeed, during this process the symmetry of the system is C_2, and the symmetric HOMO crosses the antisymmetric LUMO. Therefore, the interconversion of these two skeletal isomers cannot occur through the least motion pathway shown above. Consequently, the fluxional behavior of compounds having structure-type III shown below, in which the cluster is transformed into its mirror image through intermediate Type IV is also symmetry-forbidden.

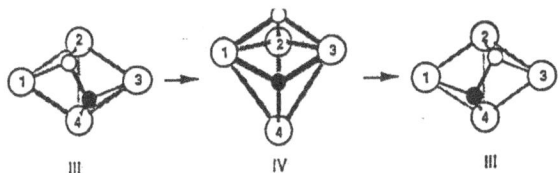

One could think of a simple octahedron as an intermediate structure, *i.e.* a simple rotation of the E-E vector, with no opening of the E-E bond. This pathway, schematized below, is also symmetry-forbidden. Moreover, the octahedron is very unstable for the count of 8 SEP's, as predicted by the Wade-Mingos rules, making it a bad potential candidate for a transition state.

In fact, the easiest symmetry-allowed pathway that we have found for this hypothetical skeletal rearrangement corresponds to a rotation of the E-E vector in the other way (see next page). The transition state structure VII looks like a basket-type one. However, the computed energy barrier is rather high (~1.9 eV). It is interesting to note that this hypothetical transition state would be thermodynamically stable for a 10-SEP count and that its core geometry is close to that of the tricyclic polyphosphane $P_6(C_5Me_5)_2$ [16] which bears 10 SEP's [14]. With a 8-SEP count, this transition state has two unoccupied non-bonding MO's rendering it Jahn-Teller unstable. Again, the transition state structure is not that presenting the highest possible connectivity, but the one with the highest connectivity, providing no antibonding orbital at all is occupied.

292

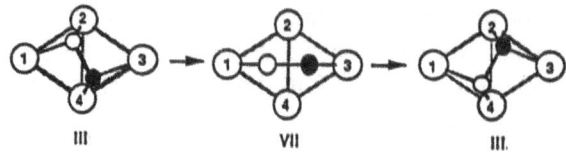

III → VII → III.

4. Cluster Core Isomerization of Mixed Platinum/Molybdenum or Platinum/Tungsten Tetanuclear Clusters: From Planar to Tetrahedral

The cluster compounds of general formula $Pt_2M_2Cp_2(CO)_6(PR_3)_2$ (M=Mo, W) have been shown by Braunstein et al to adopt two different skeletal arrangements in the solid state [17].

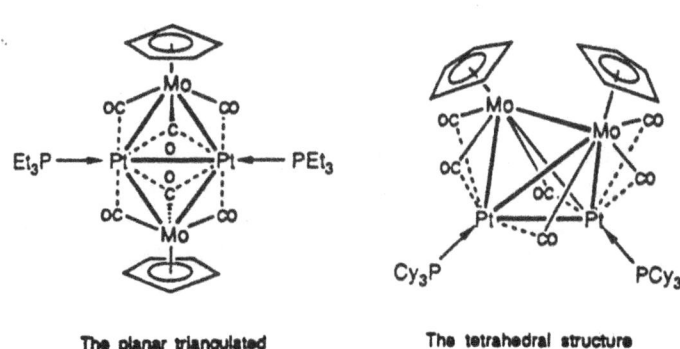

The planar triangulated
rhombohedral (PTR) structure

The tetrahedral structure

With small phosphine ligands, the planar triangulated rhombohedral (PRT) arrangement of the metal atoms is observed. This situation corresponds to an electron-precise species with five 2-electron-2-centre bonds. With bulky phosphines, the compounds crystallizes in a tetrahedral arrangement in which the Pt-Pt bond is rather long: 2.99 Å in the case of M=Mo and R=Cy [17]. A similar structure has also been reported for isoelectronic heteroplatinum compounds [18]. These 5-SEP tetrahedral clusters are electron-deficient with respect to the PSEP rules which indicate stability for 6 SEP's. In addition, these latter compounds present both forms in equilibrium in solution at room temperature [17]. The MO diagrams of the PRT and tetrahedral isomers of the $Pt_2Mo_2Cp_2(CO)_6(PH_3)_2$ model are shown below.

In the PRT case, it is possible to identify five occupied bonding MO's and five vacant antibonding MO's associated with the five metal-metal bonds. In particular the $1a_g$ level can be considered as being the σ_{Pt-Pt} orbital, and the $2a_u$ level as being the σ^*_{Pt-Pt} orbital. Similarly, $2a_g$ and $2b_u$ can be identified as σ_{Mo-Mo} and σ^*_{Mo-Mo}, respectively. Both are occupied and non-bonding, in agreement with the fact that there is no Mo-Mo bond in the PRT form. This situation is different in the tetrahedral case, since now there is a Mo-Mo bond: The σ_{Mo-Mo} level (derived from the '3a$_1$' frontier orbital of the $Mo_2Cp_2(CO)_6$ fragment) is occupied, while the σ^*_{Mo-Mo} level (derived from the '2b$_1$' frontier orbital of the $Mo_2Cp_2(CO)_6$ unit) is vacant. On the other side, both σ_{Pt-Pt} and σ^*_{Pt-Pt} (derived from the '1a$_1$' and '1b$_2$' orbitals of the $Pt_2(PH_3)_2$ fragment) are both occupied. At first sight, one could conclude that there is no Pt-Pt bond in this form. The calculated Pt-Pt overlap population indicates that there is a significant bonding character (+0.14, versus +0.20 in the PRT case). The reason lies in the mixing of the Pt-Pt bonding '2a$_1$' frontier orbital of the $Pt_2(PH_3)_2$ fragment in several occupied MO's of the cluster. This mixing is rather large due to the low energy and the diffuse character of this frontier orbital.

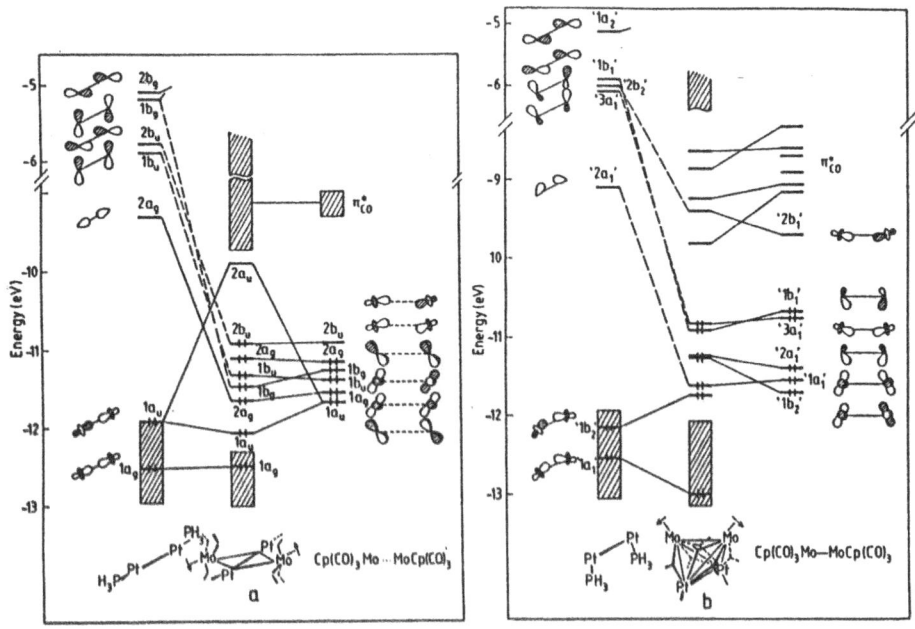

MO diagrams of the PRT (left) and tetrahedral (right) isomers of
$Pt_2Mo_2Cp_2(CO)_6(PH_3)_2$.

An important feature of this MO analysis is that σ^*_{Pt-Pt} is unoccupied in the PRT case while this is σ^*_{Mo-Mo} which is vacant in the tetrahedral case. Since the corresponding bonds are perpendicular, it could be concluded that the least-motion interconversion between both isomers is symmetry-forbidden. This is in agreement with the existence of two isomers, (two minima on the potential energy surface). This is, of course, in full disagreement with the existence of an equilibrium between both isomers observed for some of the studied compounds[17]. The reason originates from

294

two major factors. The first one is the very low symmetry of the ligand shell in the tetrahedral form which renders the process (at least formally) symmetry-allowed. The second factor comes from the mixing of the 6s- and 6p-type frontier orbitals of the $Pt_2(PH_3)_2$ unit in the occupied cluster MO's, which becomes larger as soon as the PRT form begins to bend. This mixing, which contributes significantly to the stabilization of the transition state and the tetrahedral isomer, can be related to some tendency of platinum to achieve the 18-electron configuration.

5. Conclusion

In conclusion, we hope having convinced the reader that the combination of extended Hückel calculations with various experimental data and simple electron counting rules can provide a useful understanding of the mechanisms of skeletal rearrangements.

6. References

1. Wade, K. (1976) *Adv. Inorg. Chem. Radiochem.* **18**, 1.
    Wade, K. (1981) in B.F.G. Johnson (ed), *Transition Metal Clusters*, John Wiley and Sons Publishers, New York, p 193.
    Mingos, D.M.P. (1984) *Acc. Chem. Res.* **17**, 311.
    Mingos, D.M.P. and Wales, D.J. (1990) *Introduction to Cluster Chemistry*, Prentice-Hall Publishers, Englewood Cliffs.
2. Stone, A.J. (1980) *Mol. Phys.* **41**, 1339.
    Stone, A.J. (1981) *Inorg. Chem.* **20**, 563.
    Stone, A.J. and Alberton, M. J. (1982) *Inorg. Chem.* **21**, 2297.
    Stone, A.J. (1984) *Polyhedron* **3**, 1299.
3. Hoffmann, R. (1982) *Angew. Chem. Int. Ed. Engl.* **21**, 711.
4. McPartlin, M., Eady, C.R., Jonhson, B.F.G. and Lewis, J. (1976) *J. Chem. Soc. Chem. Commun.* 883.
5. Mingos, D.M.P. and Forsyth, M.I. (1977) *J. Chem. Soc. Dalton Trans.* 610.
6. Wucherer, E., Tasi, M., Hansert, B., Powell, A.K., Garland, M.T., Halet, J.-F., Saillard, J.-Y. and Vahrenkamp, H. (1989) *Inorg. Chem.* **28**, 3564.
7. Stohrer, W.-D. and Hoffmann, R. (1972) *J. Am. Chem. Soc.* **94**, 779.
8. Jaouen, G., Marinetti, A., Saillard, J.-Y., Sayer, B.G. and McGlinchey, M.J. (1982) *Organometallics* **1**, 225.
9. Mlekuz, M., Bougeard, P., Sayer, B.G., Peng, S., McGlinchey, M.J., Marinetti, A., Saillard, J.-Y., Ben Naceur, J., Mentzen, B. and Jaouen, G. (1985) *Organometallics* **4**, 1123.
10. Deeming, A.J. (1978) *J. Organomet. Chem.* **150**, 123.
11. Schilling, B.E.R. and Hoffmann, R. (1979) *Acta Chem. Scand.* **B33**, 231.
12. Bantel, H., Hansert, B., Powell, A.K., Tasi, M. and Vahrenkamp, H. (1989) *Angew. Chem. Int. Ed. Engl.* **28**, 1059.

13. Song, J.-S., Han, S.-H., Nguyen, S.T., Geoffroy, G.L. and Rheingold, A.L. (1990) *Organometallics* **9**, 2386.
14. Kahlal, S., Halet, J.-F., Saillard, J.-Y. (1991) *New J. Chem.* **15**, 843.
15. Keller, E. and Wolters D. (1984) *Chem. Ber.* **117**, 1572.
16. Jutzi, P., Kroos, R., Müller, A. and Penk, M. (1989) *Angew. Chem. Int. Ed. Engl.* **28**, 600.
17. Braunstein, P., de Méric de Bellefon, C., Bouaoud, S.-E., Grandjean, D., Halet,
J.- F. and Saillard, J.-Y. (1991) *J. Am. Chem. Soc.* **113**, 5282.
18. Farrugia, L.J., Howards, J.A.K., Mitprachachon, P., Stone, F.G.A. and Woodward, P.J. (1981) *J. Chem. Soc. Dalton Trans.* 1134. *Ibid.* 1274.
   Braunstein, P., Dehand, J. and Nenning, J.-F. (1975) *J. Organomet. Chem.* **92**, 117.
   Bhaduri, S., Sharma, K.R., Clegg, W., Sheldrick, G.M. and Stalke, D. (1984) *J. Chem. Soc. Dalton Trans.* 2851.

SOME OLD AND NEW REDOX REACTIONS OF POLYNUCLEAR ORGANOMETALLIC COMPLEXES.

Dedicated to Professor L.F. Dahl on the occasion of his 65th birthday.

Heinrich Vahrenkamp
*Institut für Anorganische und Analytische Chemie der Universität Freiburg,
Albertstr. 21, D-79104
Freiburg, Germany.*

Abstract

The author's work on organometallic compounds with redox properties is reviewed. Electron transfer studies have ranged from exploratory electrochemistry via ESR and bonding considerations to a detailed investigation of ET catalysed CO substitutions. Preparative studies have involved metal exchange in cluster compounds, organic reactions in the ligand sphere, and the synthesis, electron transfer and mixed valence chemistry of cyanide bridged organometallic Prussian Blue analogs.

Introduction

The author's first transition metal compound was a paramagnetic organometallic cluster, $Ni_3Cp_3(\mu_3\text{-}S)_2$ [1]. Despite having obtained his Dr. degree at the University of München and having received his academic training, among others, from W. Hieber and E.O. Fischer, he first became a boron chemist under the guidance of H. Nöth, and only the postdoctoral years took him to the laboratory of L.F. Dahl in Madison, Wisconsin. There again he did not expect to do preparative organometallic chemistry, not to speak of the study of species with unpaired electrons, but to learns the art of crystal structure analysis. Things, however, went differently. In Madison, while being impregnated with the academic spirit, scientific enthusiasm, and personal warmth of Larry Dahl, the author also gained the conviction that preparation is the focal point of chemistry.

The preparation of the above-mentioned compound was not the result of this conviction but one of the triggering points leading to it. The author stumbled upon the cluster $Ni_3Cp_3(\mu_3\text{-}S)_2$ while desperately trying to do something useful in the lab and to

297

L. J. Farrugia (ed.), The Synergy Between Dynamics and Reactivity at Clusters and Surfaces, 297–316.
© 1995 Kluwer Academic Publishers.

fill the waiting time for the highly desired X-ray machines. The synthesis from [CpNi(CO)]$_2$ and S$_8$ is anything but designed, and there are no obvious reasons why its main product should be a paramagnetic organometallic compound. Together with the fact that many classical cluster syntheses are equally unpredictable it has made the author set out for a 25-year journey in search of a systematic cluster chemistry. During this journey there have been sporadic outbreaks of electron transfer work and physicochemical studies of polynuclear organometallic complexes with unpaired electrons.

This paper tries to present some highlights from the work with paramagnetic species, put older and newer work in context, and take the reader from safe and proven results to some unpublished and possibly disputable findings.

Exploratory Electrochemistry

The group of L.F. Dahl as well as our own group have come back several times to the nickel cluster 1 and to the analogous cobalt cluster 2. Based on the 18 electron rule, 1, a 53e cluster, has five excess electrons, and 2, a 50e cluster, has two excess electrons. We showed by DC voltammetry, that all those five electrons can be removed, though not reversibly, from 1, and that the two electrons can be removed reversibly from 2 [2]. Dahl was able to structurally characterize the mono- and dication of 2 [3]. Very recently this was also achieved for the monocation of 1 [4] as part of a detailed reinvestigation of its redox chemistry. We have finally been able to extend the voltammetry of 1 and 2 with modern electrochemical techniques [5]. As a result, oxidation states from +5 to -1 have now been accessed for 1 and from +2 to -1 for 2. With ultrapure dichloromethane as a solvent, reversible redox transitions could be found between the total oxidation states of -1, 0, +1, +2 and +3 for the nickel cluster. Figure 1 overleaf shows the cyclic voltammogram. The clusters 1 and 2 have thereby provided a unique example of the verification of simple electron counting schemes by electrochemical, crystallographic and molecular orbital methods.

The basic fact which governs the redox and structural chemistry of 1 and 2 is that HOMO's and LUMO's in such compounds are closely related to metal-metal bonding. Another simple test for the validity of this statement was made possible by the existence of the series of compounds 3, 4 and 5 which all adopt basically the same molecular structure [6]. In this structure electron counting requires a V-V double bond for 3, a Cr-Cr single bond for 4, and no Mn-Mn bond for 5. As this amounts to a variation of only 0.9 Å in the metal-metal bond distances [6] it was plausible to assume that the variation in the electron count between 3, 4 and 5 could also be achieved electrochemically for each of the single compounds, i.e. for instance by oxidation of 5 or by reduction of 3. This was borne out to a wide extent electrochemically [7]. Figure 2 shows on a voltage scale which of the possible electron counts between 32 (for neutral 3) and 38 (for dianionic 5) were accessible for the three compounds. Specifically the

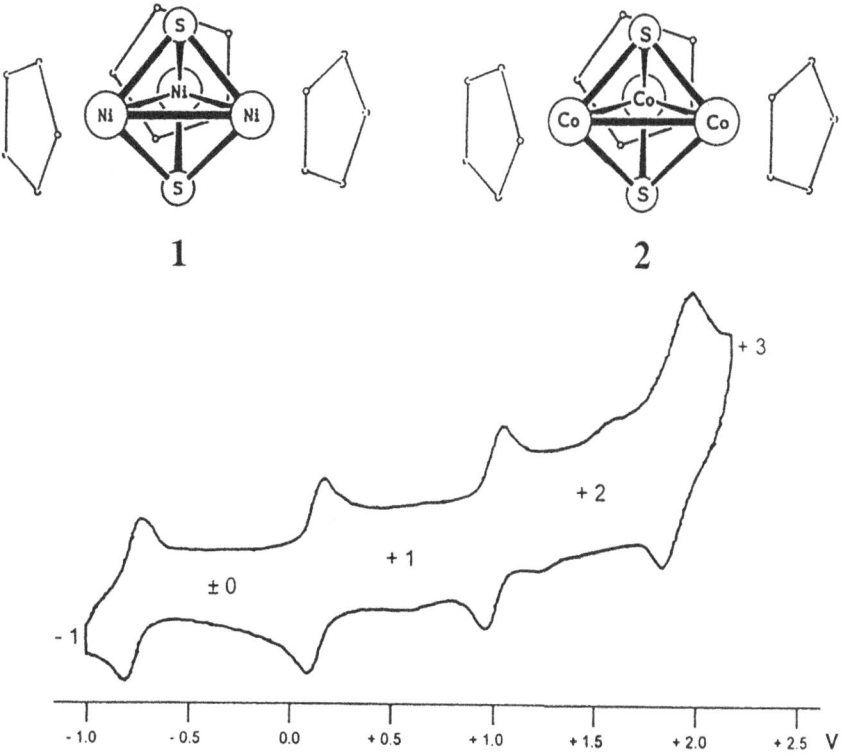

1 **2**

Fig 1. Cyclic voltammogram of **1** (in dichloromethane, room temp., scan speed 1 V/sec, E vs. Ag/AgCl)

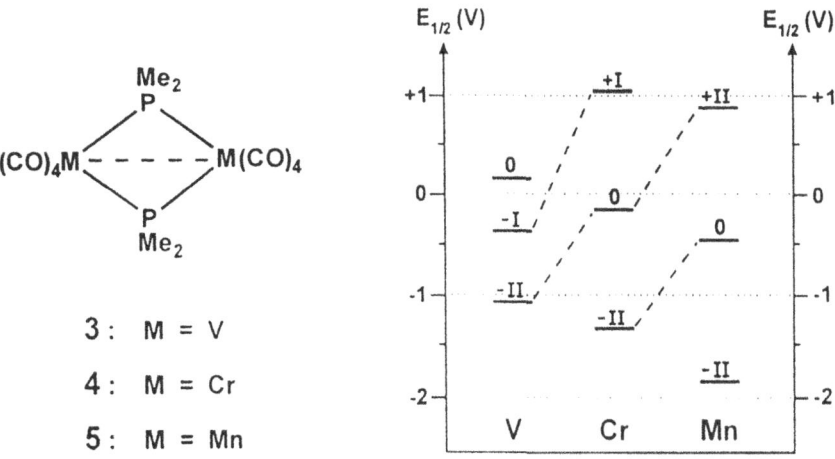

3 : M = V

4 : M = Cr

5 : M = Mn

Fig 2. Accessible redox states of **3**, **4** and **5** and isoelectronic relations between them (Oxidation state 0 arbitarily set)

ease of reduction of **3** and the reluctance of **5** towards reversible oxidations underline the tendency of the complexes to achieve the full electron count and to minimise metal-metal bonding.

That metal-metal bonding is an energetically unfavourable compromise is also the basis of the preparatively important reductive cleavage of many dinuclear metal carbonyl derivatives to form their anionic mononuclear fragments like $Mn(CO)_5^-$ etc.. During our early exploratory electrochemical work we hoped to find quantitative data on metal-metal bond strengths in such simple dinuclear species. While in some closely related cases there are linear dependencies between reduction potentials and electronic absorption energies [8], simple general relations don't exist. Our electrochemical studies, instead, led us to the observation of a preparatively important phenomenon, the existence of small quantities of the mononuclear neutral organometallic fragments in equilibrium with their dinuclear parent compounds [8,9]. Figure 3 presents the significant observations which were achieved with the cyclopentadienyl metal carbonyls **6, 7** and **8**.

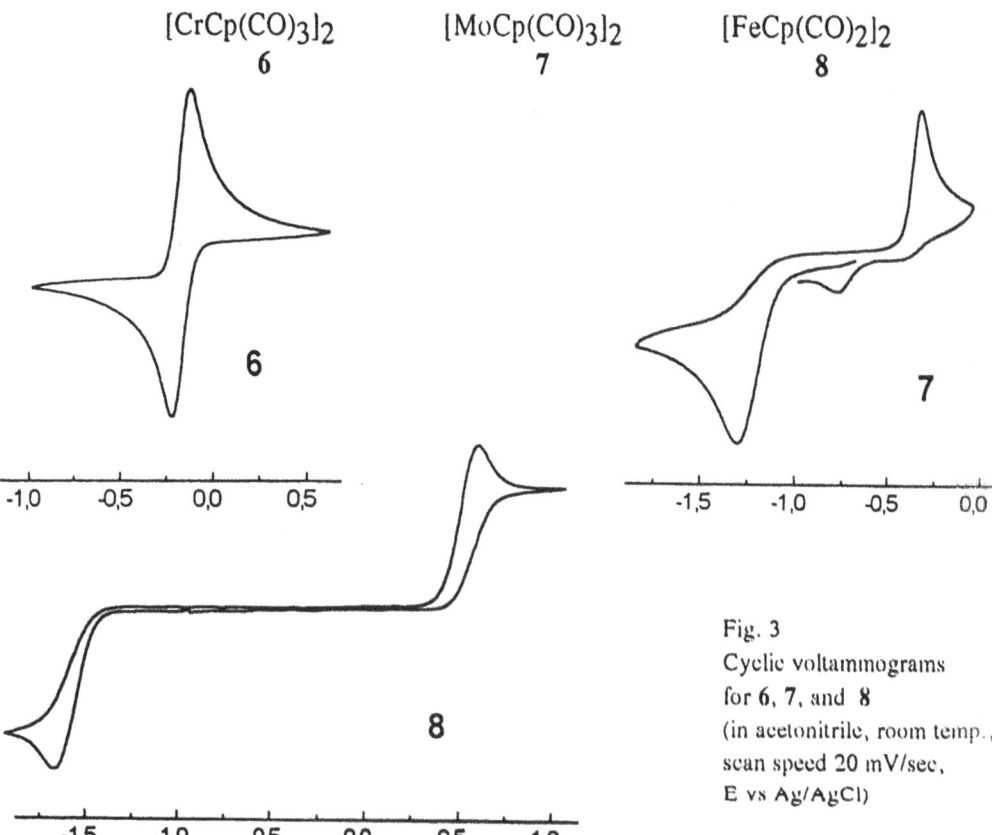

Fig. 3
Cyclic voltammograms
for **6, 7,** and **8**
(in acetonitrile, room temp.,
scan speed 20 mV/sec,
E vs Ag/AgCl)

The electrochemical reduction of a dinuclear metal carbonyl cannot be electro-chemically reversible as the oxidized and the reduced species (X_2 vs. X^-) are different chemical entities. Accordingly the potentials for reduction and oxidation must be different. The CV of **8** at slow scan speeds represents this expected behaviour and provides a deceptively simple case of chemical reversibility. The CV of **7** seems to indicate the presence of a decomposition product which is reduced at a potential near the oxidation potential of $MoCp(CO)_3^-$. The CV of **6** looks as if there is no dinuclear species in solution to be reduced and as if now the "decomposition product" is the only redox-active species. We have identified the "decomposition product" as the mononuclear fragment $MCp(CO)_3^{-}$ [9]. In the case of M = Cr the neutral $MCp(CO)_3$ is formed so quickly from its dimer that all dimer is used up in the reduction via this species. In the case of M = Mo the monomer and the dimer compete for the reduction with the dimer dominating. For the iron compound **8** the amount of monomer is unobservable in the slow scan electrochemical experiment. Our conservative estimate for the amount of $Cp(CO)_3Cr$ in equilibrium with **6** was 1%. Later this was found to be more like 10% [10], and certain ligand or Cp ring substituted derivatives of **6** were found to be fully dissociated [11]. For our own group the apparent ease of metal-metal bond breaking, i.e. the availability of unstable organometallic fragments in solution, has opened the way to some of the most satisfying preparative advances.

A Look at Bonding

In an extensive series of papers L.F. Dahl and his group have qualified and quantified the bonding situation of trinuclear organometallic clusters with triply bridging ligands, mainly by structure determinations and molecular orbital calculations. We have been able, in part, to complement these investigations by our electrochemical work and by ESR measurements for which we have enjoyed the cooperation with B.H. Robinson and B.M. Peake at the University of Otago, Dunedin, New Zealand, for many years. The basis for the measurements was again our preparative work which has provided us with various series of compounds.

One such isoelectronic series is represented by clusters **9, 10** and **11**. Neutral **9** is a stable 49e cluster with one upaired electron. Its ESR spectrum, in accordance with MO calculations, indicates that the HOMO of **9** is comprised mainly of a symmetrical combination of d orbitals from the three cobalt atoms. Each cobalt nucleus ($I = 7/2$) contributes equally to the symmetrical 22-line multiplet (Fig. 4). Noticeably a ESR- (i.e. MO-) contribution by the phosphorus atom is not present [12].

The findings for **9** were validated by the ESR data for the radical anions of **10** and **11**. The 48e clusters **10** and **11** can be reduced to stable (**10**) or short-lived (**11**) anions [12,13]. Their ESR spectra (Fig. 4.) indicate hyperfine splitting due to two (**10**) and one (**11**) cobalt nuclei, while the phosphorus atom again shows no interaction. Although the contribution of the iron and nickel atoms remains unoticeable due to their

302

lack of magnetically active nuclei, it can be deduced indirectly: The hyperfine splittings for **9, 10**·, and **11**·· are roughly equal (30 Gauss) which leads us to the conclusion that the MO contributions of the individual metal atoms (including the "invisible" Fe and Ni atoms) are also roughly equal. This means that a variation of neighbouring metal atoms

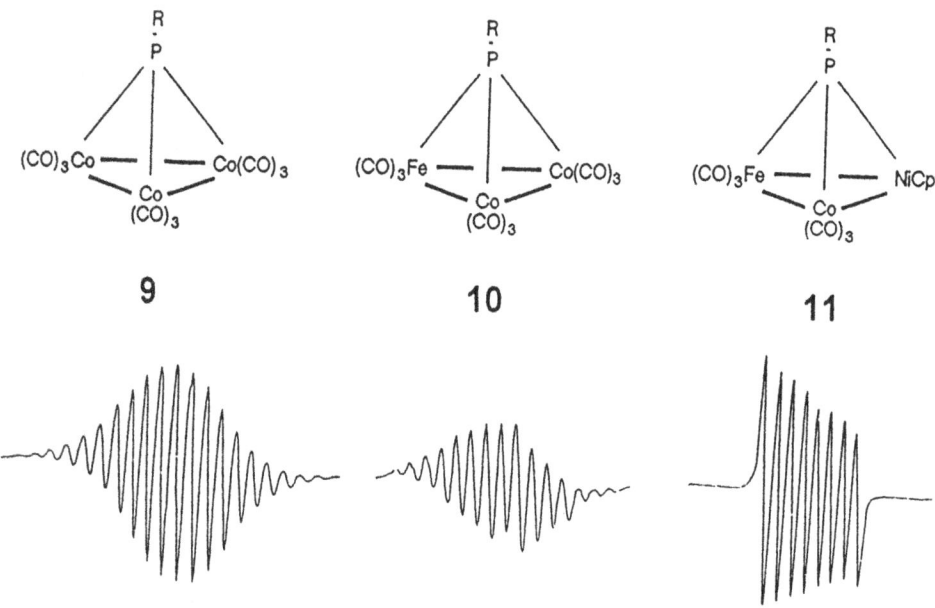

Fig 4. ESR spectra of **9, 10**·· and **11**··

in a cluster does not cause significant changes in the electronic situation, as long as the overall symmetry is maintained. Or, to put it in a phrase which resulted from our preparative experiences "all metals are equal". We have supported this phrase by many measurements [13], and there is also some theoretical justification for it [14].

Just like the ESR data the redox potentials obtained by cyclic voltammetry underline the small size of the electronic changes brought about by varying the core composition of clusters with the same overall structure. We have tested this again for clusters with the general framework **A** comprised of a metal triangle with a triply bridging main group element ligand [15]. If in this framework one element is replaced by a homologous one with the same charge (S→Se, Fe→Ru, Mo→W) the corresponding change of the first reduction potential is always less than 0.18V. This is negligible compared to the changes brought about by replacing a $M(CO)_3$ group by an isoelectronic M'Cp group, e.g. $Co(CO)_3$/NiCp, or by just replacing a metal atom by a neighbouring one which means varying the overall charge, e.g. Co/Fe⁻. Thus donor ligands (Cp, PR_3)

are shown to have much more effect in varying the overall electronic situation of a cluster than the metal atoms themselves.

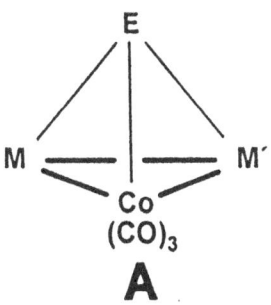

core	replacement	$\Delta E_{1/2}$ (V)
EFeCo$_2$	S / Se	+ 0.1
SMCo$_2$	Fe(CO)$_3$ / Ru(CO)$_3$	- 0.1
SFeCoM'	MoCp(CO)$_2$ / WCp(CO)$_2$	- 0.1
SFeCoM'	Co(CO)$_3$ / NiCp	- 0.5
SFeCo$_2$	Co(CO)$_3$ / Co(CO)$_2$PR$_3$	- 0.3
SFeCoM'	Co(CO)$_3$ / Fe(CO)$_3^-$	- 1.0

E = S, Se;   M = Fe(CO)$_3$, Ru(CO)$_3$

M' = Co(CO)$_3$, NiCp, MoCp(CO)$_2$, WCp(CO)$_2$, Fe(CO)$_3^-$

In contrast to the M$_3$(μ_3-E) clusters discussed above where the contribution of E to the HOMO's and LUMO's is negligible, the group of tetranuclear clusters with a M$_4$(μ_4-E)$_2$ composition provides an orbital situation with significant metal and E contributions [16]. We have contributed the saturated cluster 12 and its unsaturated counterpart 13 to this group of compounds [17]. The unsaturated nature of 13 gives rise to a rich electrochemistry. Its first reduction occurs already at positive potentials (Fig. 5), and the anionic species of 13 and all of its derivatives like 13a are stable compounds [18].

12

13

13a

A systematic ESR investigation has revealed full agreement between theory and experiment for this class of compounds with respect to the ease of reduction and the electronic nature of the reduced species [18]. Figure 6 shows the ESR spectrum of **13a·⁻** as an example. The symmetrical nature of the radical anion is obvious from the triplet signal due to hyperfine coupling with the equivalent μ_4-bridging phosphorus atoms. The magnitude of this coupling (ca. 30 Gauss) demonstrates the high contribution of phosphorus to the HOMO. The coupling due to the terminal phosphorus ligand comes in as an additional doublet splitting with ca. 15 Gauss.

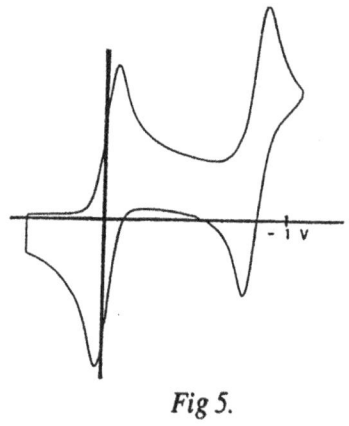

Fig 5.

CV for the reduction of **13**

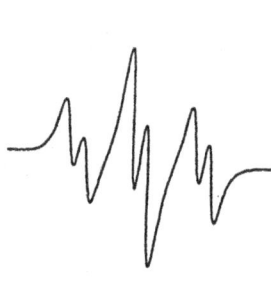

Fig 6

ESR spectrum of **13a·⁻**

It was fun to find these simple bonding statements by non-sophisticated methods and to use them to evaluate electronic interactions between several clusters connected by bridging ligands [18]. We are convinced that there are still enough clusters waiting with constitutions simple enough to lend themselves to such simple bonding investigations.

Mechanistic Intricacies

While doing exploratory electrochemistry and ESR studies on the trinuclear clusters of type **A** we were always plagued by the problem that phosphine substitution of the clusters results in ill-defined electrochemical measurements. Our collaborative partners B.H. Robinson et al. had suffered from this problem too which seemed to be related to enhanced lability towards ligand substitution and cluster fragmentation. They tried to clarify the situation by further measurements on phosphine derivatives of $RCCo_3(CO)_9$ [19] which could not yield a fully consistent interpretation, however. At this point we succeeded in attracting the attention of our Freiburg colleague J. Heinze for this kind of electrochemical problem. It was with his expertise and the efforts of his coworkers that for once one of our redox-active cluster systems was subjected to an in-depth mechanistic study [20].

It took a large number of electrochemical and chemical experiments to clarify the situation. The result of all observations and evaluations is a consistent picture of the electron transfer chemistry of **14** and its derivatives and of their electron transfer induced ligand substitutions [20]. This picture includes a re-evaluation of the stability of the radical anion of **14** and the existence of stable mono- and dianions of the phosphine-rich derivatives of **14**.

The seemingly well-established observation, based on a persistent ESR signal, that **14** forms a stable radical anion, was always inconsistent with the fact that **14** becomes very labile towards substitution upon reduction. We found that lucky circumstances indicate a reversible reduction of **14** and stability of **14**·⁻ under the conditions of the ESR experiment (Fig. 7a). These circumstances lie in the kinetic parameters of the CO dissociation from **14**·⁻ (first order) and the re-addition of CO to the resulting $MeCCo_3(CO)_8$·⁻ (second order). By lowering the concentration of **14**, the second-order addition becomes too slow, and **14**·⁻ is partially decomposed before it is reoxidized (Fig. 7b). Addition of CO to the low-concentration solution favours the

	L^1	L^2	L^3
14	CO	CO	CO
15	CO	CO	PR_3
16	CO	PR_3	PR_3
17	PR_3	PR_3	PR_3

the readdition, and **14**·⁻ is seemingly stable again on the time scale of the cyclic voltammogram (Fig. 7c). These and similar experiments have proved the rapid CO dissociation from **14**·⁻ under all conditions. Thereby it was also proved that the electron transfer-induced CO substitution (e.g. **14** → **15** → **16**) is a dissociative process. The high efficiency of this process is obvious from Fig. 7d: the cyclic voltammogram of **14** in the presence of PPh₃ barely indicates the formation of **14**·⁻ while the reduction of **15** to **15**·⁻ is the dominant feature. This is so because the initially formed small amounts of **14**·⁻ are quickly converted to **15**·⁻ via dissociative CO/PPh₃ substitution. From then on homogeneous (and not electrochemical) reduction of **14** by **15**·⁻ takes place, thereby rapidly increasing the amount of **15** and reducing the amount of **14**. All this happens in about one second while the electrochemical apparatus sweeps past the reduction potential of **14**. Further measurements of this kind gave firm support for this mechanism of the electron transfer-catalyzed formation reactions of all clusters **15, 16** and **17**.

The second most important observation in these experiments was made during the voltammetry of the tris-phosphine substituted clusters **17**. Of these the PEt₂Ph substituted compound **17a** gave the cyclic voltammogram shown in Figure 8a. There is a crossover phenomenon on the reverse sweep which calls for the unusual explanation that despite the voltage sweep towards more oxidizing potentials there is still a reductive current flow. This means that after the initial reduction a chemical follow-up reaction has produced a species with a more positive reduction potential. If the voltage is subsequently swept continuously between -1.25 and -1.85V the cyclic voltammogram of Figure 8b results. All lines of this voltammogram go through one point called an isopotential point, and finally the ideal curve of a new reversible one-electron reduction evolves.

The isopotential point, just like an isosbestic point, proves that a clean interconversion between just two chemical entities occurs. It could be shown [20] that, after the first reduction sweep accompanied by the transfer of two electrons and loss of

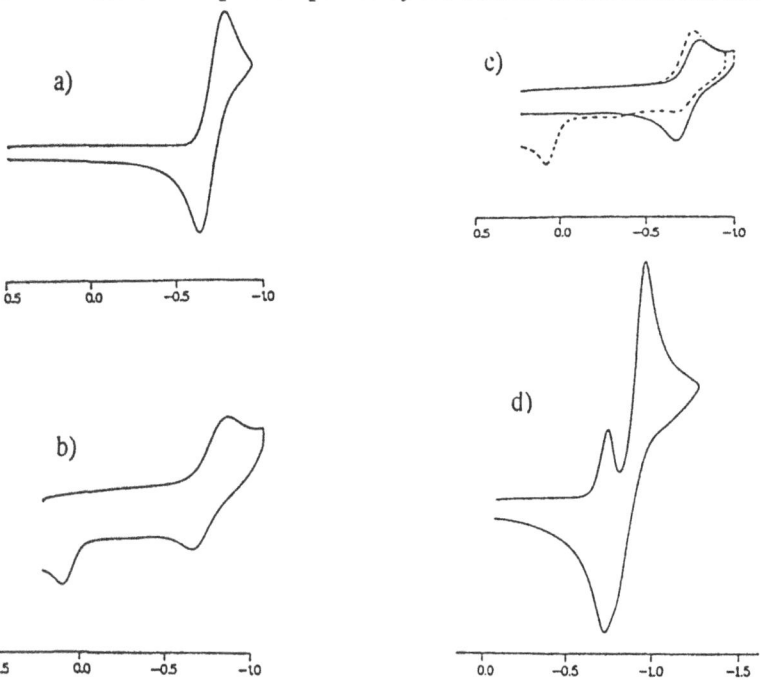

Fig 7. Cyclic voltammograms of **14** (in acetonitrile, room temp., scan speed 0.2 V/sec, E vs. Ag/AgCl,); (a) 2.4 x 10⁻³ M ; (b) 7.2 x 10⁻⁵ M ; (c) same as (b) but after addition of CO;(d) 9 x 10⁻⁴ M with equimolar amount of PPh₃

one phosphine ligand, the dianion of **17b** is formed initially from **17a**. In a diffusion-controlled reaction **17b²⁻** then transfers one electron to further incoming **17a**, generating **17b·⁻** and **17a·⁻**. **17a·⁻** in turn quickly loses one PR_3 ligand and forms **17b·⁻** as well. This occurs also after the first sweep reversal, and hence the crossover phenomenon since a reduceable species (**17b·⁻**) has been produced non-electrochemically. If the multisweep between -1.25 and -1.85V is continued long enough all **17a** near the electrode is used up and the various redox reactions leave **17b·⁻** and **17b²⁻** as the only species in solution which forms a reversible redox couple.

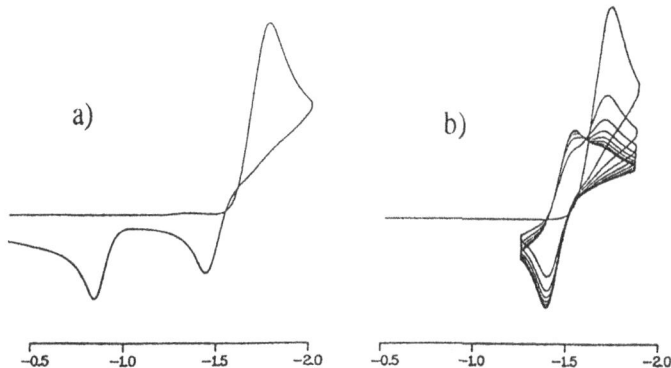

Fig 8. Cyclic voltammograms of **17a**: (a) single sweep, (b) multiple sweep between -1.25 and -1.85 V (8.0×10^{-4} M in acetonitrile, room temp., scan speed 0.4 V/sec, E vs. Ag/AgCl

$MeCCo_3(CO)_6(PEt_2Ph)_3$ **17a** $MeCCo_3(CO)_6(PEt_2Ph)_2$ **17b**

The novelty of these observations lies in the proof of a two-electron reduction of clusters **15-17** with concomitant phosphine elimination and of the observable existence of the electron and ligand deficient radical anion **17b·⁻**. Radical anions of this type are the key intermediates in electron-transfer catalyzed ligand substitutions and had been proposed before only on the basis of kinetic data or product distributions. That such a radical anion can be obtained from **17b** must have to do with the electron-richness and steric bulk of the specific phosphine ligands present. Thus, in order to find **17b·⁻** preparative and electroanalytical expertise had to be paired with a considerable amount of tenacity.

Are there Preparative Applications?

The normal result of an electron transfer reaction of an organometallic cluster is a paramagnetic odd electron species. In general this means that this species is of low stability and labile towards loss of ligands and cluster fragments. There can be only a limited range of isolable cluster radical ions which may be subjected to further reactivity studies. On the other hand the high reactivity of short-lived radical species derived from polynuclear compounds should be the basis of a rich follow-up chemistry.

M' \ M	$Cp(CO)_3Cr$	$Cp(CO)_3Mo$	$Cp(CO)_3W$	$(CO)_5Mn$	$Cp(CO)_2Fe$	$(CO)_4Co$	$Cp(CO)Ni$
$Cp(CO)Ni$	Cr – Ni	Mo – Ni	W – Ni	Mn – Ni	Fe – Ni	Co – Ni	Ni – Ni
$(CO)_4Co$	Cr – Co	Mo – Co	W – Co	Mn – Co	Fe – Co	Co – Co	
$Cp(CO)_2Fe$	Cr – Fe	Mo – Fe	W – Fe	Mn – Fe	Fe – Fe		
$(CO)_5Mn$	Cr – Mn	Mo – Mn	W – Mn	Mn – Mn			
$Cp(CO)_3W$	Cr – W	Mo – W	W – W				
$Cp(CO)_3Mo$	Cr – Mo	Mo – Mo					
$Cp(CO)_3Cr$	Cr – Cr						

Equilibrations

$$M - M + M' - M' \rightleftharpoons 2 M - M'$$

Fig 9. Equilibration reactions of dinuclear metal carbonyls

We experienced this early on during our exploratory electrochemistry work (see above). There we found the equilibrium between $[CrCp(CO)_3]_2$ and monomeric $CrCp(CO)_3$. We also observed that it is impossible to prepare pure $(CO)_3CpMo-WCp(CO)_3$ by methathetical reactions like the one between $(CO)_3CpMoCl$ and $NaWCp(CO)_3$, and that actually any sample of so-called $(CO)_3CpMo-WCp(CO)_3$ is a 1:2:1 statistical mixture of the heterodinuclear and the two homodinuclear species [8]. It was easy to conclude from this that any dinuclear metal carbonyl can deliver its mononuclear fragments as species which are ready for equilibration reactions. And it really took just a few weeks to test this hypothesis by executing all possible equilibrations shown in Figure 9. The seven homodinuclear starting complexes readily generate all 21 possible heterodinuclear compounds of which 8 were unknown at the time of our investigation. Conversely, when the pure heterodinuclear compound is dissolved it also quickly produces a mixture containing both homodinuclear compounds as equilibrium constituents. It was amazing to find that in most cases this equilibrium mixture contains the heterodinuclear compound as the clearly dominating species.

The odd-electron species involved in the above-mentioned reactions were actually not produced by a redox reaction. But their study was initiated by one, and the course of the reaction between $(CO)_3CpMoCl$ and $NaWCp(CO)_3$ (see above) indicates that an intermediate redox reaction is involved which generates the $MCp(CO)_3$ radicals. Such radicals or the polynuclear one-electron species are, in our opinion, much more important in preparative organometallic chemistry than is generally accepted today.

In our hands this proved true for the metal exchange reaction by which we have prepared the majority of our heterometallic and chiral clusters [21], and we actually had to learn that a redox reaction is the shortest way to replace one organometallic fragment in a cluster by another one. First we used indirect methods, i.e. attachment via linking ligands, to bring the new organometallic fragment in [22]. Then we made use of the existence of organometallic radicals in equilibrium with their homodinuclear precursors [23], and finally we found that anionic metal carbonyl species are the best reagents to replace $Co(CO)_3$ units as cluster constituents [24]. Reactions of this type, as exemplified by the conversion of **18** by $Na[MoCp(CO)_3]$ to **19**, require of course the entering anion to be more reducing than the leaving one which is one of the reasons why the cobalt carbonyl unit is such a good leaving group. This redox method of metal exchange has allowed us to enter the field of chiral hydridometal clusters which was important for mechanistic and stereochemical studies [25]. The method of choice here is the use of dianionic metal exchange reagents like $Fe(CO)_4^{2-}$ or $Ru(CO)_4^{2-}$ which by expulsion of $Co(CO)_4^-$ produce anionic clusters which can then be neutralized by protonation.

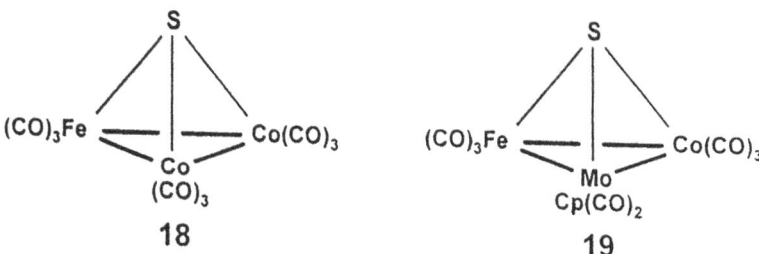

The mechanism of the redox-driven metal exchange is not clear. In our opinion however, it is a sequence of one electron transfers whereby first the cluster is turned into an anion and the organometallic reagent into a radical. Both are then odd-electron species with enhanced lability which make the successive reactions possible. An arguement in favour of this mechanistic proposal comes from the observation that our type of metal exchange reaction can be catalyzed by such typical ETC reagents as benzophenone ketyl [26].

ETC (electron transfer catalysis) is a subject which in cluster chemistry, just like in other branches of chemistry, probably has more importance than presently known.

As mentioned above, we have worked it out in detail for the phosphine substitution in $RCCO_3(CO)_9$. We also have used it in the way developed by Bruce and Nicholson [27] for many ligand substitutions. But we and others have yet to put their energies to an exploitation of the further possibilities offered by ETC.

Another area of cluster redox chemistry which has yet to be exploited is that of clusters as redox reagents. This is no wonder, as it is still widely unknown how valuable the ferricinium ion or cobaltocene are as clean one-electron oxidizing and reducing agents respectively. One of the advantages of ferricinium ion or cobaltocene is that they become diamagnetic after the redox process. This advantage is present in all of the paramagnetic clusters mentioned in this review. A multitude of clusters offer a second advantage in that they can span the full range of redox potentials, including the ability of clusters like **2** or **13** to be reducing even at positive potentials. This should mean that there are applications as redox reagents even though the synthesis of clusters is cumbersome and the available quantities are bound to be small.

$$20^{\pm} + PhCH_2Br \rightarrow 20 + \tfrac{1}{2}\,(PhCH_2)_2 + Br^-$$

20

We encountered a possible application of this kind when trying to establish organic reactions of cluster-bound nitriles [28]. During these studies we observed that the cluster **20** forms a rather stable radical anion. It was then hoped that organic electrophiles might attack this anion such that the nitrile ligand is derivatized on the cluster. Preparative attempts in this direction met with little success, however. Instead the normal course of events was a redox reaction between the cluster and the reagent which regenerates neutral **20** and leads to a Wurtz-type coupling of the organic reagent. While being disappointing for the specific case, this reaction may be of general value for less trivial cases of organic electron transfer chemistry where the usual redox reagents don't work.

The basis of the attempted reactions between the radical anion of **20** and organic electrophiles was the assumption that the well established electron-transfer properties of the clusters should open an entry to an electron transfer chemistry of substrates attached to clusters. Thereby the redox equivalent would be transferred from a cheap redox reagent via the cluster to the substrate, making the cluster the equivalent of a catalytically active surface and of an electrode at the same time in terms of the cluster-surface analogy. We have made various attempts to find realizations of this hypothesis but we haven't found a simple straightforward one yet. In general, lability of the clusters complicates the course of the reactions, as exemplified by the interconversion of the alkyne ligand in cluster **21** [29]. Clusters of type **21** can be converted by cobaltocene to solutions of their radical anions. These, upon treatment with alkylating agents and after work-up, produce the dinuclear complexes **22**. The new organic ligand in **22** is composed of the original alkyne, CO, and the electrophile R'. Thus the nucleophilicity resulting from reduction of the cluster **21** has become effective in its organic ligand sphere but only after extensive rearrangements: one cobalt carbonyl unit has been lost, one CO has been inserted, and the point of attack of the electrophile is this CO and not the original alkyne.

21   **22**

The question "Are there preparative applications?" finds two answers in our work. As long as cluster synthesis and the handling of organometallic fragments are involved, redox chemistry has brought significant advances. As long as new organic transformations in the ligand sphere of clusters are involved, redox chemistry is still more an attractive prospect, rather than a practical success.

An Organometallic Prussian Blue Story

Another question to be asked with respect to the redox properties of polynuclear organometallic complexes is "Are there physical applications?". To answer this question one might look for optical or electrical properties of the species with unpaired electrons. We, like others [30,31] are doing this with cyanide bridged oligonuclear compounds.

Although the cyanide ligand is one of the best-known ligands in classical coordination chemistry, its organometallic chemistry is rather underdeveloped, and its high potential to support electron transfer and mixed valence which is so evident in Prussian Blue or Krogmann's salts has yet to be exploited in coordination chemistry, classical as well as organometallic. This is amazing if one realizes how many cyanide complexes are known which all can act as ligands towards other metals through the cyanide nitrogen atoms.

We found out that it is particularly easy to link organometallic units by cyanide ligands starting from organometallic cyanide complexes. Thus, using $Cp(CO)_2Mn-CN^-$ and $Cp(CO)_2Fe-CN$ as ligands, the four isoelectronic compounds 23a-d were prepared in a straightforward way [32,33]. They all undergo reversible one-electron oxidations accompanied by distinct colour changes, thereby supporting the idea that the cyanide bridge is a good transducer and mediator of electronic effects. Specifically 23d which in the oxidised form contains Fe(II) and Fe(III) can be called a minimal molecular analogue of Prussian Blue. Due to this we are calling our preparative and electron transfer work with cyanide bridged complexes "Organometallic Prussian Blue Chemistry".

$$Cp(CO)_2Mn-CN-Fe(CO)_2Cp \qquad Cp(CO)_2Fe-CN-Mn(CO)_2Cp$$

23a 23b

$$[Cp(CO)_2Mn-CN-Mn(CO)_2Cp]^- \qquad [Cp(CO)_2Fe-CN-Fe(CO)_2Cp]^+$$

23c 23d

Two questions which have to be tackled with this type of compounds are those of isomerism and mixed valence. 23a (red) and 23b (orange) are isomers which are easily distinguished by their colours and spectra. For a related pair of isomers, 24a and 24b, crystal structure determinations have revealed that the molecular shapes are almost undistinguishable. The two compounds are isomorphous and the molecules differ only marginally in their M-C and M-N distances [33]. Thus the Cr-(CN)-Fe units and the donor/acceptor properties of the ligands level off the electronic imbalances within the two compounds. This statement is equally true for the cations 24a' and 24b' of which 24a' can be isolated and crystallized as the BF_4 salt [34]. The structure of 24a' is again practically identical to that of 24a and 24b. Spectroscopic data of 24a' including a strong intervalence transfer band in the near IR indicate that 24a' is a mixed valence complex of class II with a high degree of delocalization, once more underlining the analogy with Prussian Blue [34].

$$(CO)_5Cr-CN-Fe(dppe)Cp \quad \underset{+e}{\overset{-e}{\rightleftharpoons}} \quad [(CO)_5Cr-CN-Fe(dppe)Cp]^+$$

24a 24a'

$$Cp(dppe)Fe-CN-Cr(CO)_5 \quad \underset{+e}{\overset{-e}{\rightleftharpoons}} \quad [Cp(dppe)Fe-CN-Cr(CO)_5]^+$$

24b 24b'

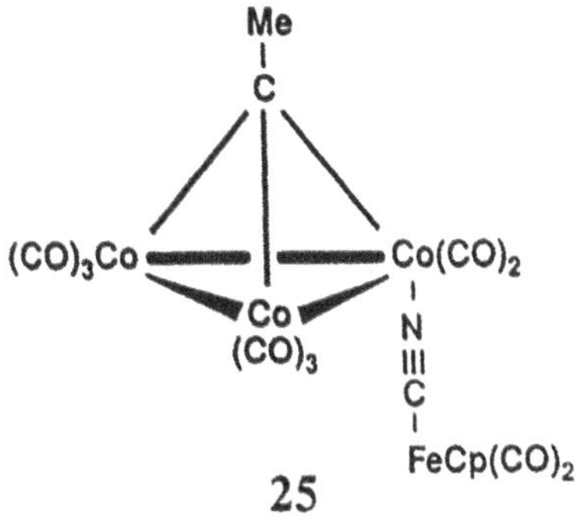

25

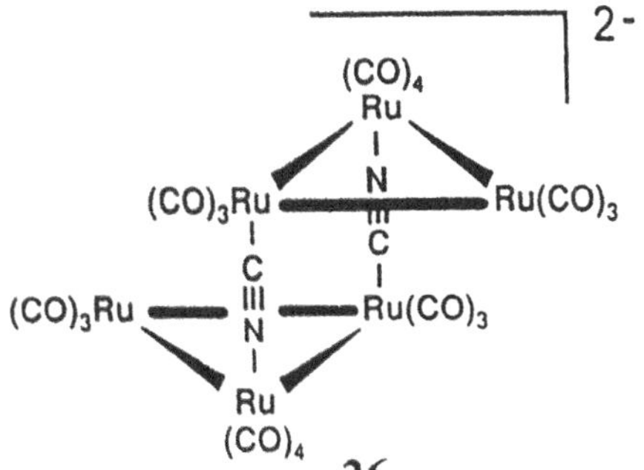

26

Clusters linked to other organometallic units by cyanide bridges are equally easy to prepare. Two examples are **25** [33] and **26** [35]. The electrochemistry of this type of compounds is generally complex as long as the clusters bear mainly carbonyl ligands. This has to do with the enhanced lability of such clusters towards ligand substitution after a single electron transfer as was discussed above for the $RCCo_3(CO)_9$ derivatives. For CO free clusters connected to organometallic cyanides, however, the cluster as well as the external organometallic units can get involved in reversible redox steps [33,36].

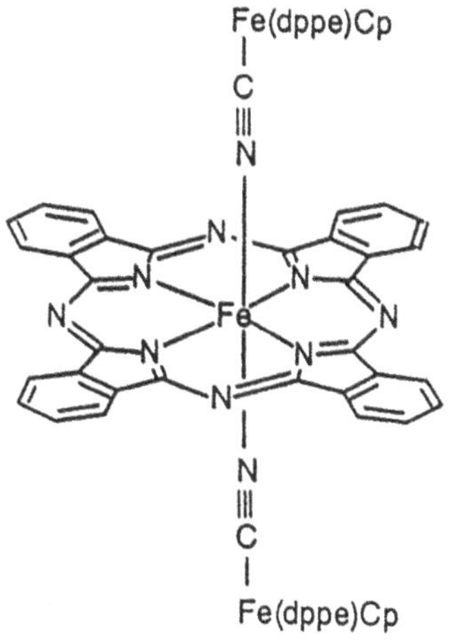

27

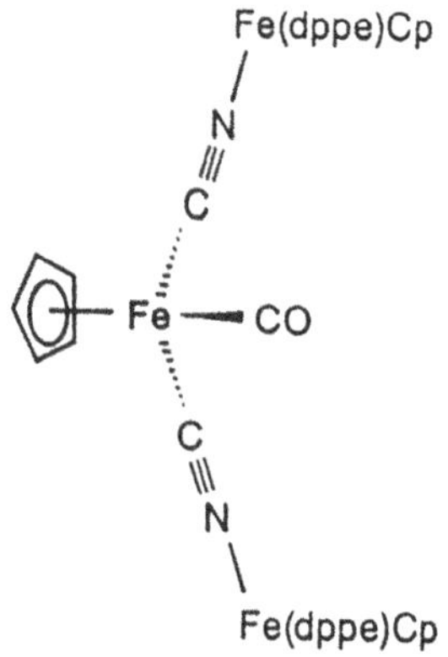

28

The analogy with Prussian Blue can be extended to oligonuclear organometallic complexes containing chains of metal atoms linked by cyanide ions. The simpler ones of these complexes again pose no synthetic problems. We have used several approaches to prepare trinuclear systems of which **27** and **28** stress the anlogy by containing only

314

iron as a metal. **27** is diamagnetic if it contains Fe(II) in the phthalocyanine ring. Upon exposure to air its green colour turns deep blue demonstrating the colour effects of the haeme-like ring system and/or the mixed valence situation.

Complexes like **27** and **28** exhibit a very rich electrochemistry. Figure 10 shows a CV of the manganese analog of **27** as an example. It contains a total of six electrochemical one-electron transitions interrelating the total charges of the complex between -2 and +4. Colour changes and electron transfers indicates a development towards electric conductivity which was actually achieved in polymers containing stacked (phthalocyanine)Fe-CN units after partial oxidation [37]. Our investigations in this field are just in the beginning, and we have yet to apply the methodology of mixed valence and charge transfer studies to understand the level of electronic mobility in the new complexes.

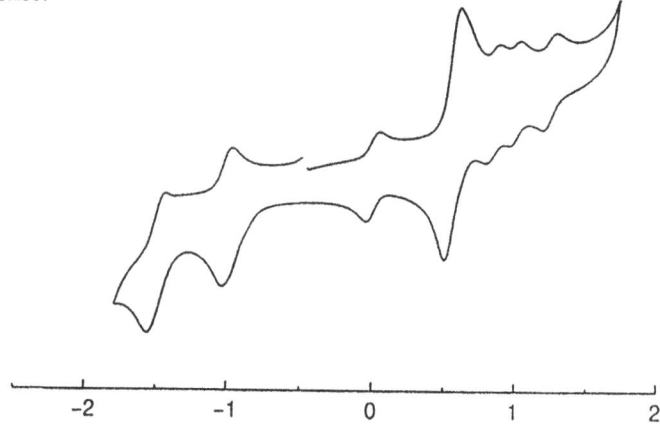

Fig 10. Cyclic voltammogram of the analog of **27** with a central Mn atom (in benzene/ acetonitrile 2:1, room temp., scan speed 0.5 V/sec, E vs. Ag/AgCl. The large wave results from a decomposition product).

Where next?

This account of our electron transfer work wouldn't be complete if it would hide our major failures. We have been dreaming of linking clusters directly or via conducting ligands to polymers with electric conductivity. We have been dreaming of isolating salts containing paramagnetic cluster ions as electrical conductors. We have hoped to find catalytic applications of odd electron organometallic compounds. We didn't, but almost all competitors didn't either. So the challenge remains, and we are looking for new ways to succeed.

As we move on, our philosophy remains the same. First we shall be making compounds, then we shall try to learn about them. There are many clusters and oligonuclear complexes whose redox chemistry waits to be evaluated. In our opinion combinations of classical and organometallic complexes promise a rich harvest. There are many "conducting" ligands other than the cyanide ligand waiting to be used to link complex fragments. Preparative chemistry is the key to possible advances, and we shall continue applying and propagating it.

Acknowledgement

I am indebted to the colleagues who taught me the practical art of electrochemistry and electron spin resonance, H. Baumgärtel (Berlin) and B.M. Peake (Dunedin). In our electron transfer work I have enjoyed collaboration over many years with B.H. Robinson (Dunedin) and J. Heinze (Freiburg). T. Madach and J.S. Field have set milestones in my research group. The German Science Foundation and the State of Baden-Württemberg have provided a financial basis, and the Graduiertenkolleg "Systeme mit ungepaarten Elektronen" provides a stimulating interdisciplinary atmosphere.

All this wouldn't have been written had it not been for Lawrence F. Dahl to whom this account is dedicated and who in the most critical moments of my academic career gave me the faith and strength to continue. He has given faith and strength to many people, and he deserves to be honoured for it.

References

1.   H. Vahrenkamp, V.A. Uchtman and L.F. Dahl, *J. Am. Chem. Soc.*, 1968, **90**, 3272.
2.   T. Madach and H. Vahrenkamp, *Chem. Ber.* 1981, **114**, 505.
3.   C.R. Pulliam, J.B. Thoden, A.M. Stacy, B. Spencer, M.H. Englert and L.F. Dahl, *J. Am. Chem. Soc.* 1991, **113**, 7398.
4.   T.E. North, J.B. Thoden, B. Spencer and L.F. Dahl, *Organometallics*, 1993, **12**, 1299.
5.   J. Heinze, H. Vahrenkamp and coworkers, unpublished.
6.   H. Vahrenkamp, *Chem. Ber.*, 1978, **111**, 3472.
7.   T. Madach and H. Vahrenkamp, *Chem. Ber.*, 1981, **114**, 513.
8.   T. Madach and H. Vahrenkamp, *Z. Naturforsch.* 1979, **34b**, 573.
9.   T. Madach and H. Vahrenkamp, *Z. Naturforsch*, 1978, **33b**, 1301.
10.   S. J. McLain, *J. Am. Chem. Soc.* 1988, **110**, 643.
11.   cf. K.A.E. O'Callaghan, S.J. Brown, J.A. Page and M.C. Baird, *Organometallics*, 1991, **10**, 3119.
12.   H. Beurich, T. Madach, F. Richter and H. Vahrenkamp, *Angew. Chem.* 1979, **91**, 751; *Angew. Chem. Int. Ed. Engl.* 1979, **18**, 690.

13. P.N. Lindsay, B.M. Peake, B.H. Robinson, J. Simpson, U. Honrath, H. Vahrenkamp and A.M. Bond, *Organometallics* 1984, **3**, 413.

14. R. Fisel and R. Hoffmann, cited in B.M. Peake, P.H. Rieger, B.H. Robinson and J. Simpson, *Inorg. Chem.* 1981, **20**, 2540.

15. U. Honrath and H. Vahrenkamp, *Z. Naturforsch*, 1984, **39b**, 545.

16. J.F. Halet, R. Hoffmann and J.-Y. Saillard, *Inorg. Chem.* 1985, **24**, 1695; J.F. Halet and J.-Y. Saillard, *Nouv. J. Chim.* 1987, **11**, 315.

17. T. Jaeger, S. Aime and H. Vahrenkamp, *Organometallics*, 1986, **5**, 245.

18. J.T. Jaeger, J.S. Field, D. Collison, G.P. Speck, B.M. Peake, J. Hähnle and H. Vahrenkamp, *Organometallics*, 1988, **7**, 1753.

19. A.J. Downard, B.H. Robinson and J. Simpson, *Organometallics*, 1986, **5**, 1122; 1986, **5**, 1132; 1986, **5**, 1140.

20. K. Hinkelmann, J. Heinze, H.T. Schacht, J.S. Field and H. Vahrenkamp, *J. Am. Chem. Soc.* 1989, **111**, 5078.

21. H. Vahrenkamp, *Comments Inorg. Chem*, 1985, **4**, 253.

22. H. Beurich and H. Vahrenkamp, *Chem. Ber*, 1982, **115**, 2385.

23. H. Beurich, R. Blumhofer and H. Vahrenkamp, *Chem. Ber.* 1982, **115**, 2409.

24. R. Blumhofer, K. Fischer and H. Vahrenkamp, *Chem. Ber*, 1986, **119**, 194.

25. D. Mani, H.T. Schacht, A.K. Powell and H. Vahrenkamp, *Chem. Ber*, 1989, **122**, 2245.

26. U. Honrath and H. Vahrenkamp, *Z. Naturforsch*, 1984, **39b**, 559.

27. M.I. Bruce, J.G. Matisons and B.K. Nicholson, *J. Organomet. Chem.*, 1983, **247**, 321.

28. P. Suter, Dissertation, Universität Freiburg, 1994.

29. H. Bantel, A.K. Powell and H. Vahrenkamp, *Chem. Ber.*, 1990, **123**, 1607.

30. F.L. Atkinson, A. Christofides, N.G. Connelly, H.J. Lawson, A.C. Lyons, A.G. Orpen, G.M. Rosair and G.H. Worth, *J. Chem. Soc., Dalton Trans.*, 1993, **1441**, and references therein.

31. F. Scandola, R. Argazzi, C.A. Bignozzi, C. Chiorboli, M.T. Indelli and N.A. Rampi, *Coord. Chem. Rev.*, 1993, **125**, 283.

32. B. Oswald, A.K. Powell, F. Rashwan, J. Heinze and H. Vahrenkamp, *Chem. Ber.*, 1990, **123**, 243.

33. N. Zhu, A. Geiß and H. Vahrenkamp, unpublished.

34. N. Zhu and H. Vahrenkamp, *J. Organomet. Chem.*, 1994, **472**, C5.

35. G. Lavigne, N. Lugan and J.J. Bonnet, *J. Chem. Soc., Chem. Commun.* 1987, 957.

36. D.H. Johnston, C.L. Stern and D.F. Shriver, *Inorg. Chem.* 1993, **32**, 5170.

37. J. Metz and M. Hanack, *J. Am. Chem. Soc.* 1983, **105**, 828.

ARENE CLUSTER COMPOUNDS

Brian F.G. Johnson,
Crum Brown Professor of Chemistry,
Department of Chemistry,
The University of Edinburgh,
West Mains Road,
Edinburgh,
EH9 3JJ, U.K.

Abstract

The synthesis, isolation and characterisation of a wide and diverse range of arene clusters based on nuclearities of three to eight metal atoms are described. Their use as models for the chemisorption of benzene on the macromolecular surface will be considered and matters relevant to the chemistry on the surface such as bonding type, mobility and reactivity will be discussed. Certain of these compounds readily undergo photoisomerisation to produce benzyne-derivatives by the cleavage of two C-H arene bonds. This process may be tuned by varying the nature of the other ligands present. In the solid, many of the arene clusters exhibit strong arene-arene interactions which are 'graphitic-like' leading to the formation of chains, ribbons and snakes. Use of bridging bifunctional arenes such as Ph-CH=CH-CH=CH-CH=CH-Ph leads to useful monomeric precursors to arene-cluster polymers.

1. Introduction

Our interest in this area was originally stimulated [1] by the chance discovery of the arene hexaruthenium derivatives $Ru_6C(CO)_{14}(C_6H_{6-n}Me_n$ (n = 0,1,2,3) and the subsequent synthesis of the *bis*(benzene) derivative $Ru_6C(CO)_{11}(\eta^6\text{-}C_6H_6)(\mu_3:\eta^2:\eta^2:\eta^2\text{-}C_6H_6)$ [2]. The observation that in this latter interesting compound the benzene molecules adopted not only the familiar η^6-terminal bonding mode found in *e.g.* $Cr(\eta^6\text{-}C_6H_6)_2$, but also the then highly unusual μ_3-face-capping mode, led us to investigate the chemistry of arene-clusters in fuller detail.

Over the recent past we have developed a systematic and diverse chemistry of these materials, not only based on Ru_6C units, but also for a variety of clusters containing from three to eight metal atoms.

317

L. J. Farrugia (ed.), The Synergy Between Dynamics and Reactivity at Clusters and Surfaces, 317–333.
© 1995 *Kluwer Academic Publishers.*

For us, basically *four* patterns of interest have emerged, *viz.*,

(i)   the cluster/surface analogy,

(ii)   the photochemistry of arene cluster and C-H bond cleavage reactions,

(iii)   graphitic-like interactions in the solid-state leading to chains, snakes and ribbons and

(iv)   the generation of monomer and dimer precursors to conducting polymers.

These themes are not independent; they rely heavily on one another. Nevertheless, each provides an excellent research area in its own right and undoubtedly other related areas of interest will emerge.

2. The Cluster/Surface Analogy

The idea behind these investigations emerged naturally from our studies of $Ru_6C(CO)_{11}(\eta^6-C_6H_6)(\mu_3:\eta^2:\eta^2:\eta^2-C_6H_6)$. Our discovery of this compound corresponded with reports by Somorjai [3] and others of the interaction of benzene with the Rh(111) surface both in the presence of differing amounts of CO and in its absence. It seemed apparent to us that a clear relationship existed between our benzene/carbonyl cluster and the benzene/carbonyl/surface interaction (*Figure 1*). The feeling that our systems, which were so easily studied by the usual vibrational (IR and R) spectroscopic techniques, NMR spectroscopy and diffraction methods (X-ray and neutron), might

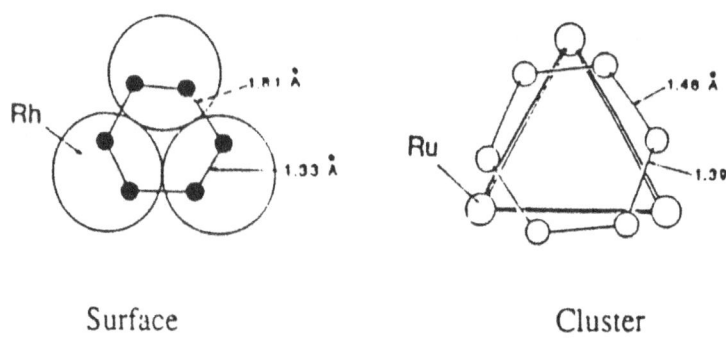

Surface   Cluster

Figure 1

enable a clearer understanding of the surface chemistry, led us to contemplate the synthesis of simple model compounds. In the limiting case, the single-crystal metallic surface may be regarded as being composed of approximately close-packed planar arrays of spheres extending infinitely in two dimensions. A consequence of this long-range periodicity is that local atomic and electronic structure will be influenced by more distant atoms in both the surface and the bulk of the metal. Hence, CO absorption on one site will influence the characteristics of sites adjacent (or even removed from) it. Recent advances in the dynamical theory of low energy diffraction (LEED) by ordered absorbate overlayers has led to the structural characterisations of several metal surface-benzene complexes by Somorjai, van Hove and co-workers [3]. Optimised structures of the surface complexes Rh(111) - 3 x 3 - C_6H_6 + 2CO (a), Rh(111) - C{2(3)$^{\frac{1}{2}}$ x 4}rect-C_6H_6 + CO (b), Rh(111) - {2(3)$^{\frac{1}{2}}$ x 3}rect-2C_6H_6 (c) and Pt(111) -{2(3)$^{\frac{1}{2}}$ x 4}rect-2C_6H_6 + 4CO (d) are illustrated in *Figure 2*. In each overlayer, benzene is chemisorbed intact and lies parallel to the close-packed metal surface. In (I) and (II) adsorption occurs at a three-fold h.c.p.-type site - clearly related to the μ_3-bonding mode adopted in $Ru_6C(CO)_{11}(\eta^6\text{-}C_6H_6)(\mu_3{:}\eta^2{:}\eta^2{:}\eta^2\text{-}C_6H_6)$ with an expanded C_6-ring showing in-plane Kekule distortions: C-C bond distances alternate between 1.46(15) and 1.58(15) Å in (I) and between 1.31(15) and 1.81(15) Å in (II), the short bonds lying above single atoms while the long C-C bonds form bridges linking pairs of metal atoms. Although the CO-free structure (III) has not been analysed in detail, the principal features are known; the superlattice contains two benzene molecules *per* unit cell occupying two-fold bridging sites. Bridge sites occupancy has been substantiated in the ordered Pt(111) structure (IV), with the probable distortion in molecular symmetry to C_{2v}. For us, the challenge to produce model cluster systems which correspond to these observations was raised.

The metal skeletons of high nuclearity carbonyl cluster compounds are frequently structurally comparable to fragments of bulk metallic lattice, *e.g.* $[Rh_{13}(CO)_{24}H_3]^{2-}$, $[Os_{10}C(CO)_{24}]^{2-}$ and $[Rh_{14}(CO)_{25}]^{4-}$ may be recognised as fragments of h.c.p., c.c.p. and b.c.c. structures respectively [4]. The series of osmium clusters based on four, ten, twenty and thirty-five metal atoms, $H_4Os_4(CO)_{12}$, $[Os_{10}C(CO)_{24}]^{2-}$, $[Os_{20}(CO)_{40}]^{2-}$ and $Os_{35}(CO)_{56}^{2-}$ follow the c.c.p. growth sequence precisely. Significantly, in the dianion $[Os_{20}(CO)_{40}]^{2-}$, the face of the cluster corresponds directly to the metal (111) surface. Smaller carbonyl clusters are typically deltahedra and many of their metal core configurations may be regarded as microscopic fragments of common close-packed lattices. However, the role of the ligand sphere in stabilising these bulk-like geometries should be appreciated, as should the case in which they undergo geometrical transformation as the overall electron count is modified, *e.g.* monocapped trigonal bipyramid in $Os_6(CO)_{18}$ (84e) to octahedral in $[Os_6(CO)_{18}]^{2-}$ (86e).

Structural comparison between the chemisorbed state and metal clusters are best made when there is a correspondence in the metal, ligands, ligand coverage of the cluster or surface and in the crystallography. Detailed structural analysis of the gas-metal interface is currently only feasible for metal crystals having well-defined flat

320

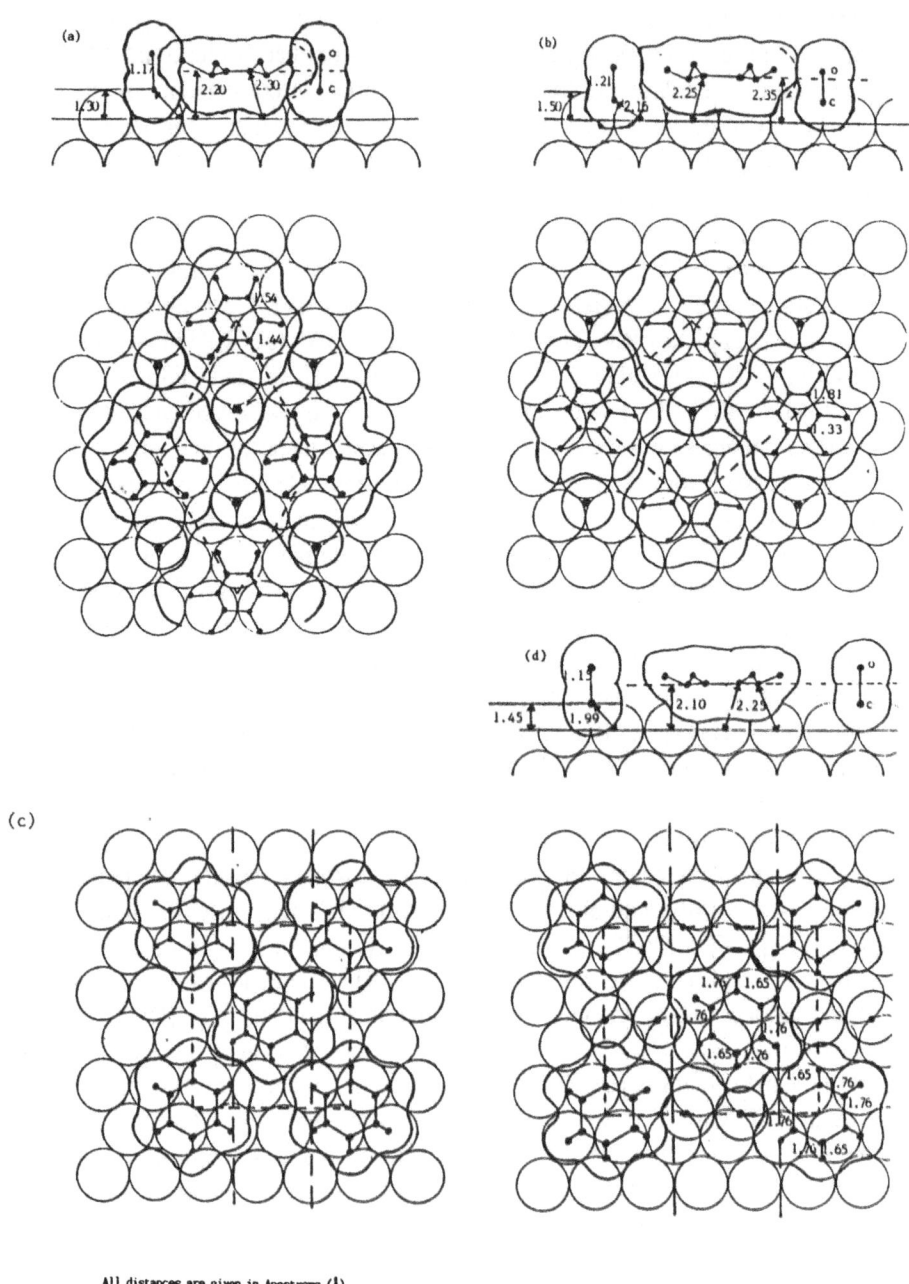

All distances are given in Angstroms (Å)

Figure 2

surfaces. For chemisorption on these low Miller index surface planes, the metal-ligand and metal-metal connectivity is respectively smaller and larger than for clusters, although on higher planes the metal coordination number decreases at irregular features such as step and kink sites.

Thus, the challenges to the molecular chemist may be described as:

(i)   the production of systems containing benzene in bonding modes related to those supposed to exist on the surface

(ii)   to investigate the change in the electronic character and bonding mode of the arene as the number and type of metal atoms changes,

(iii)   to monitor related changes in bonding as a function of ligand type,

(iv)   to explore the migratory patterns that exist on the cluster surface and to relate them to motion on the bulk.

In this paper some of our efforts made in response to these challenges are reported.

2.1 The Model Compounds $[Ru_3(CO)_9(\mu_3:\eta^2:\eta^2:\eta^2C_6H_6)]$ and $[Os_3(CO)_9(\mu_3:\eta^2:\eta^2:\eta^2-C_6H_6)]$

The simplest model of the [111] surface is the metal triangulo-arrangement found in the simplest clusters $Ru_3(CO)_{12}$ and $Os_3(CO)_{12}$. In this work we took advantage of this fact to explore routes to benzene/carbonyl models. We were lucky that our first attempted syntheses of the title compounds proceeded moderately easily and in favourable yields. The best routes to the ruthenium and osmium compounds $M_3(CO)_9(\mu_3:\eta^2:\eta^2:\eta^2-C_6H_6)$ are, however, different. For the ruthenium compound, we find that treatment of $Ru_3(CO)_{12}$ in dichloromethane with Me_3NO in the presence of excess 1,3-cyclohexadiene, leads to the required product in a single stage process [5].The corresponding osmium derivative may be prepared in a related fashion (*see Scheme 1*), but is best prepared from the very reactive unsaturated cluster $(\mu_2-H)_2Os_3(CO)_{10}$ (*see Scheme 2*) [2]. The molecular structures of the two compounds are, as expected, very similar [2,6]. In each case the central metal triangle supports the μ_3-bonded benzene and nine CO ligands. Because of the greater precision of the ruthenium structure, the C-H vectors are seen to bend away from the C_6 plane and Ru_3 unit, in agreement with the Somorjai model. In addition, in both cases, the C_6-ring shows the expected Kekule distortion. Clearly these compounds are good model systems. Detailed vibrational analysis of both these cluster and surface absorbed systems is currently under investigation.

    Many other 'model' systems have now been prepared and detailed synthetic routes to clusters based on M_4, M_5, M_6, M_7 and M_8 units have been devised. The Ru_6C system provides a good example of the synthetic routes employed and also the versatility of the benzene, which may bond to produce a variety of geometric forms (*Scheme 3*) [7].

322

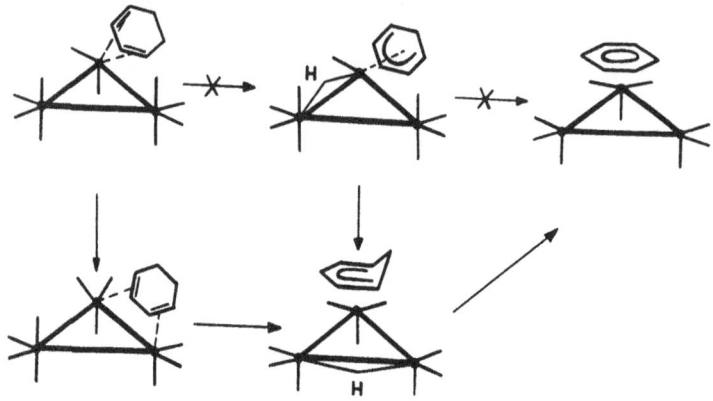

Scheme 1

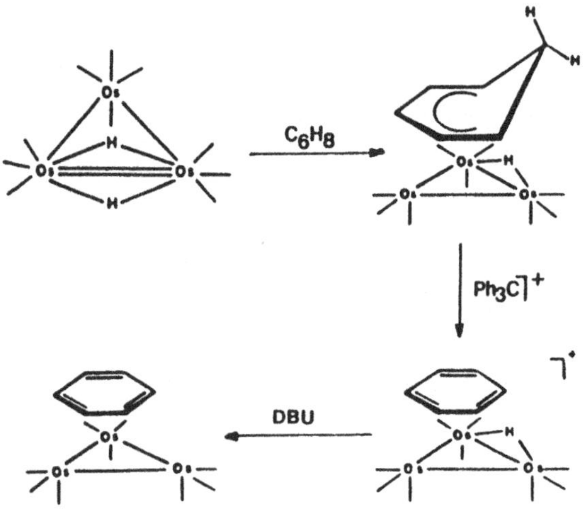

Scheme 2

One of our primary objectives in this aspect of the work has been to increase the number of benzene or related molecules supported by the cluster unit - leading ultimately to a system containing only benzene on the surface - a goal we've yet to meet!

In our synthetic programme essentially the same synthetic approach has been employed for all systems under investigation. The initial cluster is activated by oxidative removal of CO as CO_2 by reaction with Me_3NO, followed by addition of cyclohexa-1,3-diene which usually bonds first through a single double bond and then through the 1,3-diene unit. By this means the complexes containing the six membered ring bonded to central cluster units have been isolated and characterised.

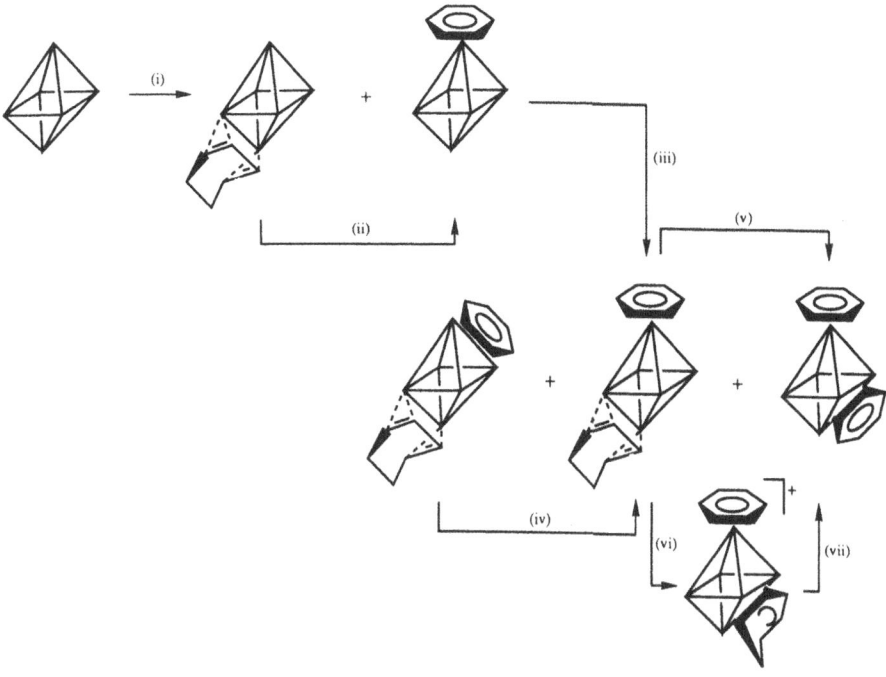

Scheme 3. Reagents and conditions; (i) Me_3NO/C_6H_8, (ii) Me_3NO, (iii) Me_3NO/C_6H_8, (iv) hexane Δ, (v) Me_3NO, (vi) $[Ph_3C][BF_4]$, (vii) DBU

The bonding mode adopted by the cyclohexadiene unit in this scheme is *not* always the same [5]. For the Os_3 system we have good evidence to suggest that two forms of the compound $Os_3(CO)_{10}(C_6H_8)$ exist. One, for which the molecular structure has been established by single crystal X-ray analysis, with the diene bonded to *one* osmium atom. This compound is reluctant to undergo conversion either to the dienyl derivative by single C-H bond cleavage, or to the benzene derivative *via* the cleavage of two C-H bonds. In contrast, the second isomer, which we suppose contains the diene bridge-bonding to *two* osmium atoms, readily undergoes conversion directly to $Os_3(CO)_9(\mu_3-\eta^2:\eta^2:\eta^2-C_6H_6)$. This observation is significant, indicating as it does that C-H cleavage, leading to face-bridging arene, is assisted by multiple interactions with the cluster

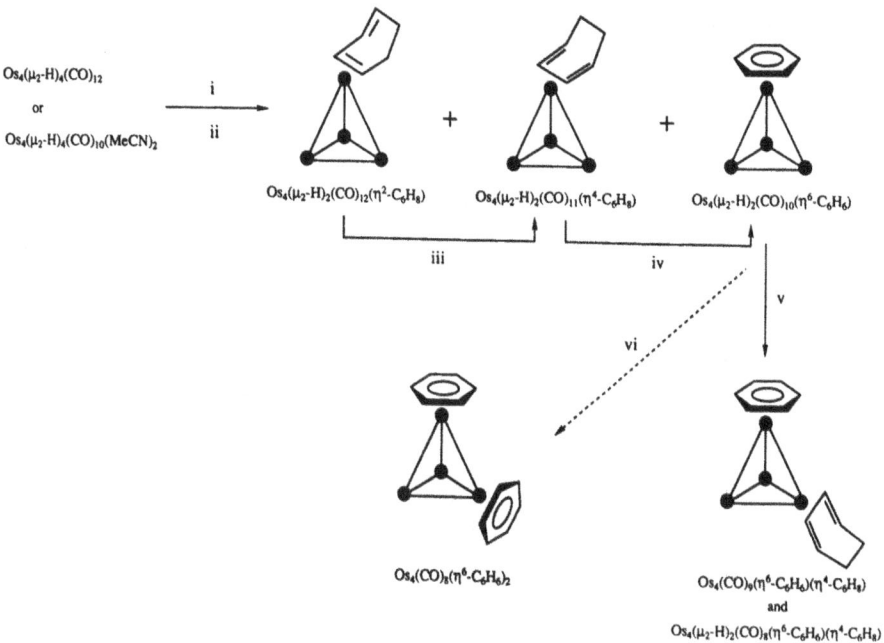

Scheme 4. (i) 3.2 equiv $Me_3NO/CH_2Cl_2/1,3-C_6H_8$, (ii)$CH_2Cl_2/1,3-C_6H_8$, (iii) and (iv) Δ hexane, (v) 3.2 equiv. $Me_3NO/CH_2Cl_2/1,3-C_6H_8$, (vi) 3.2 equiv. Me_3NO/CH_2Cl_2/benzene.

surface in this case. However, this cannot be the whole story, since in our studies of the Os_4 systems (*Scheme 4*) we have been able to isolate and characterise the complete range of intermediate compounds leading up to the final benzene derivative [8]. It is important to note, however, that here the hexadiene molecule bonds to only *one* osmium atom and benzene apparently adopts only the η^6-terminal bonding mode. Similar behaviour is noted with the Co_4 systems where again only the η^6-mode for benzene is observed [9]. A feature of these Co_4 and Os_4 tetrahedral systems is the relative ease with which the terminally bonded arene undergoes exchange with uncoordinated arene.

For both the Os_4 and Co_4 systems this exchange provides a highly convenient method of producing different arene-cluster derivatives. However, this behaviour is not general and systems based on the Ru_6C unit are reluctant to undergo exchange.

In most other systems (see e.g. Scheme 3), the diene is found to bridge one edge of the cluster polyhedron and on treatment of these complexes with Me_3NO, or on pyrolysis, smooth conversion to the appropriate benzene derivative is observed [7].

For the Ru_6C series of compounds, benzene (and other arenes) adopt both μ_3- and μ_1- bonding configurations [2]. With the compound $Ru_6C(CO)_{14}$(arene), apart from the derivative derived from cyclophane (see below), all arenes adopt the terminal bonding mode and no evidence for face-bridging has been found [10]. This is not the case for the bis-benzene derivative $Ru_6C(CO)_{11}(C_6H_6)_2$, which may be easily prepared from $Ru_6C(CO)_{17}$ using $Me_3NO/1,3-C_6H_8$ in a sequential manner. Here several different isomeric forms have been isolated.

Significantly, these forms undergo exchange [11] but barriers to isomer interconversion are relatively high. The mechanism by which isomerisation occurs has not been established, but we consider that the most likely route involves an intermediate in which the C_6-ring spans a polyhedral edge and is coordinated through two 'allyl' fragments within the C_6-ring. Although direct examples for this mode have not been found for the simple benzene derivatives, a compound containing benzene bound in this manner has been observed by others in $(CpRh)_2(\mu_2-\eta^3:\eta^3-C_6H_6)$[12] and we have fully characterised a related cyclophane derivative, $[Ru_2(CO)_6(\mu_2-\eta^3:\eta^3C_{16}H_{16})]$ (Figure 3) [13].

In general, the benzene compounds described so far cannot be prepared directly from the reaction of benzene with the appropriate cluster. The Co_4 system provides a rare example, where direct reaction does occur with either $Co_2(CO)_8$ or $Co_4(CO)_{12}$ to produce $(C_6H_6)Co_4(CO)_9$ [9]. Interestingly, we have noted that in the reaction of $Ru_3(CO)_{12}$ with $Me_3NO/1,3-C_6H_8$ the presence of benzene greatly enhances the reaction rate, although we have not established the role of the benzene [5]. Benzyne, rather than benzene, derivatives of Os_3, e.g. $(\mu_2-H)_2Os_3(CO)_9(C_6H_4)$, are formed directly on reaction of the activated cluster $Os_3(CO)_{10}(MeCN)_2$ with C_6H_6; no evidence has been found that $Os_3(CO)_9(C_6H_6)$ is an intermediate in this reaction, but we have discovered that on irradiation or prolonged heating $Os_3(CO)_9(C_6H_6)$ is converted to the same benzyne derivative (see below).

2.2. Electronic Characters and Change in Bonding Mode of Arene.

A number of interesting features have emerged but we remain a long way from truly understanding the factors which govern the bonding mode adopted by arenes on the cluster surface. Apart from $Ru_2(CO)_6(\mu_2-\eta^2:\eta^2-C_{16}H_{16})$ we have no direct evidence for the bis-allyl bonding mode (Figure 3) although there is limited evidence to suggest the formation of unstable edge-bridged intermediates in the μ_3- to η^6- conversion process.

The η^6-terminal bonding mode is commonly observed for all cluster nuclearities

326

and seems to be independent of the metal-metal connectivities within the clusters, being bound to metal atoms with connectivity (M-M) of *two* in $M_3(CO)_7(R_2C_2)(C_6H_6)$, [16] *three* in $H_2Os_4(CO)_{10}(C_6H_6)$, [8] $Co_4(CO)_9(C_6H_6)$ [9] and $Ru_5C(CO)_{12}(C_6H_6)$ [17] (isomer **1**) and *four* in $Ru_5C(CO)_{12}(C_6H_6)$ [17] (isomer **2**) and $Ru_6C(CO)_{14}(C_6H_6)$ [17]. Migration of arene over the cluster surface to produce different isomeric forms may be induced chemically [16] or thermally [17,18]. At present there appears to be no correlation between the number of carbonyl groups also present and the arene bonding

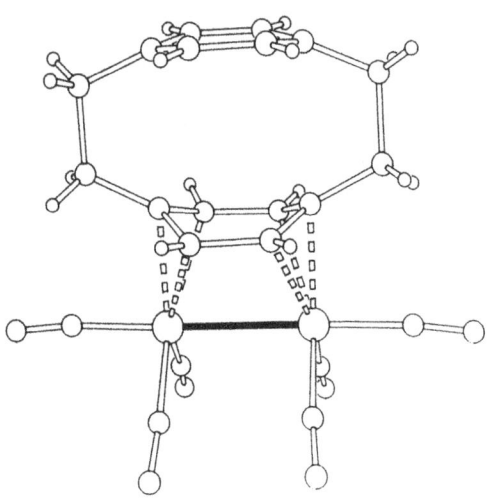

Figure 3. Molecular structure of $Ru_2(CO)_6(\mu-\eta^2:\eta^2-C_{16}H_{16})$

type. For systems containing more than one arene moiety, as in $Ru_6C(CO)_{12}(C_6H_6)_2$, we detect a slight tendency for face-bonding to be preferred, but this may be more apparent than real!

One exciting development has arisen from our observation that the C-H chemical shift in the NMR spectra of the range of η^6-bonded benzene derivatives $Ru_5C(CO)_{12}(\eta^6-C_6H_6)$ ($\delta 5.93$), $Ru_6C(CO)_{14}(\eta^6-C_6H_6)$ ($\delta 5.56$), $H_4Ru_8(CO)_{18}(\eta^6-C_6H_6)$ ($\delta 5.40$) undergo a systematic variation as a function of the number of metals in the cluster unit [19]. Although great care should always be exercised in any correlation of this type, it is reasonable to assume that the shift reflects the change in electron density available for bonding on the cluster surface and, in any event appears to be a more sensitive probe than *e.g.* the CO stretching vibration in carbonyl clusters.

A similar variation is noted for the ^1H resonances of the two rings in face-capping derivatives of cyclophane (*see below*) [20]. In this case the degeneracy of the two rings is removed on coordination and the chemical shift observed at $\delta 6.51$ for the parent is seen to separate. As the number of metals increases from 3 to 6 the separation also increases; for $Ru_3(CO)_9(\mu_3-\eta^2:\eta^2:\eta^2-C_{16}H_{16})$ the values are $\delta 3.76$ and 7.22 ppm ($\Delta = 3.46$) and for $Ru_6C(CO)_{14}(\mu_3-\eta^2:\eta^2:\eta^2-C_{16}H_{16})$ the values are $\delta 3.43$ and 7.44 ppm ($\Delta = 4.01$). In this case the synergic bonding between the bonded ring and the cluster in turn affects the synergic bonding between the two rings through their p_π interactions. This is a particularly sensitive probe of the cluster surface and it would be interesting to examine the interactions of cyclophane with the bulk surface.

2.3 Bonding as a Function of Ligand Type.

Our investigations here have been limited to date. Undoubtedly the presence of CO on the cluster surface affects the nature and type of bonding mode adopted by the arene. In one key experiment we have observed that reaction of the triangulo-clusters $M_3(CO)_9(C_6H_6)$ (M = Ru or Os) with alkynes R_2C_2 leads to the displacement of the benzene to a terminally bonded mode and the concomitant bond formation of the alkyne in the familiar $(2\sigma,-\pi)$ mode across the triangular face to form the complex $M_3(CO)_7(R_2C_2)(C_6H_6)$ [16]. For the series of compounds $Os_3(CO)_8(L)(C_6H_6)$ (L = C_2H_4, PR_3, *etc*) the barrier of the (C_6H_6) to rotation is found to vary but not by very much [21]. At present we are studying the variation in the ^1H chemical shift for a series of compounds in which a CO ligand is replaced by a series of phosphine ligands of different basicity.

2.4. Migratory Patterns of the Arene Ligand over the Cluster Surface.

As we have commented previously, a range of different isomeric forms have been observed for the series of *bis*-substituted compounds $Ru_6C(CO)_{11}(C_6H_6)_2$ [7] (*Scheme* 3) and the *mono*-substituted system $Ru_5C(CO)_{12}(C_6H_6)$ [17]. In each case almost every possible isomeric form is observed (except the *bis* - μ_3 - form!) for $Ru_6C(CO)_{11}(C_6H_6)_2$. Interconversion is readily monitored spectroscopically but at no stage have stable forms been observed in which the arene straddles the edge. Nevertheless, we believe this to be the most likely intermediate (*see above*).

3. The Photochemistry of Arene Clusters and C-H Cleavage Reactions.

We have observed that on irradiation of solutions of $[Os_3(CO)_9(C_6H_6)]$ in toluene, the benzyne derivative $[H_2Os_3(CO)_9(C_6H_4]$ is produced. Interest in this reaction centres around the activation of the two adjacent ring C-H bonds. Although the mechanism by which the photochemical reaction occurs has not been fully established, the accepted view is that the Os-Os bond cleavage ($\sigma \to \sigma^*$) occurs as the initial step. Although related work has not been carried out for other triosmium arene compounds containing C_6H_5Me, $C_6H_4Me_2$, $C_6H_3Me_3$, *etc*., we have established that the direct reaction of a wide range of arenes with $Os_3(CO)_{10}(MeCN)_2$ generates the appropriate benzyne

328

derivatives [22]. No evidence for the intermediate formation of μ_3- arene intermediate compounds has been found and it is probable that reaction takes place in these instances *via* a μ_2- bonded intermediate. Evidence for a reaction of this kind has been found in our related studies of the reaction of cyclohexene with $H_4Os_4(CO)_{12}$ or $[H_4Os_4(CO)_{10}(MeCN)_2]$. Here, totally unexpectedly, cleavage of one of the 'olefin' C-H bonds to generate the cyclohexaenyl derivatives $H_3Os_4(CO)_9(C_6H_9)$, which undergoes a second C-H bond cleavage on heating to form the 'yne' derivatives $[H_2Os_4(CO)_9-C_6H_8)]$, in which the C_6H_8 moiety spans the tetrahedral Os_4-face. Related studies of the corresponding ruthenium systems have led to the isolation of a series of yne-like complexes based on the $Ru_4(CO)_{12}(C_6H_8)$ unit but, in this case, presumably because of the weaker Ru-Ru bonds, the yne-unit straddles a butterfly Ru_4 arrangement [23]. Again, olefinic C-H bond cleavage has occurred in preference to methylene CH_2 bond cleavage.

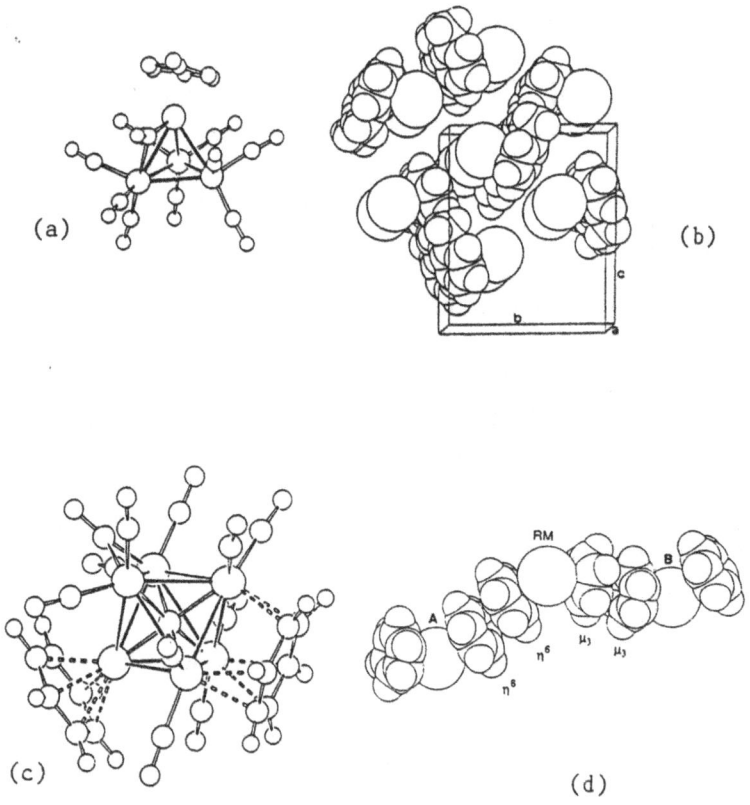

Figure 4 (a) Molecular structure and (b) packing of $H_2Os_4(CO)_{10}(\eta\text{-}C_6H_6)$. (c) molecular structure and (d) packing of $Ru_6C(CO)_{11}(\eta\text{-}C_6H_6)(\mu_3\text{-}\eta^2{:}\eta^2{:}\eta^2\text{-}C_6H_6)$

4. Graphitic-like Interactions in the Solid State.

Many of the compounds reported in this article have been the subject of detailed analysis by X-ray diffraction studies on single crystals. Several points of interest emerge. For several of the *mono*-substituted derivatives, based on Os_4 and Ru_6C units with simple arenes, the organic rings are observed to interlock to produce layers of organic substrate and layers of the cluster units (*Figure 4*) [24]. In the *bis*-substituted derivatives of Ru_6C the various different isomeric forms give rise to different crystallographic packing arrangements. Thus, in *trans* $Ru_6C(CO)_{11}(\eta^6-C_6H_3Me_3)_2$ and *trans*-$Ru_6C(CO)_{11}(\eta^6-C_6H_6)_2$, the rings from adjacent molecules in the crystal are superimposed *albeit* with some displacement and the inter-ring distances are compatible with those observed in graphite. In these examples, interaction between the rings occurs throughout the solid, leading to molecular 'chains'. In the related system *cis*-$Ru_6C(CO)_{11}(\eta^6-C_6H_6)_2$, the interactions are similar but because of the distribution of the arenes about the Ru_6C cluster unit the 'graphitic-like' interactions lead to molecular 'snakes' throughout the solid. For the system $Ru_6C(CO)_{11}(\eta^6-C_6H_6)(\mu_3-C_6H_6)$, the η^6- rings on adjacent molecules in the solid interact in a graphitic-like manner as do the adjacent μ_3- ring systems. This leads to molecular 'ribbons'. These interactions are of fundamental importance, particularly since they provide a mechanism or route by which charge transfer may occur throughout the lattice in well designated pathways. In an effort to extend the studies of this phenomenon we are currently investigating the formation of arene clusters containing much large polycyclic systems. It is our intention to produce organometallic cluster materials in which ring-ring interactions are maximised. To date we have prepared $Ru_3(CO)_9(C_{12}H_{10})$ which shows the desired effect [25].

5. Monomer and Dimer Precursors to Conducting Polymers.

Following on from these studies (*see Section 3*) we have synthesised a whole range of cyclophane cluster derivatives [20]. At the onset of these studies it was our intention to produce monomers with one cyclophane bonded to the central unit and then dimers in which the cyclophane moiety is bonded simultaneously to two separate cluster units and thereby serving as a bridge allowing each separate cluster unit to exchange electron density. Ultimately, our goal is to generate polymeric chains of such materials which hopefully will serve as conductors or semi-conductors. We are also interested in producing clusters containing two cyclophane units which pack in chains, snakes or ribbons as with the simpler arenes, but through which charge transfer may be mediated. Examples of cyclophane clusters we have prepared and characterised are shown in *Figure 5*. Several features appear. First, the cyclophane ligand shows a marked tendency to function as a face cap rather than the η^6-terminal group. However, this is not always the case and our attempts to prepare $Ru_6C(CO)_{11}(C_{16}H_{16})_2$, the first cluster with two face-capping rings, have resulted in the η^6-, μ_3-complex. Secondly, as commented above, the ^1H NMR of the coordinated $C_{16}H_{16}$ ligand shows highly significant ^1H chemical shift values which appear to be highly sensitive to the number of metal atoms within the cluster. Finally, we have observed that in the compound $Ru_6C(CO)_{12}(\mu_3\text{-}C_{16}H_{16})(\mu_2\text{-}C_6H_8)$, it is the C- atoms of the coordination ring that sit directly above the metal atoms of the triangular face and not, as in all other cases, the C=C bond. Significantly, major changes in the configuration of the cyclophane rings are also observed.

In related studies we have prepared derivatives of linked arene systems such as $Ph(CH_2)_nPh$ (n = 0, 1 2) and PhCH=CH-CH=CH-CH=CH-CH-CHPh [26]. Again, our objective is simply to produce cluster systems linked by an organo bridge and ultimately to produce organometallic cluster materials. The molecular structures of the series of compounds $Ru_6C(CO)_{14}(Ph(CH_2)_nPh)$ (n=0,1,2) have been established and two are shown in *Figure 6*.

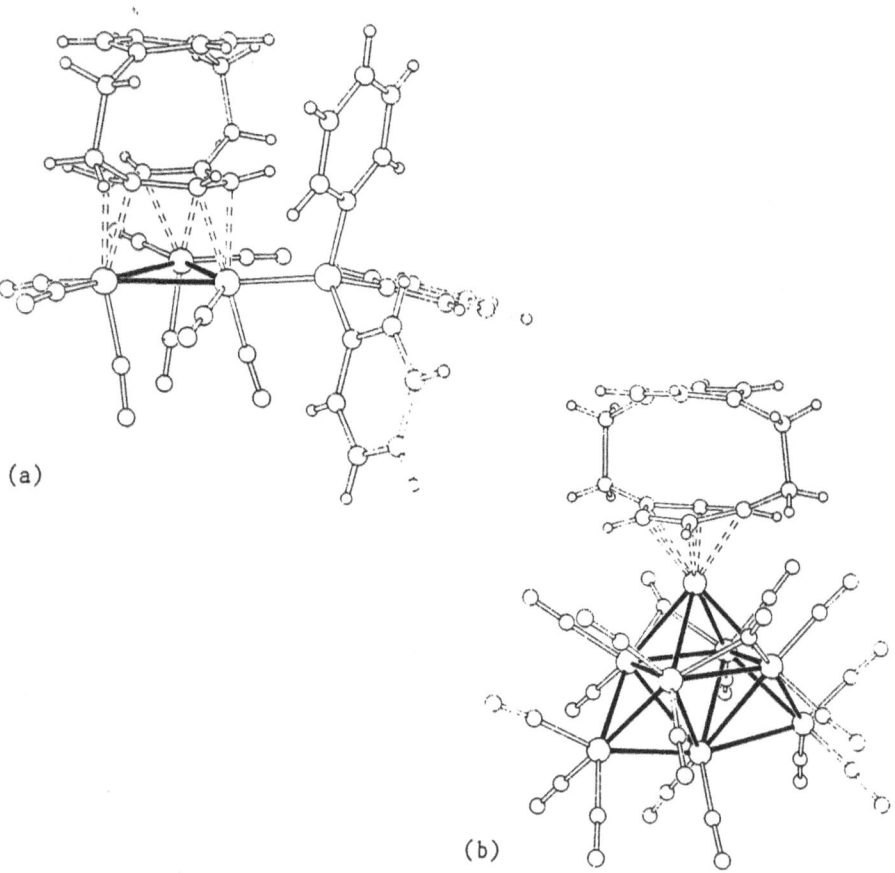

(a)

(b)

Figure 5 Molecular structure of (a) $Ru_3(CO)_8(\mu_3\text{-}\eta^2\text{:}\eta^2\text{:}\eta^2\text{-}C_{16}H_{16})$ and (b) $H_4Ru_8(CO)_{18}(\eta\text{-}C_{16}H_{16})$.

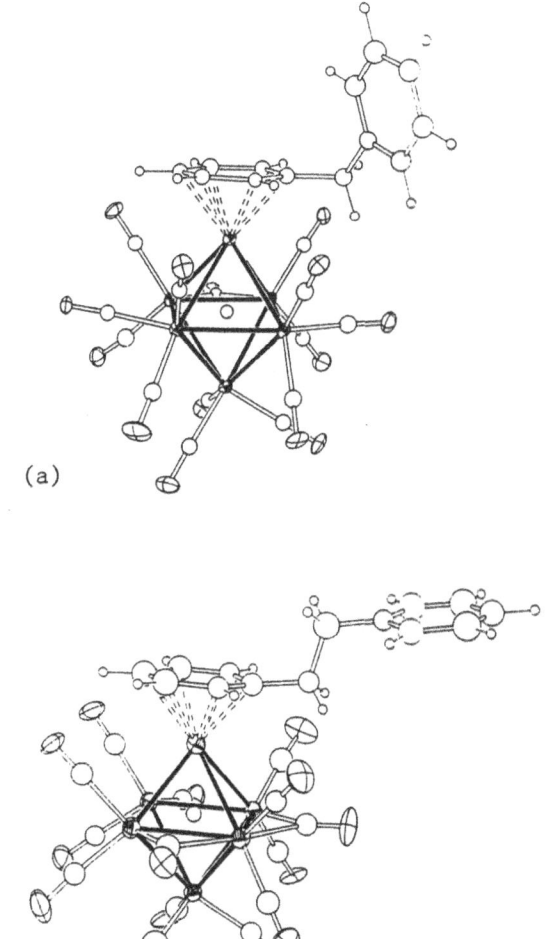

Figure 6 Molecular structure of (a) Ru$_6$C(CO)$_{14}$(C$_6$H$_5$.CH$_2$.C$_6$H$_5$) and (b) Ru$_6$C(CO)$_{14}$(C$_6$H$_5$.C$_2$H$_4$.C$_6$H$_5$)

332

6. References.

1a.   Johnson, B.F.G., Johnston, R.D. and Lewis, J. (1967) , *J. Chem. Soc., Chem. Comm.* 1057.

1b.   Johnson, B.F.G., Johnston, R.D. and Lewis, J. (1968) *J. Chem. Soc. (A)* 2865 - 2868

2.   Gomez-Sal, M.P, Johnson, B.F.G., Lewis, J, Raithby, P.R. and Wright, A.H. (1985) . *J. Chem. Soc. Chem. Comm.* 1682-1683.

3.   Somorjai, G.A. (1990) *J. Phys. Chem.* **94**, 1013-1023.

4.   Muetterties, E.L., Rhodin, T.N., Band, E., Brucker, C and Pretzer, H. (1979) , *Chem. Rev.* **79**, 91-137 and references therein.

5.   Blake, A.J., Dyson, P.J., Johnson, B.F.G., Martin, C.M., Nairn, J.G.M., Parisini, E. and Lewis, J. (1993) *J. Chem. Soc. Dalton Trans.* 981-984.

6a.   Johnson, B.F.G., Lewis, J., Martinelli, M., Wright, A.H., Braga, D and Grepioni, F. (1990) *J. Chem. Soc., Chem. Comm.* 364-365.

6b.   Braga, D., Grepioni, F., Johnson, B.F.G., Lewis, J., Housecroft, C.E. and Martinelli, M. (1991) *Organometallics* **10**, 1260-1268.

7.   Dyson, P.J., Johnson, B.F.G., Lewis, J., Martinelli, M., Braga, D., and Grepioni, F.(1993). *J. Am. Chem. Soc.* **115**, 9062-9068.

8.   Johnson, B.F.G., Blake, A.J., Martin, C.M., Braga, D., Parisini, E. and Chen, H. (in press) *J. Chem. Soc., Dalton Trans.*

9.   Khand, I.V., Knox, G.R., Pauson, P.L. and Watts, W.E. (1973) *J. Chem. Soc., Perkin I* 975-977.

10.   Braga, D., Grepioni, F., Parisini, E., Dyson, P.J., Blake, A.J. and Johnson, B.F.G. (1993) *J. Chem. Soc., Dalton Trans.* 2951-2957.

11.   Dyson, P.J., Johnson, B.F.G., Reed, D., Braga, D., Grepioni, F. and Parisini, E. (1993) *J. Chem. Soc., Dalton Trans.* 2817-2825.

12.   Müller, J., Gaede, P.E. and Qiao, K. (1993) , *Angew. Chem. Int. Ed. Engl.* **32**, 1697-1699.

13.   Blake, A.J., Dyson, P.J., Johnson, B.F.G. and Martin, C.M. (in press), , *J. Chem. Soc., Chem. Comm.*

14.   Gallop, M.A., Johnson, B.F.G., Lewis, J., McCamley, A. and Perutz, R.N. (1988) *J. Chem. Soc., Chem. Comm.* 1071-1072.

15.   Dyson, P.J. and Johnson, B.F.G. The University of Edinburgh. Unpublished results.

16a.   Braga, D., Grepioni, F., Johnson, B.F.G., Lewis, J., Martinelli, M. and Gallop, M.A. (1990) *J. Chem. Soc., Chem. Comm.* 53-55.

16b.   Braga, D., Grepioni, F., Johnson, B.F.G., Parisini, E., Martinelli, M., Gallop, M.A. and Lewis, J. (1992), *J. Chem. Soc., Dalton Trans.* 807-812.

17.   Braga, D., Grepioni, F., Sabatino, P., Dyson, P.J., Johnson, B.F.G., Lewis, J., Bailey, P.J., Raithby, P.R. and Stalke, D. (1993) *J. Chem. Soc., Dalton Trans.* 985-992.

18. Braga, D., Grepioni, F., Prisini, E., Dyson, P.J., Johnson, B.F.G., Reed, D., Shephard, D.S., Bailey, P.J. and Lewis, J. (1993) *J. Organomet. Chem.* **462**, 301-308.

19. Johnson, B.F.G., Martin, C.M., Braga, D., Grepioni, F. and Parisini, E., (1994) *J. Chem. Soc., Chem. Comm.* **10**, 1253-1254.

20. Dyson, P.J., Johnson, B.F.G., Martin, C.M., Blake, A.J., Braga, D. and Parisini, E. (1994) *Organometallics*, **13**, 2113-2117.

21. Gallop, M.A., Gomez-Sal, M.P., Raithby, P.R. and Wright, A.H. (1992) *J. Am. Chem. Soc.* **114**, 2502-2510.

22. Johnson, B.F.G., Nairn, J.G.M., Brown, D.B., Lewis, J., Gallop, M.A. and Parker, D.G. (in preparation)

23. Martin, C.M. and Johnson, B.F.G. The University of Edinburgh. Unpublished results.

24. Braga, D., Grepioni, F., Dyson, P.J. and Johnson, B.F.G. (1992) *Journal of Cluster Science*, **3**, 297-311 and references therein.

25. Nairn, J.G.M. and Johnson, B.F.G. The University of Edinburgh. Unpublished results.

26. Shephard, D.S. and Johnson, B.F.G. The University of Edinburgh. Unpublished results.

SILVER AND GOLD CLUSTERS STABILIZED BY Fe(CO)$_4$ LIGANDS

Francesca Calderoni, Maria Carmela Iapalucci, Giuliano Longoni and
Umberto Testoni
University of Bologna
Dipartimento di Chimica Fisica ed Inorganica, Viale del Risorgimento 4
40136, Bologna, Italy

1. Summary

The reaction pathway of the oxidation of the [Fe(CO)$_4$]$^{2-}$ dianion with Ag(I) and Au(I) salts to Fe(0) species, namely Fe(CO)$_5$, Fe$_3$(CO)$_{12}$ and iron metal, involves the intermediate formation of several Ag-Fe and Au-Fe clusters, which may be selectively obtained by using substoichiometric molar ratios (comprised in the 0.33-1.75 range) between the reagents. On the basis of isolobal analogies, all the compounds which have been structurally characterized so far may be viewed as clusters of Group 11 elements stabilized by d8-Fe(CO)$_4$ or bidentate Fe$_2$(CO)$_8$ ligands, in their several possible carbonyl stereochemistries. Rationalization of the miscellaneous steps of the above reactions enabled the synthesis of related heterometallic Cu-Ag and Au-Ag clusters. Their spectroscopic and structural behaviour are compared with those of previously reported Cu, Pd and Pt related compounds, as well as the corresponding Cu, Ag and Au cluster derivatives stabilized by d9-Co(CO)$_4$ groups.

2. Introduction

A large number of Fe-M (where M is either a transition metal or a post-transition element) clusters, containing either C$_{3v}$ or C$_{2v}$-Fe(CO)$_4$ groups is known [1-6]. A few structurally characterized examples for Group 10-15 elements, which may be assumed to be representative of the miscellaneous bonding behaviour of the Fe(CO)$_4$ fragment, are collected in TABLES 1, 2 and 3, together with their references.

During the chemical characterization of some Fe-P and Fe-In anionic clusters [7,8], we observed that their oxidation with AgBF$_4$ could produce trace amounts of an identical paramagnetic by-product showing a complex EPR pattern due to hyperfine couplings with several nuclei [9]. It appeared probable, therefore, that a common Ag-Fe derivative could be at hand and we began a systematic exploration of the reactions of the iron carbonyl anions with silver salts. As a result of that original aim, we have isolated and characterized a wide series of new Ag-Fe clusters [9-12], and more recently we have extended our investigations to Au-Fe [13,14] and tri-metallic Ag-Cu-Fe and Ag-Au-Fe derivatives.

L. J. Farrugia (ed.), The Synergy Between Dynamics and Reactivity at Clusters and Surfaces, 335–349.
© 1995 *Kluwer Academic Publishers.*

TABLE 1. Some Representative Fe-M Clusters Containing Terminal C_{3v} $Fe(CO)_4$ Ligands

Compound	Reference
Linear 2-fold Coordination	
$[M\{Fe(CO)_4\}_2]^{2-}$ (M=Zn,Cd,Pb)	15,16
Trigonal-planar 3-fold Coordination	
$[In\{Fe(CO)_4\}_3]^{3-}$	8
$[M\{Fe(CO)_4\}_3]^{2-}$ (M=Sn,Pb)	17,18
Homoleptic Tetrahedral Coordination	
$[Sb\{Fe(CO)_4\}_4]^{3-}$	19
$[Bi\{Fe(CO)_4\}_4]^{3-}$	20
Heteroleptic Tetrahedral Coordination	
$[Cl_2Sn\{Fe(CO)_4\}_2]^{2-}$	21
$[Bu^iBi\{Fe(CO)_4\}_3]^{2-}$	22
$[HAs\{Fe(CO)_4\}_3]^{2-}$	23

TABLE 2. Some Representative Clusters of Elements of Group 10-14 Stabilized by Bridging C_{2v} μ_2-$Fe(CO)_4$ Ligands

Compound	Reference
Dinuclear	
$[Pt_2\{\mu_2\text{-}Fe(CO)_4\}(CO)\{P(OPh)_3\}_3]$	24
$[Au_2\{\mu_2\text{-}Fe(CO)_4\}(PPh_3)_2]$	25
$[Au_2\{\mu_2\text{-}Fe(CO)_4\}(dppm)]_2$	26
$[In_2\{\mu_2\text{-}Fe(CO)_4\}_2Br_4]^{2-}$	8
$[Sn_2\{\mu_2\text{-}Fe(CO)_4\}_2\{Fe(CO)_4\}_2Br_2]^{2-}$	27
Trinuclear	
$[Pt_3\{\mu_2\text{-}Fe(CO)_4\}_3(CO)_3]^{n-}$ (n=0,1,2)	28,29,30
$[Cu_3\{\mu_2\text{-}Fe(CO)_4\}_3]^{3-}$	31
Tetranuclear	
$[Cd_4\{\mu_2\text{-}Fe(CO)_4\}_4(Me_2CO)_2]$	32
Hexanuclear	
$[Pt_6\{\mu_2\text{-}Fe(CO)_4\}_4(CO)_6]^{2-}$	28

TABLE 3. Some Representative Clusters of Elements of Group 10 and 11, Stabilized by Bridging C_{2v}, C_{3v} or T_d μ_3-$Fe(CO)_4$ Ligands

Compound	Reference
C_{2v}-$Fe(CO)_4$	
$[Cu_5\{\mu_2\text{-}Fe(CO)_4\}_2\{\mu_3\text{-}Fe(CO)_4\}_2]^{3-}$	31
C_{3v}-$Fe(CO)_4$	
$[Cu_6\{\mu_3\text{-}Fe(CO)_4\}_4]^{2-}$	33
$[Ag_6\{\mu_3\text{-}Fe(CO)_4\}_3\{HC(PPh_2)_3\}]$	34
T_d-$Fe(CO)_4$	
$[HPd_6\{\mu_3\text{-}Fe(CO)_4\}_6]^{3-}$	35

3. Results and Discussion

3.1. THE INITIAL STEPS OF THE OXIDATION OF $[Fe(CO)_4]^{2-}$ WITH Ag^+ AND Au^+ IONS.

The reaction of the $[Fe(CO)_4]^{2-}$ dianion with Cu^+, Ag^+ and Au^+ salts has been investigated mainly in THF as solvent, using molar ratios between the reagents in the range 0.33-2. Whereas the counteranions of Cu^+ and Ag^+ do not play any significant role (the anions Cl^-, NO_3^- and BF_4^- have been routinely employed), the nature of the Au(I) reagent is more critical, owing to its often ready disproportionation to Au(III) salts and gold metal. After examining complexes of AuCl such as Au(CO)Cl, $Au(SEt_2)Cl$, and also $[AuBr_2]^-$ salts as potential starting materials, only the latter has been routinely employed owing to its better performance.

The three Group 11 ions display significant differences even in the early stages of reaction. Thus, the first product detectable by IR in the reaction of $[Fe(CO)_4]^{2-}$ with both Cu^+ and Ag^+ is the $[M\{Fe(CO)_4\}_2]^{3-}$ trianion. This is quantitatively obtained according to equation (1), upon addition of 0.5 equivalents of M^+ [11]. In contrast, as is evident from IR monitoring, the corresponding reaction with $[AuBr_2]^-$ salts is more complicated, and probably involves the intermediate formation of $[Au\{Fe(CO)_4\}_{3-x} Br_x]^{(5-x)-}$ species. Of these $[Au\{Fe(CO)_4\}_3]^{5-}$ separates out upon addition of ca. 0.3 equivalents as a white microcrystalline precipitate from the reaction solution (equation 2), owing to the low solubility in THF of its sodium salt. The further addition of 0.2 equivalent of $[AuBr_2]^-$ to the above suspension results in the formation of the corresponding trianion $[Au\{Fe(CO)_4\}_2]^{3-}$ according to reaction (3):

$$2 [Fe(CO)_4]^{2-} + M^+ \rightarrow [M\{Fe(CO)_4\}_2]^{3-} \quad (M=Cu,Ag) \qquad (1)$$

$$3 [Fe(CO)_4]^{2-} + [AuBr_2]^- \rightarrow [Au\{Fe(CO)_4\}_3]^{5-} + 2 Br^- \qquad (2)$$

$$2 [Au\{Fe(CO)_4\}_3]^{5-} + [AuBr_2]^- \rightarrow 3[Au\{Fe(CO)_4\}_2]^{3-} + 2 Br^- \qquad (3)$$

The sodium salts of the $[M\{Fe(CO)_4\}_2]^{3-}$ (M=Cu,Ag,Au) trianions have been precipitated from their THF solutions by addition of 15-crown-5 [11,14]. All these salts have elemental analyses in keeping with the given formulations and their infared spectra are collected in TABLE 4. Although their X ray structures were not determined, it appears rather likely that the $[M\{Fe(CO)_4\}_2]^{3-}$ (M=Cu,Ag,Au) trianions adopt a linear structure, such as that displayed by the isoelectronic $[Hg\{Fe(CO)_4\}_2]^{2-}$ [16] and $[M\{Co(CO)_4\}_2]^-$ (M=Cu,Au) [36,37]. A trigonal planar arrangement is suggested for $[Au\{Fe(CO)_4\}_3]^{5-}$, similar to that shown by the isoelectronic $[In\{Fe(CO)_4\}_3]^{3-}$ [8] and $[M\{Fe(CO)_4\}_3]^{2-}$ (M=Sn,Pb) [17,18] anions.

The solid state structures of the above clusters contain $Fe(CO)_4$ fragments which display a C_{3v} carbonyl stereochemistry. The $Fe(CO)_4$ fragment (which is formally derived from the trigonal bipyramidal $Fe(CO)_5$ by loss of an axial ligand) becomes, on addition of two electrons as negative charges, a 2-electron σ-donor analogous to Cl^-. Thus, the $[Au\{Fe(CO)_4\}_3]^{5-}$ and $[M\{Fe(CO)_4\}_2]^{3-}$ (M=Cu,Ag,Au) may be considered as the counterparts of the $[MX_3]^{2-}$ and $[MX_2]^-$ complexes.

3.1.1. The $[M\{Fe(CO)_4\}_2]^{3-}$ (M=Cu,Ag,Au) Trianions as Ligands.

As previously noted, the bonding capability of a $Fe(CO)_4$ fragment changes as a function of the carbonyl stereochemistry [34,38,39]. When the carbonyls approach a C_{2v} or a D_{4h} stereochemistry (respectively derived from a cis or trans coordinately-divacant octahedron), the $Fe(CO)_4$ fragments in the species $[M\{Fe(CO)_4\}_2]^{n-}$ (M=Cu,Ag,Au,Zn,Cd and Hg) have available for bonding a further filled σ orbital. Therefore, the above linear clusters may be respectively considered as potential analogs of bidentate ligands such as $Ph_2P-CH_2-PPh_2$ (dppm) and 1,4-dicyanobenzene. It appeared, therefore, possible that further addition of M^+ ions to $[M\{Fe(CO)_4\}_2]^{3-}$ (M=Cu,Ag,Au) could result either in cyclic clusters $[M_4\{\mu_2-Fe(CO)_4\}_4]^{4-}$ such as $[Cd_4\{\mu_2-Fe(CO)_4\}_4]$ [32], zig-zag chain polymers $[M_n\{\mu_2-Fe(CO)_4\}_n]^{n-}$ or $[M_n\{\mu_2-Fe(CO)_4\}_{n+1}]^{(n+2)-}$ such as found for one of the two modifications of $[Cu\{Co(CO)_4\}]_n$ [40,41], as well as linear chain polymers, as partially exemplified by the linear $[Cu_2\{\mu_2-Fe(CO)_4\}(PPh_3)_4]$ [33].

3.2. THE INTERMEDIATE STEPS OF THE OXIDATION OF $[Fe(CO)_4]^{2-}$ WITH Ag^+ AND Au^+ IONS.

Partially in keeping with the above suggestions, both $[Ag\{Fe(CO)_4\}_2]^{3-}$ and $[Au\{Fe(CO)_4\}_2]^{3-}$ react with one equivalent of Ag^+ and Au^+, respectively, to afford quantitatively the new $[Ag_4\{\mu_2-Fe(CO)_4\}_4]^{4-}$ and $[Au_4\{\mu_2-Fe(CO)_4\}_4]^{4-}$ cyclic clusters [11,14]. The two possible cross reactions gave rise to the identical product $[Ag_2Au_2\{\mu_2-Fe(CO)_4\}_4]^{4-}$. It should be noted that reaction (4) is at variance with the corresponding reaction of $[Cu\{Fe(CO)_4\}_2]^{3-}$ with Cu^+, which only affords the previously reported [31] $[Cu_3\{\mu_2-Fe(CO)_4\}_3]^{3-}$ trimer.

$$2\ [M\{Fe(CO)_4\}_2]^{3-} + 2\ M^+ \rightarrow [M_4\{\mu_2-Fe(CO)_4\}_4]^{4-}\ (M=Ag,Au) \qquad (4)$$

The nature of the products from the above reactions has been ascertained by full X ray structural investigations or by molecular weight determinations following unit cell and density measurements of miscellaneous crystals [11,14]. The structure of $[Ag_4\{\mu_2-Fe(CO)_4\}_4]^{4-}$ is shown in *Figure 1*. Interestingly, the unit cell of the corresponding $[Au_4\{\mu_2-Fe(CO)_4\}_4]^{4-}$ salt contains two structural modifications of the tetraanion. The first is very close to that of $[Ag_4\{\mu_2-Fe(CO)_4\}_4]^{4-}$, while the second represents a snapshot of the deformation depicted by sketch (B) in *Figure 2*, owing to a significant elongation of two opposed Au-Au interatomic contacts.

The presence in the unit cell of two structural modifications of the $[Au_4\{\mu_2-Fe(CO)_4\}_4]^{4-}$ molecular ion, as well as the two distinct sets of Au-Au and Au-Fe interatomic separations found in its second modification, suggests that the M-M bonds are probably weak in comparison to the M-Fe interactions. Thus, the loss in M-M bond energy upon elongation may apparently be intramolecularly compensated by a concomitant shortening of the energetically more favourable M-Fe bonds. Nevertheless, the inward displacement of the M atoms systematically observed in both $[M_4\{\mu_2\text{-Fe(CO)}_4\}_4]^{4-}$ (M=Ag,Au) and $[M_4\{\mu_2\text{-Co(CO)}_4\}_4]$ (M=Cu,Ag) [36,41], and which is absent in $[Cd_4\{\mu_2\text{-Fe(CO)}_4\}_4]\cdot 2Me_2CO$ [32], can be taken as a first hint of the presence of an attractive component. Secondly, in all the above cyclic clusters the M-M interatomic parameters are only slightly longer than the corresponding ones in the bulk metals.

The possible factors concerning the origin of a M-M bonding component in Group 11 derivatives have been thoroughly discussed [3,6,11,42,43], including the reasons why gold often displays shorter M-M contacts than silver [44,45]. We shall comment here only on the diversity of molecular complexity of the above $\{[MFe(CO)_4]^-\}_n$ (M=Ag and Au, n=4; M=Cu, n=3) congeners. The adoption of an octa-ring structure by $[M_4\{\mu_2\text{-Fe(CO)}_4\}_4]^{4-}$ (M=Ag,Au), *versus* a hexa-ring for $[Cu_3\{\mu_2\text{-Fe(CO)}_4\}_3]^{3-}$, could arise from the fact that the former allows a better linearity of the Fe-M-Fe axis as the M-M/Fe-M bond ratio increasingly departs from 1. Indeed, while the Cu-Cu/Fe-Cu ratio in $[Cu_3\{\mu_2\text{-Fe(CO)}_4\}_3]^{3-}$ is 1.07 [31], the corresponding ratio in the octa-ring $[M_4\{\mu_2\text{-Fe(CO)}_4\}_4]^{n-}$ (M=Ag and Au, n=4; M=Cd, n=0) and $[M_4\{\mu_2\text{-Co(CO)}_4\}_4]$ (M=Cu,Ag) is in the range 1.11-1.28.

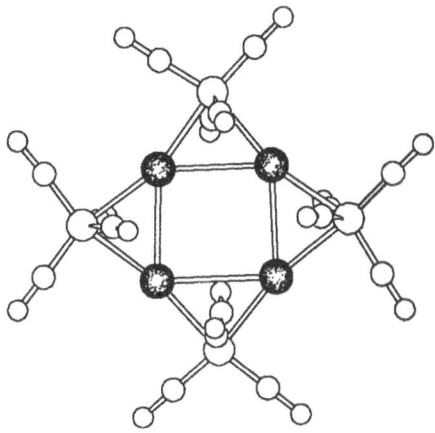

Figure 1. X ray Structure of $[Ag_4\{\mu_2\text{-Fe(CO)}_4\}_4]^{4-}$ [11]

In all the above reactions there is no unambiguous evidence for the formation of zig-zag or linear chain oligomeric intermediates with formulae as suggested above, prior to the final 1:1 molar ratio between the $[Fe(CO)_4]^{2-}$ and M^+ starting materials being fulfilled. Although alternate zig-zag or linear chain polymeric structures could be favoured by solubility considerations because of the progressive possibility of separation into the solid state as the charge and/or the molecular complexity increases, it is conceivable that the formation of linear chains is thermodynamically disfavoured because of the partial or complete sacrifice of the enthalpic contribution arising from the weak, but nevertheless significant, M-M interactions.

It was thought possible that oligomerization of $[M\{Fe(CO)_4\}_2]^{3-}$ trianionic species with Group 12 dications could be more feasible. Indeed, $[Ag\{Fe(CO)_4\}_2]^{3-}$ reacts with one equivalent of Cd^{2+} in a two-steps reaction, which affords the $[Ag_2Cd\{Fe(CO)_4\}_4]^{4-}$ tetraanion after addition of 0.5 equivalents, and the $[Ag_2Cd_2\{Fe(CO)_4\}_4]^{2-}$ dianion as the final product. Unfortunately, attempts to override the problems arising from the similarity of X ray scattering power of the adjacent metals Ag and Cd, by adopting alternate Ag-Hg and Au-Cd combinations have been so far proved unsuccessful.

3.2.1 The $[M_4\{\mu_2\text{-}Fe(CO)_4\}_4]^{4-}$ Clusters as Ligands.

In the $[M_4\{\mu_2\text{-}Fe(CO)_4\}_4]^{4-}$ clusters the $Fe(CO)_4$ moieties have not yet come to the end of their bonding capability. Indeed, a distortion of the metal frame according to the sketches (A) and (B) of *Figure 2*, will transform these clusters into potential 8-crown-4 and 8-crown-2 ligands. This primarily arises from the fact that by widening the M-Fe-M angle from 90 up to 144°, a filled in-plane σ-orbital directed toward the centre of the ring will become available on such an $Fe(CO)_4$ fragment, which has a geometry formally derived from a pentagonal bipyramid by removal of three adjacent equatorial sites. Such an $Fe(CO)_4$ fragment is therefore a potential 4-electron donor, and this may probably compensate for the loss of four or two M-M interactions, respectively. Of the two envisioned distortions, the one depicted in (B) seems the most probable since it does not require the sacrifice of the Fe-M-Fe bond linearity. It is indeed rewarding that the second crystalline modification of $[Au_4\{\mu_2\text{-}Fe(CO)_4\}_4]^{4-}$ almost perfectly conforms to sketch (B) of *Figure 2*.

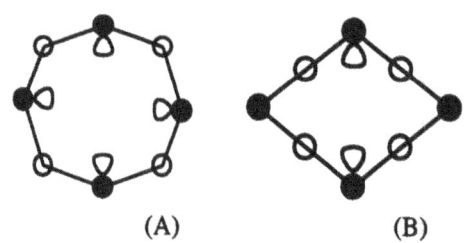

(A)   (B)

Figure 2.
Possible distortions of the $[M_4\{\mu_2\text{-}Fe(CO)_4\}_4]^{4-}$
clusters which make available in-plane σ-orbitals
directed toward the centre (Filled circles represent
the $Fe(CO)_4$ fragments; open circles indicate the
Ag or Au atoms).

In keeping with this analysis of the $[M_4\{\mu_2\text{-Fe(CO)}_4\}_4]^{4-}$ clusters, these react quantitatively with one equivalent of M'^+ to give $[M_4(\mu_6\text{-M'})\{Fe(CO)_4\}_4]^{3-}$ crown complexes, according to equation (5). The compounds so far isolated are reported in TABLE 4, while the X ray structure of $[Ag_4(\mu_6\text{-Ag})\{\mu_2\text{-Fe(CO)}_4\}_2\{\mu_3\text{-Fe(CO)}_4\}_2]^{3-}$ is shown in *Figure 3*.

$$[M_4\{\mu_2\text{-Fe(CO)}_4\}_4]^{4-} \quad + \quad M'^+ \rightarrow \quad [M_4(\mu_6\text{-M'})\{Fe(CO)_4\}_4]^{3-} \qquad (M=Ag,Au;$$
$M'=Cu,Ag,Au)$ (5)

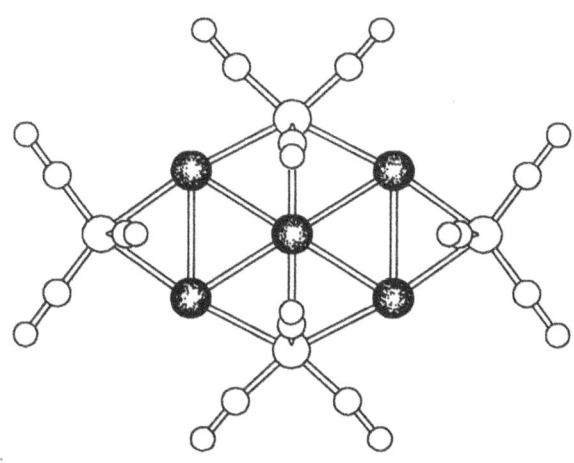

Figure 3. X ray structure of $[Ag_5\{Fe(CO)_4\}_4]^{3-}$ [11]

Completely identical $[Cu_5\{Fe(CO)_4\}_4]^{3-}$ derivatives were previously isolated from the reaction of $[Cu_3\{Fe(CO)_4\}_3]^{3-}$ with CuCl and structurally characterized [31]. The remaining species $[M_4(\mu_6\text{-M'})\{Fe(CO)_4\}_4]^{3-}$ (M=Ag, M'=Cu,Au; M=Au, M'=Au) were not characterized by X ray diffraction, in view of their very similar IR patterns (see TABLE 4). Owing to this, we cannot yet assume that the M' atom has really centred the crown in all cases.

In contrast, all attempts to use the $[Ag_4\{Fe(CO)_4\}_4]^{4-}$ cluster as a crown ligand for dications such as Pt^{2+} and Sn^{2+} have been so far proved unsuccessful owing to competitive formation of the known $[Pt\{\eta^2\text{-Fe}_2(CO)_8\}_2]^{2-}$ [35] or mixtures of $[Sn\{\eta^2\text{-Fe}_2(CO)_8\}_2]$ [46] and $[Sn\{\eta^2\text{-Fe}_2(CO)_8\}\{Fe(CO)_4\}_2]^{2-}$ [47], rather than $[Ag_4(\mu_6\text{-M'})\{Fe(CO)_4\}_4]^{2-}$ (M'=Pt,Sn) derivatives.

3.2.2. *The $[M_4(\mu_6\text{-M'})\{Fe(CO)_4\}_4]^{3-}$ Clusters as Ligands.*
In the $[M_4(\mu_6\text{-M'})\{Fe(CO)_4\}_4]^{3-}$ clusters the two $\mu_3\text{-Fe(CO)}_4$ fragments came to an end of their bonding capability, but the two $\mu_2\text{-Fe(CO)}_4$ can still make a third filled σ-orbital available upon rearrangement of the carbonyl stereochemistry from C_{2v} to C_{3v}

and tilting of the carbonyl *umbrella* outside the plane of the metal skeleton. Therefore, they remain potentially bidentate ligands. Two different reaction pathways may be reasonably envisioned to occur on addition of M^+ ions. As depicted in sketch (B) of *Figure 4*, upon bending of the $[M_5\{Fe(CO)_4\}_4]^{3-}$ cluster along the shortest diagonal, the two edge-fused M_3Fe_3 moieties give rise to an incipient M_5 square pyramid, with two bridged edges and faces. Each of the two edge-bridging $Fe(CO)_4$ moieties, upon adoption of a C_{3v} stereochemistry of the carbonyl groups

TABLE 4. Infrared Carbonyl Absorptions of the Heterometallic Cu, Ag and Au Clusters containing $Fe(CO)_4$ or $Fe_2(CO)_8$ fragments.

Compound	n_{CO} (cm-1) in CH_3CN
$[Au\{Fe(CO)_4\}_3]^{5-}$	1820w,1796m,1721s
$[Cu\{Fe(CO)_4\}_2]^{3-}$	1916w,1884m,1807s
$[Ag\{Fe(CO)_4\}_2]^{3-}$	1913w,1889m,1795s
$[Au\{Fe(CO)_4\}_2]^{3-}$	1915w,1889m,1809s
$[Ag_4\{\mu_2\text{-}Fe(CO)_4\}_4]^{4-}$	1926ms,1851s
$[Au_4\{\mu_2\text{-}Fe(CO)_4\}_4]^{4-}$	1931ms,1867s
$[Ag_2Au_2\{\mu_2\text{-}Fe(CO)_4\}_4]^{4-}$	1928ms,1862s
$[Ag_2Cd\{Fe(CO)_4\}_4]^{4-}$	1937ms,1848s
$[Ag_2Cd_2\{Fe(CO)_4\}_4]^{2-}$	1993ms,1909s
$[Ag_5\{\mu_2\text{-}Fe(CO)_4\}_2\{m_3\text{-}Fe(CO)_4\}_2]^{3-}$	1949ms,1879s
$[Ag_4Cu\{Fe(CO)_4\}_4]^{3-}$	1945ms,1891s
$[Ag_4Au\{Fe(CO)_4\}_4]^{3-}$	1950ms,1882s
$[Au_5\{Fe(CO)_4\}_4]^{3-}$	1946ms,1881s
$[Ag_6\{Fe(CO)_4\}_4]^{2-}$	1970s,1885m
$[Ag_5Au\{Fe(CO)_4\}_4]^{2-}$	1972s,1884m
$[Ag_4Au_2\{Fe(CO)_4\}_4]^{2-}$	1973s,1893m
$[Ag_{13}\{\mu_3\text{-}Fe(CO)_4\}_8]^{4-}$	1980s,1905m
$[Ag_{13}\{\mu_3\text{-}Fe(CO)_4\}_8]^{3-}$	2000s,1917m
$[Au\{\mu_2\text{-}Fe_2(CO)_8\}_2]^{-}$	2015s,1990m,1957mw
$[Au\{\mu_2\text{-}Fe_2(CO)_8\}_2]^{2-}$	1984s,1945ms,1770m
$[Au\{\mu_2\text{-}Fe_2(CO)_8\}_2]^{3-}$	1979m,1931s,1766m

adoption of a C_{3v} stereochemistry of the carbonyl groups, will have an inner-pointing filled orbital available for binding an entrant M^+ ion. It should be noted that the implicit transformation of the $\mu_3\text{-}Fe(CO)_4$ fragment only requires a remixing of the orbital of the fragment and does not change their occupation. In both bonding situations represented in sketches (A) (the coordinately trivacant pentagonal pyramid) and (B) (the coordinately trivacant capped octahedron), the $\mu_3\text{-}Fe(CO)_4$ fragment behaves as a 4-electron donor [11,34,39]. As a result, an M_6 octahedral cluster capped on four non-adjacent faces by $Fe(CO)_4$ moieties is one of the likely products. This possibility has been realised in the reaction of $[Cu_5\{Fe(CO)_4\}_4]^{3-}$ with CuCl, which gave $[Cu_6\{\mu_3\text{-}Fe(CO)_4\}_4]^{2-}$ [33].

On the other hand, on maintaining planarity of the $[M_5\{Fe(CO)_4\}_4]^{3-}$ metal skeleton, polymeric species such as $[\{M^+\}_n\{M_5\{Fe(CO)_4\}_4\}_n]^{2n-}$, could be expected. For sake of clarity, only the first step of the polymer growth is depicted in sketch (C) of *Figure 4*.

Experimentally the reaction of $[Ag_5\{Fe(CO)_4\}_4]^{3-}$ with one equivalent of Ag^+ or Au^+ is greatly influenced by the counterion of the former and has not yet been fully clarified. On working with alkali counterions the purported $[Ag_5M\{Fe(CO)_4\}_4]^{2-}$ (M=Ag,Au), analogous to the known $[Cu_6\{\mu_3\text{-}Fe(CO)_4\}_4]^{2-}$ [33], is quantitatively obtained by reaction (6).

As shown in TABLE 4, the carbonyl absorption frequencies and their overall

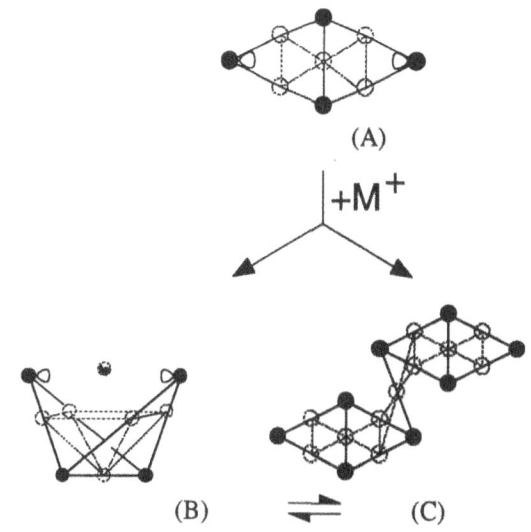

Figure 4.
The $[M_5\{Fe(CO)_4\}_4]^{3-}$ clusters as potential bidentate ligands (full circles represent the $Fe(CO)_4$ groups and open circles indicate the Group 11 metal atoms).

$$[Ag_5\{Fe(CO)_4\}_4]^{3-} + M^+ \rightarrow [Ag_5M\{Fe(CO)_4\}_4]^{2-} \text{ (M=Ag,Au)} \qquad (6)$$

pattern, which points out a change in the carbonyl stereochemistry, are in keeping with such a formulation. However, all attempts to grow crystals suitable for X ray diffraction studies have been so far unsuccessful, both for the alkali and the tetrasubstituted ammonium or phosphonium salts. In particular, by metathesis of the sodium salts with quaternary ammonium or phosphonium halides, amorphous precipitates, completely insoluble in most organic solvents, were invariably obtained. Identically behaving materials were also directly obtained from reaction of $[Ag_5\{Fe(CO)_4\}_4]^{3-}$, as quaternary ammonium or phosphonium salts, and $AgBF_4$. The latter amorphous insoluble material shows a solid state IR with absorption frequencies similar to that of $[Ag_6\{Fe(CO)_4\}_4]^{2-}$, but differing in the pattern which consists of two equally intense broad absorptions. Furthermore, it slowly dissolves in solvents such as DMSO, or in most solvents in the presence of amines, to give $[Ag_6\{Fe(CO)_4\}_4]^{2-}$. Although the data available at present are insufficient to safely conclude that path (C) is at hand, these are at least in keeping with a possible equilibrium between the product (B) and (C) of *Figure 4*. Analogous results have been obtained on reacting either $[Ag_5\{Fe(CO)_4\}_4]^{3-}$ or $[Ag_4Au\{Fe(CO)_4\}_4]^{3-}$ with one equivalent of $Au(SEt_2)Cl$.

The corresponding reaction of $[Au_5\{Fe(CO)_4\}_4]^{3-}$ with gold(I) derivatives probably follows a similar course. However, so far we only have spectroscopic evidence for the formation of a species which might be $[Au_6\{Fe(CO)_4\}_4]^{2-}$ and have no evidence for the formation of an insoluble polymeric species such as that observed in the corresponding reaction with Ag^+.

3.3. THE FINAL STEPS OF THE OXIDATION OF $[Fe(CO)_4]^{2-}$ WITH Ag^+ and Au^+ IONS.

For $M^+/[Fe(CO)_4]^{2-}$ molar ratios ≥ 1.5, the course of the reaction of $[Fe(CO)_4]^{2-}$ with Ag^+ begin to differ significantly from that with Au^+ [11-14]. Thus, the further addition of 0.5 equivalents of Ag^+ to $[Ag_6\{Fe(CO)_4\}_4]^{2-}$ results into quantitative formation of $[Ag_{13}\{Fe(CO)_4\}_8]^{3-}$, according to equation (7).

$$2\,[Ag_6\{Fe(CO)_4\}_4]^{2-} + Ag^+ \rightarrow [Ag_{13}\{Fe(CO)_4\}_8]^{3-} \qquad (7)$$

The corresponding $[Ag_{13}\{Fe(CO)_4\}_8]^{4-}$ tetraanion has been obtained by reduction of the initially obtained $[Ag_{13}\{Fe(CO)_4\}_8]^{3-}$ with alkali metal or $[Fe(CO)_4]^{2-}$.

In contrast, 1-1.5 equivalents of $[AuBr_2]^-$ are necessary in order to quantitatively convert the presumed $[Au_6\{Fe(CO)_4\}_4]^{2-}$ dianion into mixtures of an as yet uncharacterized brown derivative and $[Au\{\eta^2\text{-}Fe_2(CO)_8\}_2]^-$, along with some gold metal. The dark green $[Au\{\eta^2\text{-}Fe_2(CO)_8\}_2]^-$ monoanion has been more conveniently obtained from the reaction of $[Fe_3(CO)_{11}]^{2-}$ or $[Fe_2(CO)_8]^{2-}$ with $[AuCl_4]^-$ salts, and has been fully characterized [13]. As shown in *Figure 5*, the structure of this compound is very similar to that displayed by the isoelectronic $[M\{\eta^2\text{-}Fe_2(CO)_8\}_2]^{2-}$ (M=Pd,Pt) dianion [35]. The major differences can be found in the stereochemistry of the axial carbonyl groups. In the latter, a pair of non-consecutive axial carbonyls above and below the metal plane is tilted toward the central atom to give rise to a tetrahedral coordination of C atoms around both Pd and Pt, while the remaining axial carbonyls are tilted away. In contrast, the axial carbonyls of $[Au\{\eta^2\text{-}Fe_2(CO)_8\}_2]^-$ are all tilted toward the central gold atom, although this occurs to a much lesser extent than in the above complexes.

These different structural features probably also have some chemical relevance. Thus, the $[Au\{\eta^2\text{-}Fe_2(CO)_8\}_2]^-$ and the $[M\{\eta^2\text{-}Fe_2(CO)_8\}_2]^{2-}$ (M=Pd,Pt) anions can be formally described as arising respectively from d^8 square-planar Au^{3+} or Pd^{2+} and

Pt^{2+} ions coordinated by two bidentate $[Fe_2(CO)_8]^{2-}$ 4-electron ligands, analogous for instance to a diphosphine. All bond theories (Crystal Field, Valence Bond and MO theories), as well as EHMO calculations performed on a model compound, agree in suggesting the possible presence of a low-lying empty orbital mainly centred on the unique Group 10 or 11 atom. In accord with this prediction, $[Au\{\eta^2\text{-}Fe_2(CO)_8\}_2]^-$ can be reversibly reduced to the corresponding $Au\{Fe_2(CO)_8\}_2]^{2-}$ and $[Au\{Fe_2(CO)_8\}_2]^{3-}$ derivatives, both chemically and electrochemically (equation (8)).

$$[Au\{Fe_2(CO)_8\}_2]^- \overset{E°=-0.73\ V}{\rightleftharpoons} [Au\{Fe_2(CO)_8\}_2]^{2-} \overset{E°=-0.93\ V}{\rightleftharpoons} [Au\{Fe_2(CO)_8\}_2]^{3-} \quad (8)$$

As shown in TABLE 4, the infrared spectra of both these products show absorptions in the region of bridging carbonyls. In particular, $[Au\{Fe_2(CO)_8\}_2]^{2-}$ shows an infrared spectrum almost superimposable on that of $[Cd\{Fe_2(CO)_8\}_2]^{2-}$. This latter complex was obtained quantitatively from the reaction of $CdCl_2$ with $[Fe_2(CO)_8]^{2-}$ [48]. It appears likely therefore that the above mentioned carbonyl bridges span the Fe-Fe, rather than Au-Fe bonds, as it occurs for instance in $[Au\{\eta^2\text{-}Fe_2(CO)_8\}(PPh_3)]^-$ [49]. Unfortunately, all attempts to crystallize these compounds have been hampered by their sensitivity to humidity and oxygen.

Significantly, the $[M\{\eta^2\text{-}Fe_2(CO)_8\}_2]^{2-}$ (M=Pd,Pt) clusters do not display reversible redox behaviour analogous to that reported in equation (8) [13].

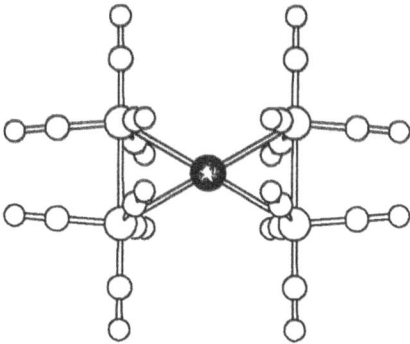

Figure 5. *X Ray Structure of* $[Au\{\eta^2\text{-}Fe_2(CO)_8\}_2]^-$ [13]

The $[Ag_{13}\{\mu_3\text{-}Fe(CO)_4\}_8]^{n-}$ (n=3,4) anions show identical structures and only that of the tetraanion is shown (see *Figure 6*). Once one considers that a C_{3v} $\mu_3\text{-}Fe(CO)_4$ fragment is isolobal with a μ_3-S atom, the $[Ag_{12}(\mu_{12}\text{-}Ag)\{\mu_3\text{-}Fe(CO)_4\}_8]^{3-}$ trianion can be considered as an Ag^+ cryptate of the $[Ag_{12}\{\mu_3\text{-}Fe(CO)_4\}_8]^{4-}$ cryptand, related to the recently characterized $[Au_{12}(\mu_{12}\text{-}Na)(\mu_3\text{-}S)_8]^{3-}$ [50]. Consequently, $[Ag_{13}\{\mu_3\text{-}Fe(CO)_4\}_8]^{4-}$ formally represents the corresponding cryptate of a neutral odd-electron Ag atom. Accordingly, this species shows a complex EPR spectrum pattern, which suggests that the unpaired electron is at least delocalized over all silver atoms. However, the values of the coupling constants with the interstitial silver isotopes, which are ca. 45 times greater than that with the peripheral silver atoms, signify a greater contribution of the 5s orbital of the interstitial atom to the SOMO. Incidentally, the observed EPR signal is coincident with those observed during the oxidation with $AgBF_4$ of the Fe-P and Fe-In anionic clusters cited in the introduction. Therefore, the initial aim of this work has been fully accomplished.

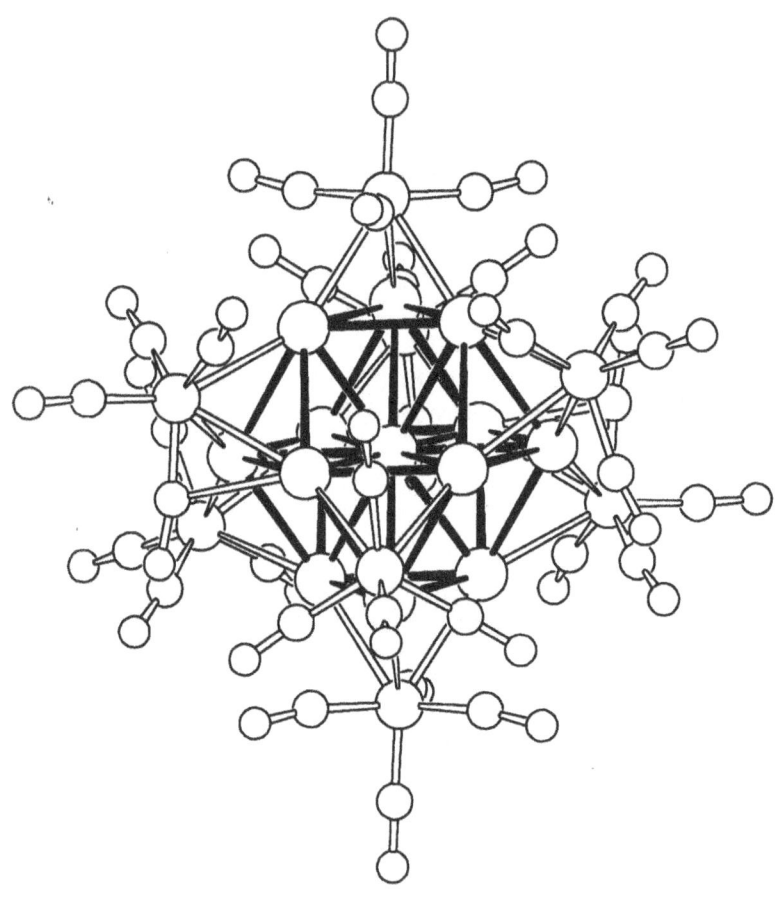

Figure 6.   *X Ray structure of* $[Ag_{12}(\mu_{12}\text{-}Ag)\{\mu_3\text{-}Fe(CO)_4\}_8]^{4-}$ [10]

4. Conclusions

As shown by the formal equations (9) and (10), the synthesis of $[Ag_{13}\{\mu_3\text{-}Fe(CO)_4\}_8]^{3-}$ requires an overall $Ag^+/[Fe(CO)_4]^{2-}$ molar ratio of 1.625, whereas that of $[Au\{\eta^2\text{-}Fe_2(CO)_8\}_2]^-$ indicates a $[AuBr_2]^-/[Fe(CO)_4]^{2-}$ molar ratio of 1.75 as necessary. Both these stoichiometric ratios are in good agreement with experiments.

$$8 \, [Fe(CO)_4]^{2-} + 13 \, Ag^+ \rightarrow [Ag_{13}\{\mu_3\text{-}Fe(CO)_4\}_8]^{3-} \tag{9}$$

$$4 \, [Fe(CO)_4]^{2-} + 7 \, [AuBr_2]^- \rightarrow [Au\{\eta^2\text{-}Fe_2(CO)_8\}_2]^- + 6 \, Au + 14 \, Br^- \tag{10}$$

On increasing the above molar ratios up to 2, both the $[Ag_{13}\{\mu_3\text{-}Fe(CO)_4\}_8]^{3-}$ and $[Au\{\eta^2\text{-}Fe_2(CO)_8\}_2]^-$ anions progressively disappear from their solutions, and variable mixtures of $Fe(CO)_5$, $Fe_3(CO)_{12}$, Fe and Ag or Au metals are respectively obtained.

It can be concluded, therefore, that the oxidation of $[Fe(CO)_4]^{2-}$ with both Ag(I) and Au(I) occurs through a particular inner sphere mechanism [51], which requires the intermediate redox condensation of $[Fe(CO)_4]^{2-}$ with Ag^+ or Au^+ ions to give several bimetallic clusters with a sequence which has been almost completely rationalized. Irrespective of the number of metal atoms belonging to the cluster core and the associated $Fe(CO)_4$ fragments, the resulting compounds display chemical stability if the ratio between the number of negative charges and the number of iron atoms is in the range 1.66-0.25. A decrease of this ratio up to the lowest limit corresponds to the sudden complete segregation of silver, while the formation of a formal Au(III) derivative such as $[Au\{\eta^2\text{-}Fe_2(CO)_8\}_2]^-$ provides a way to keep the bimetallic cluster path still open. A further decrease in this ratio beyond that limit eventually results in segregation of gold metal even in the latter case.

The occurrence of the above mechanism in the chemistry of transition metal carbonyls, which is probably more general than commonly thought, was originally recognized by P.Chini and since then is often referred as a redox condensation mechanism or reaction [52].

5. References

1. Shriver, D.F., Kaesz, H.D., and Adams, R.D. Eds. (1990) *The Chemistry of Metal Cluster Complexes*, VCH Publishers, Inc, New York.
2. Salter, I.D. (1989) Heteronuclear Cluster Chemistry of Copper, Silver and Gold, *Adv.Organomet.Chem.* **29**, 249-343.
3. Hall, K.P. and Mingos, D.M.P., (1984) Homo- and Heteronuclear Cluster Compounds of Gold, *Prog.Inorg.Chem.*, **32**, 237-325.

4. Whitmire, K.H. (1988) The Interface of Main Group and Transition Metal Cluster Chemistry, *J.Coord.Chem.* **17**, 95-203.

5. Hermann, W.A., (1986) *Angew.Chem.Int.Ed.Engl.*, **25**, 56-74

6. Mingos, D.M.P., and Watson, M.J., (1992) Heteronuclear Gold Cluster Compounds, *Adv.Inorg.Chem.*, **39**, 327-399.

7. Canè, M., Iapalucci, M.C., Longoni, G., Demartin, F., and Grossi, L. (1991) *Materials: Chemistry and Physics* **29**, 395-404.

8. Albano, V.G., Canè, M., Iapalucci, M.C., Longoni, G., and Monari, M. (1991) *J.Organomet.Chem.* **407**, C9-C12.

9. Albano, V.G., Azzaroni, F., Iapalucci, M.C., Longoni, G., Monari, M., and Mulley, S. (1992) *Materials Research Society Symposium Proceedings* **272**, 115-125.

10. Albano, V.G., Grossi, L., Longoni, G., Monari, M., Mulley, S., and Sironi, A. (1992) *J. Amer.Chem.Soc.* **114**, 5708-5713.

11. Albano, V.G., Azzaroni, F., Iapalucci, M.C., Longoni, G., Monari, M., Mulley, S., Proserpio, D.M., and Sironi, A. (1994) *Inorg.Chem. in press.*

12. Albano, V.G., Calderoni, F., Iapalucci, M.C., Longoni, G., Monari, M., and Zanello , P. (1994) *to be submitted.*

13. Albano, V.G., Aureli, R., Iapalucci, M.C., Laschi, F., Longoni, G., and Zanello, P. (1993) *J. Chem.Soc., Chem.Comm.*, 1501-1502.

14. Albano, V.G., Calderoni, F., Iapalucci, M.C., Longoni, G., Monari, M., *to be submitted.*

15. Sosinsky, B.A., Shong, R.G., Fitzgerald, B.J., Noren, N., and O'Rourke, C. (1983) *Inorg.Chem.* **22**, 3124-3129.

16. Alvarez, S., Ferrer, M., Reina, R., Rossel, O., Seco, M., and Solans, X. (1989) *J. Organomet.Chem.* **377**, 291-303.

17. Cassidy, J.M. and Whitmire, K.H. (1989) *Inorg.Chem.* **28**, 2494-2496.

18. Cassidy, J.M., Whitmire, K.H., and Kook, A.M. (1993) *J. Organomet.Chem.* **456**, 61-70.

19. Luo, S. and Whitmire, K.H. (1989) *Inorg.Chem.* **28**, 1424-1431.

20. Churchill, M.R., Fettinger, J.C., Whitmire, K.H., and Lagrone C.B. (1986) *J. Organomet.Chem.* **303**, 99-109.

21. Ferrer, M., Rossel, O., Seco, M., Solans, X., and Gòmez, M. (1990) *J.Organomet.Chem.* **381**, 183-189.

22. Shieh, M., Liou, Y., Peng, S.-M., and Lee, G.-H. (1993) *Inorg.Chem.* **32**, 2212-2214.

23. Bachman, R.E., Miller, S.K., and Whitmire, K.H. (1994) *Inorg.Chem.* **33** 2075-2076.

24. Albano, V.G., Ciani, G., Bruce, M.I., Shaw, G., and Stone, F.G.A., (1972) *J.Organomet.Chem.*, **42**, C99-C101.

25. Albano, V.G., Monari, M., Iapalucci, M.C., and Longoni, G. (1993) *Inorg.Chim. Acta* **213**, 183-190.

26. Briant, C.E., Hall, K.P., and Mingos, D.M.P. (1983) *J.Chem.Soc.,Chem.Comm.*, 843-845.

27. Calderoni, F., Demartin, F., Iapalucci, M.C., Longoni, G., and Soverini, M., *to be submitted.*

28. Longoni, G., Manassero, M., and Sansoni, M. (1980) *J. Amer.Chem.Soc.* **102**, 7973-7974.

29. Della Pergola, R., Garlaschelli, L., Mealli, C., Proserpio, D.M., and Zanello, P. (1990) *J. Cluster Science* **1**, 93-106.

30. Adams, R. D., Chen, G., and Wang, J.-G. (1989) *Polyhedron* **8**, 2521-2530.

31. Doyle, G., Eriksen, K.A., and Van Engen, D. (1986) *J. Amer.Chem.Soc.* **108**,445-451.

32. Ernst, R.D., Marks, T.J., and Ibers, J.A. (1977) *J. Amer.Chem.Soc.* **99**, 2090-2098.

33. Doyle, G., Eriksen, K.A., and Van Engen, D. (1985) *J. Amer.Chem.Soc.* **107**, 7914-7920.

34. Briant, C.E., Smith, R.G., and Mingos, D.M.P. (1984) *J.Chem.Soc.,Chem. Comm.*, 586-588.

35. Longoni, G., Manassero, M., and Sansoni, M. (1980) *J. Amer.Chem.Soc.* **102**, 3242-3244.

36. Fuchs, R., and Klüfers, P., (1991) *Z.Naturforsch.*, **46b**, 507-518

37. Usòn, R., Laguna, A., Laguna, M., Jones, P.G., and Sheldrick, G. (1981) *J.Chem.Soc. Dalton Trans.*, 366-370.

38. Albright, T.A., Burdett, J.K., and Whangbo, M-H. (1985) *Orbital Interactions in Chemistry*, J.Wiley and Sons, New York.

39. Hoffmann, R. (1982) *Angew.Chem.Int.Ed.Engl.*, **21**, 711-726

40. Klüfers, P. (1984) *Angew.Chem.Int.Ed.Engl.*, **23**, 307-308

41. Klüfers, P. (1985) *Angew.Chem.Int.Ed.Engl.* **24**, 70-71.

42. Merz, K.M., and Hoffmann, R. (1988) *Inorg.Chem.*, **27**, 2120-2127

43. Buhl,M.L., Long, G., and Doyle, G. (1993) *J.Organomet.Chem.*, **461**, 187-199

44. Pitzer, K.S. (1979) *Acc.Chem.Res.*, **12**, 271-276

45. Pyykko, P. and Desclaux, J-P. (1979) *Acc.Chem.Res.*, **12**, 276-281

46. Lindley, P.F. and Woodward, P. (1967) *J.Chem.Soc. A*, 382-392.

47. Whitmire, K.H., Lagrone, C.B., Churchill, M.R., Fettinger, J.C., and Robinson, B.H. (1987) *Inorg.Chem.* **26**, 3491-3499.

48. Di Cori, C. (1992) *Thesis, University of Bologna*

49. Rossell, O., Seco, M., and Jones, P. (1990) *Inorg.Chem.*, **29**, 348-350

50. Huang, S-P., and Kanatzidis, M.G. (1992) *Angew.Chem.Int.Ed.Engl.*, **31**,787-789

51. Basolo, F. and Pearson, R.G. (1967) *Mechanisms of Inorganic Reactions*, J.Wiley and Sons, New York

52. Chini, P., Longoni, G., and Albano, V.G. (1976) *Adv.Organomet.Chem.*, **14**, 285-344

REACTIONS OF SILYLALKYNES WITH TRIOSMIUM AND TRIRUTHENIUM CLUSTERS.

A.A. KORIDZE
A.N. Nesmeyanov Institute of Organoelement
Compounds, Russian Academy of Sciences,
Vavilov str. 28, 117813 Moscow, Russian Federation.

Abstract

Our continuing interest in transformations of heteroatom-substituted alkynes on transition metal clusters [1,2] has prompted us to investigate reactions of silylalkynes with triruthenium and triosmium clusters.

It should be noted that reactions of silylalkynes with metal clusters have earlier been studied by several groups but these alkynes did not reveal any unusual reactivity. We observed, however, unusual transformations of silylalkynes on triruthenium and triosmium clusters, metal complexes bearing novel organic ligands, and novel rearrangements. These findings are the subject of the present report.

1. Reactions of silylalkynes with Ru$_3$ clusters.

The μ-(O-C)-bridged carboxamido cluster Ru$_3$H(μ-C=CNMe$_2$)(CO)$_{10}$ reacts with the excess of Me$_3$SiC≡CH in hexane at the room temperature to give the mononuclear colourless complex Ru[η^5-C$_5$H$_2$(SiMe$_3$)$_2$OH](η1-COMe)(CO)$_2$ (**1**) as a single product. Complex **1** was characterised by X-ray crystallography [3].

1

L. J. Farrugia (ed.), The Synergy Between Dynamics and Reactivity at Clusters and Surfaces, 351–360.
© 1995 *Kluwer Academic Publishers.*

Complex **1** is a rare example of a stable organometallic compound with the hydroxycyclopentadienyl ligand. Its unusual stability result from intramolecular hydrogen bonding between the hydroxyl group and the oxygen atom of the acetyl group.

Complex **1** was also obtained as a minor product in the reaction of $Ru_3(CO)_{12}$ with $Me_3SiC\equiv CH$ in hot hexane; the acetylide cluster $Ru_3H(\mu_3-\eta^2-C\equiv CSiMe_3)(CO)_9$ is a major product [4].

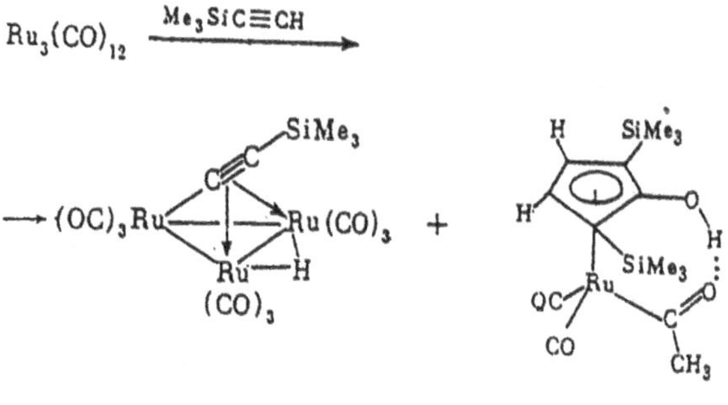

The formation of **1** in the reaction of a silylalkyne with $Ru_3(CO)_{12}$ indicates that the source of the σ-acetyl group in **1** is an alkyne molecule subjected to the desilylation and subsequent hydration of the hydrocarbyl ligands (by traces of moisture during the chromatography on silica gel) (Scheme I).

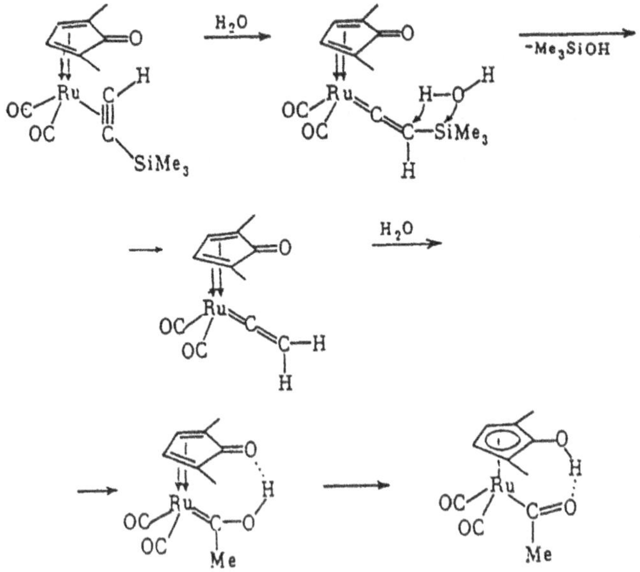

Scheme 1

The precursor of **1** might be a ruthenium complex of the η^4-cyclopentadienone ligand $(HC_2SiMe_3)_2CO$. Cyclopentadienone derivatives of metal carbonyls are known to be frequently formed in reactions of an excess of alkynes with metal carbonyls. Thus three alkyne molecules are consumed for the formation of **1**: two of them and one CO group undergo a cyclic coupling reaction, affording a pentanuclear carbocycle, and the third forms the σ-acetyl ligand.

A more interesting and unusual linear coupling of two alkynes takes place in the reaction of $Ru_3(CO)_{12}$ with an excess of $Me_3SiC\equiv CMe$ in hot hexane [4]. Two isomeric hydride complexes, **2** and **3**, are formed in this reaction in equal yields. The structure of **2** was established by an X-ray diffraction study. Molecule **2** (Fig. 1) consists of the Ru_3 triangle coordinated to eight terminal CO groups, a bridging hydride ligand, and a hydrocarbyl ligand formed by the coupling of two-alkyne molecules.

2   3

The organic ligand in **2** is bound to three metal atoms *via* a five carbon atom chain, which is formed from four acetylenic and one methyl carbon atoms provided by two initial alkyne molecules. Under the reaction conditions, one of these undergoes a dehydrogenation of the methyl group and the 1,2 shift of the Me_3Si group.

This hydrocarbyl ligand is involved into the complicated interaction with the metallic triangle *via* the carbon atoms C(9), C(10), C(11), C(12) and C(13). The C(9) atom forms a σ-bond with Ru(1) whereas the C(9), C(10) and C(11) atoms and C(11), C(12) and C(13) atoms form a delocalized π-bond with the Ru(2) and Ru(3) atoms, respectively. The C(9)-C(10) fragment bonded to Ru(1) and Ru(2) may be regarded as σ,π-alkenyl ligand, while the C(1)C(11)C(12) group represents a π-allyl system coordinated to the Ru(3) atom. The dihedral angle between the C(9)C(10)C(11) and C(11)C(12)C(13) planes is 43.0°. At the same time, the C(10), C(11), C(12) atoms form a distorted allene-unit, whose central atom is simultaneously coordinated to the Ru(2) and Ru(3) atoms, which results in its strongly distorted pseudo-tetrahedral environment; the C(10)C(11)C(12) angle being 142.5(3)°. The ^{1}H NMR spectrum of **2** (C_6D_6, δ -18.07 (1H), 0.18 (9H), 0.32 (9H), 1.97 (1H), 2.13 (3H) and 8.13 (1H) is consistent with the X-ray structure.

The ^{1}H NMR spectrum of **3** in a C_6D_6 solution shows resonances at δ -17.42

(1H), 0.18 (9H), 0.26 (9H), 1.83 (3H, d, J=6.4 Hz), 2.94 (1H, q, J=6.4 Hz) and 8.87 (1H). In an effort to shed some light on the formation of **2** and **3**, we have performed a more detailed study of trimethylsilylpropyne dimerization on the Os_3 cluster.

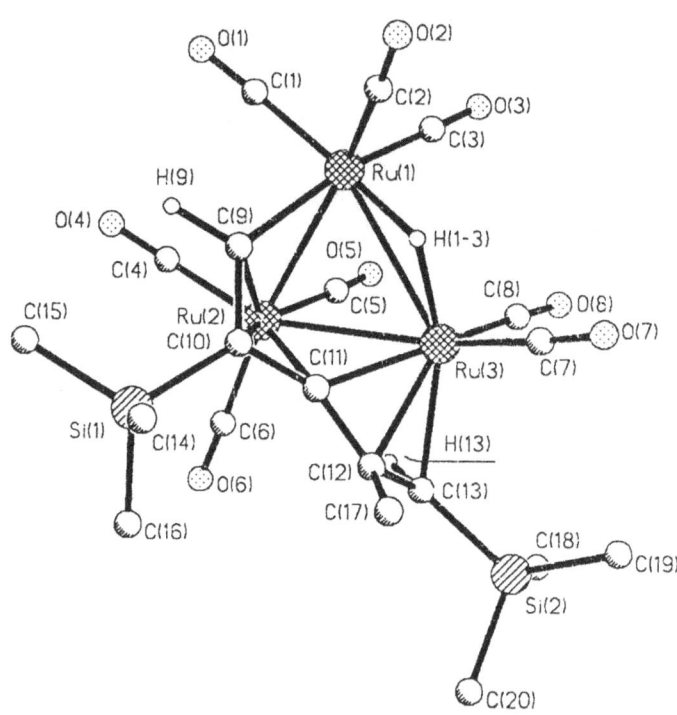

Figure 1. The molecular structure of **2**

Selected distances (Å)

Ru(1)-Ru(2) 2.788(1)	Ru(3)-C(12) 2.228(3)	Ru(1)-Ru(3) 2.945(1)
Ru(3)-C(13) 2.249(3)	Ru(2)-Ru(3) 2.918(1)	Ru(3)-H(1-3) 1.77(4)
Ru(1)-C(9) 2.082(3)	C(9)-C(10) 1.413(5)	Ru(1)-H(1-3) 1.65(5)
C(10)-C(11) 1.413(4)	Ru(2)-C(9) 2.237(3)	C(11)-C(12) 1.417(5)
Ru(2)-C(10) 2.271(3)	C(12)-C(13) 1.433(4)	Ru(2)-C(11) 2.151(4)
Ru(3)-C(11) 2.063(3)		

2. Transformation of $Me_3SiC \equiv CMe$ on an Os_3 cluster.

The reaction of $Me_3SiC \equiv CMe$ with the activated triosmium complex $Os_3(CO)_{10}$-$(C_8H_{14})_2$ gives $Os_3(\mu_3\text{-}Me_3SiC_2Me)(\mu\text{-}CO)(CO)_9$ (4), In warm methanol, compound 4 is readily desilylated, and the subsequent decarbonylation and oxidative addition of the alkyne C-H bond to the Os_3 core leads to $Os_3H(\mu_3\text{-}\eta^2\text{-}C \equiv CMe)(CO)_9$ (5),

The thermolysis of 4 in refluxing hexane leads to the allenyl compound $Os_3H\{\mu_3\text{-}C(SiMe_3)C=C=CH_2\}(CO)_9$ (6). All attempts to isomerize compound 6 into the $2\sigma,\pi$-allyl complex $Os_3H\{\mu_3\text{-}C(SiMe_3)CHCH\}(CO)_9$ were unsuccessful; in refluxing heptane or octane only 5 was formed [5]. Compound 4 adds trimethylsilylpropyne in refluxing hexane, yielding initially the red compound $Os_3\{C_4(SiMe_3)(Me)(SiMe_3)(Me)\}(CO)_n$ (7) that is then converted into the yellow compound $Os_3H\{\mu_3\text{-}C(SiMe_3)C(Me)CC(SiMe_3)\text{-}CH_2\}(CO)_8$ (8).

The reaction of $Me_3SiC \equiv CMe$ with the vinyl compound $Os_3H(\mu\text{-}\eta^2\text{-}CH=CH_2\text{-}(CO)_{10}$ proceeds analogously affording 7 and then 8. Evidently, in this case the reaction between the vinyl compound and the excess of $Me_3SiC \equiv CMe$ proceeds *via* the intermediate formation of 4. Compounds 7 and 8 were isolated by chromatography on silica gel, characterized by IR and ^1H NMR spectra, and for compound 8, a single crystal X-ray diffraction study has also been carried out.

The IR spectrum of 7 shows υ_{CO} bands at 2084s, 2044vs, 2036vs,sh, 2010vs, 1992s, 1976s and 1958w cm^{-1}. The ^1H NMR spectrum of compound 7 in a CD_2Cl_2 solution at 20°C contains two pairs of resonances belonging to the non-equivalent Me_3Si and Me groups at δ 0.29, 0.44, 1.92 and 2.06 ppm; the resonance of the Me group at 1.92 ppm is very broad. On cooling the solution to -60°C, this resonance becomes narrow but no other changes are detected in the spectrum. Unfortunately, our attempts to obtain a single crystal of 7 suitable for X-ray analysis were unsuccessful. Therefore, the structure of 7 and the nature of dynamic processes in this compound are still unclear.

The ^1H NMR spectrum of 8 in C_6D_6 shows resonances at δ -17.04 (1H, d, J=0.8 Hz), 0.06 (9H), 0.45 (9H), 1.63 (1H, dd, J=0.8, ^1J=1.4 Hz), 2.58 (3H) and 3.07 (1H, J=1.4 Hz). According to the X-ray diffraction study (Fig. 2), molecule 8 involves the triosmium triangle and the hydrocarbon chain bound to the metallic core in the same fashion as we have observed earlier for the ruthenium compound 2.

The reaction of $Me_3SiC \equiv CMe$ with 4 under more severe conditions in refluxing heptane gave the complex $Os_3H\{\mu_3\text{-}CHC(SiMe_3)CC(Me)CH(SiMe_3)\}(CO)_8$ (9) along with 8. According to the ^1H NMR spectra, compound 9 (δ -18.08 (1H), 0.17 (9H), 0.34 (9H), 1.78 (1H), 2.32 (3H) and 8.53 (1H)) and the ruthenium complex 2 have identical structures.

Unlike 4, the allenyl compound 6 does not react further with $Me_3SiC \equiv CMe$ in refluxing hexane, and 8 and 9 are obtained in low yields only after prolonged heating of 6 and the alkyne in refluxing heptane. Compound 6 cannot therefore be considered as an intermediate in the formation of the products containing dimerized alkynes.

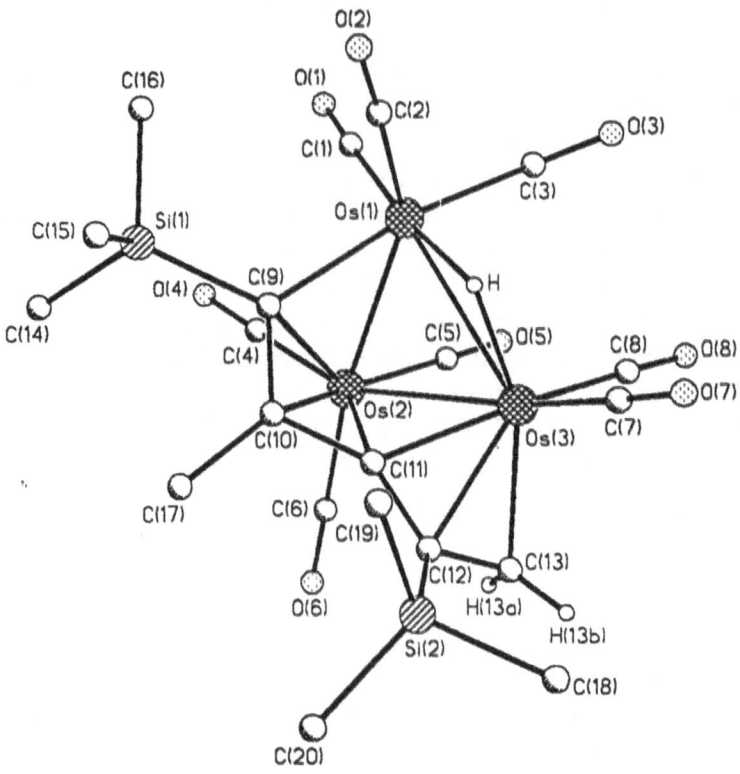

Figure 2 The molecular structure of **8**

The isolation of **8** in reactions of Me$_3$SiC≡CMe with triosmium clusters and the comparison of the organic ligand structures in **8** and **9** show that these ligands may be interconverted by a formal 1,5 shift of the hydrogen atoms, *i.e.* **8** may be a precursor of compound **9**. Indeed, we found that when heated in refluxing heptane for 7 hours, compound **8** yields a mixture of **8** and **9** in the ratio of 6:1.

This rearrangement of the organic ligand including the interchange of hydrogen atoms between the metallic core and the hydrocarbon ligand (accompanied by the internuclear migration of the CO group) may be considered as homogeneous model of rearrangements proceeding on metal surfaces.

In conclusion we have shown that the dimerization of Me$_3$SiC≡CMe on Ru$_3$ and Os$_3$ clusters proceeds with the formation of the compounds M$_3$H{μ$_3$-CHC(SiMe$_3$)-CC(Me)CH(SiMe$_3$)}(CO)$_8$ and includes such intermediates as M$_3$(μ$_3$-Me$_3$SiC$_2$Me)-(μ-CO)(CO)$_9$, Os$_3${C$_4$(SiMe$_3$)(Me)(SiMe$_3$)(Me)}(CO)$_n$, and M$_3$H{μ$_3$-C(SiMe$_3$)C(Me)CC(SiMe$_3$)CH$_2$}(CO)$_8$.

Regarding the isomerism of ruthenium complexes **2** and **3**, we observe that the

$$8 \qquad\qquad 9$$

isomers may be formed by "head-to-tail" or "head-to-head" coupling of two hydrocarbon ligands. It is remarkable that the osmium analogue of the ruthenium compound **3** has not been observed in reactions studied.

In order to elucidate the coordination mode of organic ligand(s) in **7** and related compounds, we studied mixed-alkyne clusters in which $Me_3SiC \equiv CMe$ is one of the alkynes.

3. Formation of mixed alkyne Os₃ clusters.

The alkyne complex $Os_3(\mu_3\text{-}PhC_2Ph)(CO)_{10}$ reacts with $Me_3SiC \equiv CMe$ in hot hexane to form yellow $Os_3H\{\mu_3\text{-}C(Ph)C(Ph)CC(SiMe_3)(CH_2\}(CO)_8$ (**10**) as a single product.

Under the same conditions, $Os_3(\mu_3\text{-}FcC_2CH=CHFc)(\mu\text{-}CO)(CO)_9$ (Fc = ferrocenyl) containing an enyne ligand reacts with $Me_3SiC \equiv CMe$ to give isomeric compounds $Os_3H\{\mu_3\text{-}FcC(CH=CHFc)CC(SiMe_3)CH_2\}(CO)_8$ (**11**) and $Os_3H[\mu_3\text{-}(FcCH=CH)CC(Fc)CC(SiMe_3)CH_2](CO)_8$ (**12**). According to spectral data, compounds **10, 11,** and **12** have a similar structure to that established for **8**.

No intermediates related to the red compound **7** were detected in these reactions. It is well known that organotriosmium species with dimerized alkyne ligands formed at the initial stage of the reaction contain the osmacyclopentadiene fragment. All the triosmium clusters with such a fragment which have been characterized so far have the structure **B**. An alternative structure **A** had been initially suggested for the complex with the dimerized diphenylacetylene ligand [6], but it has been shown in a later X-ray diffraction study [7] that this complex has the structure **B**, $Os_3\{\mu\text{-}2\eta^1\text{-}\eta^4\text{-}C_4Ph_4\}(CO)_9$ (**13**).

$$A \qquad\qquad B$$

358

In the reaction of Os$_3$(μ_3-Me$_3$SiC$_2$Me)(μ-CO)(CO)$_9$ (**4**) with PhC≡CH, we obtained the red triosmium compound Os$_3${μ_3-2η^1-2η^2-C(SiMe$_3$)C(Me)C(H)C(Ph)}CO)$_9$ (**14**) having the structure **B** [8]. The IR spectrum of **14** (υ_{CO} 2085m, 2049vs, 2026vw, 2011s, 1996w and 1984 cm^{-1}) is distinctly different from that of **13** and related complexes with a side-on coordinated diene fragment.

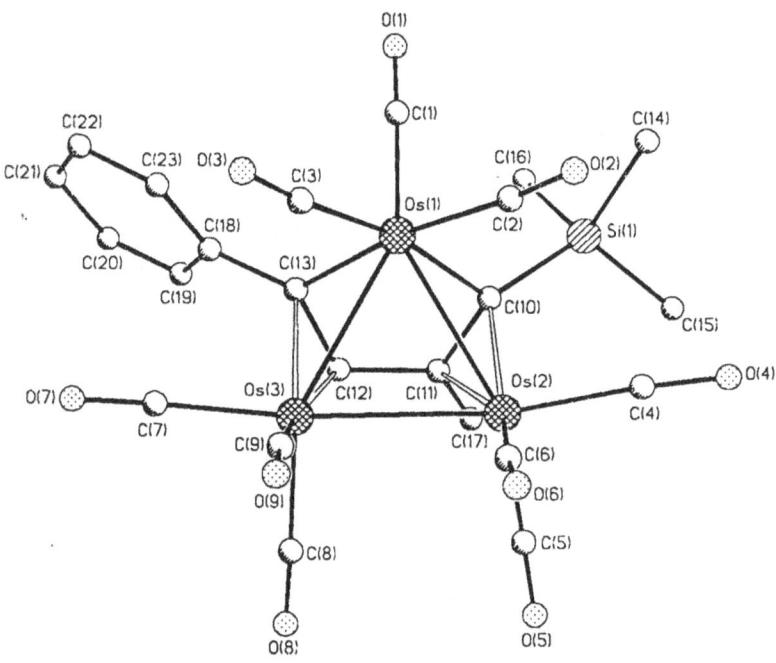

Figure 3. The molecular structure of **14**

Selected distances (Å):

Os(1)-Os(2) 2.811(1)	Os(2)-C(11) 2.42(2)
Os(1)-Os(3) 2.811(1)	Os(3)-C(12) 2.25(1)
Os(2)-Os(3) 2.825(1)	Os(3)-C(13) 2.22(1)
Os(1)-C(10) 2.19(10)	C(10)-C(11) 1.44(2)
Os(1)-C(13) 2.12(2)	C(11)-C(12) 1.46(2)
Os(2)-C(10) 2.18(1)	C(12)-C(13) 1.48(2)

Fig. 3 shows the molecular structure of compound **14**. In contrast to all previously structurally characterized trimetal clusters with the metallacyclopentadiene fragment, the hydrocarbyl ligand in **14** is directly bound to all three metal atoms. The osmacyclopentadiene ring has an envelope conformation, with the folding angle along the C(10)...C(13) line being equal to 28.8°.

The other products obtained in the reaction of **4** with PhC≡CH are the yellow hydride species $Os_3H\{\mu_3\text{-}C(SiMe_3)C(Me)C(H)C(C_6H_4)\}(CO)_8$ (**15**), the colourless $Os_2\{\mu\text{-}2\eta^1\text{-}\eta^4\text{-}C(SiMe_3)C(Me)C(H)C(Ph)\}(CO)_6$ (**16**), and the amethyst-violet $Os_3\{\mu\text{-}2\eta^1\text{-}\eta^4\text{-}C(SiMe_3)C(Me)C(Ph)CH\}(CO)_9$ (**17**). When **14** was heated in refluxing benzene, **15** and **16** were formed. Although the yellow hydride with the *ortho*-metallated phenyl group does not react with CO to form $Os_3\{\mu\text{-}2\eta^1\text{-}\eta^4\text{-}C(SiMe_3)C(Me)C(H)C(Ph)\}(CO)_9$ (**18**) with the side-on coordinated hydrocarbon unit, it does slowly react with PPh₃ at room temperature affording the phosphine derivative of **18**, the violet-black compound $Os_3\{\mu\text{-}2\eta^1\text{-}\eta^4\text{-}C(SiMe_3)C(Me)C(H)C(Ph)\}\text{-}(CO)_8(PPh_3)$ (**19**) (Scheme II).

Scheme II

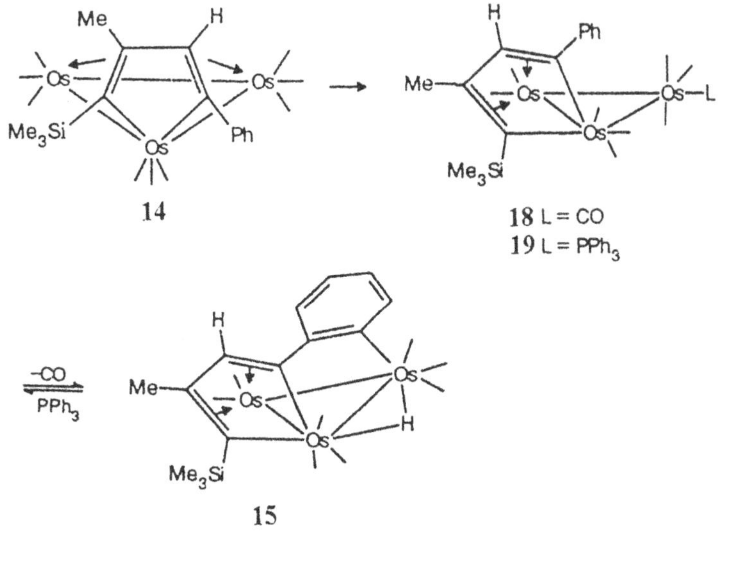

14

18 L = CO
19 L = PPh₃

15

It still remains unclear as to whether the observed **14** → **15** transformation involves the decarbonylation of complex **14** with the subsequent face-on → side-on rearrangement accompanied by *ortho*-metallation of the phenyl group or, alternatively, this reaction proceeds through the formation of complex, **18**, an unstable intermediate with a side-on osmacyclopentadiene coordination.

In conclusion it should be noted that the study of reactions between silylalkynes and Ru_3 and Os_3 clusters has provided valuable information on routes of alkyne dimerization, on novel types of complexes formed, and on novel rearrangements. A numbers of questions have arisen during this research which still have no answers, thus stimulating further development of this field of knowledge.

Acknowledgement. This work was supported by the Russian Foundation of Basic Research (Grant No. 94-03-08167)

References.

1. Koridze, A.A., Kizas, O.A., Kolobova, N.E., Vinogradova, V.N., Ustynyuk, N.A., Petrovskii, P.V., Yanovsky, A.I. and Struchkov, Yu.T. (1984) *J. Chemical Society, Chemical Communications*, 1158-1159.

2. Koridze, A.A. and Kizas, O.A. (1989), , *Metalloorganicheskaya Khimiya* **2**, 165-176.

3. Koridze, A.A., Efremidze, T.T., Struchkov, Yu.T. and Yanovsky, A.I. (1988) *Metalloorganicheskaya Khimiya* **1**, 826-830.

4. Koridze, A.A., Astakhova, N.M., Yanovsky, A.I. and Struchkov, Yu.T. (1992) , *ibid*, 886-893.

5. Koridze, A.A., Astakhova, N.M., Dolgushin, F.M., Yanovsky, A.I. and Struchkov. Yu.T. (1993) , *Izvestiya Akademii Nauk, Seriya Khimicheskaya*, 2011-2012.

6. Gambino, O., Vaglio, G.A., Ferrari, R.P. and Cetini, G. (1971) , *J. Organometallic Chemistry*, **30**, 381-385.

7. Ferraris, G. and Gervasio, G. (1974) , *J. Chemical Society, Dalton Transactions*, 1813-1817.

8. Koridze, A.A., Astakhova, N.M., Dolgushin, F.M., Yanovsky, A.I., Struchkov, Yu.T. and Petrovskii, P.V. (1994). *Izvestiya Axademii Nauk, Seriya Khimicheskaya*, 766-767.

INDEX

The manufacturer's authorised representative in the EU is Springer
Nature Customer Service Centre GmbH, Europaplatz 3, 69115 Heidelberg,
Germany. If you have any concerns regarding our products, please
contact ProductSafety@springernature.com

Printed and bound by CPI Group (UK) Ltd, Croydon, CR0 4YY
24/04/2026
02096308-0010